Marine Biology

Dr. K.P.Biswas, M.Sc.,Ph.D., D.F.Sc. (Bom), E.F. (West Germany), F.Z.S., F.A.B.S. (Kolkata), Former Joint Director Fisheries (L-1), Government of Orissa, Director of Fisheries, Andaman and Nicobar Islands and at present Faculty Member of Marine Science Department, University of Calcutta, West Bengal University of Animal and Fishery Sciences has been associated with fish, fishery and marine science for more than fifty two years.

His latest books, *"Advancement in Fish, Fisheries and Technology"* and *"Advances in Fishing Technology"* have been published in 2012.

Marine Biology

K.P. Biswas

2013
Daya Publishing House®
A Division of
Astral International Pvt. Ltd.
New Delhi – 110 002

Published by : **Daya Publishing House®**
A Division of
Astral International Pvt. Ltd.
– ISO 9001:2008 Certified Company –
4760-61/23, Ansari Road, Darya Ganj
New Delhi-110 002
Ph. 011-43549197, 23278134
E-mail: info@astralint.com
Website: www.astralint.com

Laser Typesetting : **Classic Computer Services**
Delhi - 110 035

Printed at : **Salasar Imaging Systems**
Delhi - 110 035

PRINTED IN INDIA

— Dedication —

Dedicated to Mrs. Manju Biswas
for her encouragement

Acknowledgement

The author deeply acknowledge the help of Dr. N.A. Talwar in computer setting, while preparing the manuscript of the book.

K.P. Biswas

Contents

Preface

According to a report attributed to Patricia Miloslavich of Universidad, Simon, Boliver, Venezuela, the co-scientist associated with census of marine life on a global basis, in which more than 300 scientists participated for last ten years. There are more than 2,30,000 species in world oceans.

The results show that around a fifth of the world's marine species are crustaceans, such as, crabs, shrimps, lobsters, krills and barnacles. Added mollusks (squids and octopuses) and fish (including sharks) accounts for up to half the species in the seas. The endangered species, often used in conservation campaigning, such as, whales, sea lions, turtles and sea birds account for less than two percent.

The survey also revealed areas of concern for conservationists. In every region, they have got the same story of major collapse of what were usually very abundant fish stocks or crabs or crustaceans that are now only 5 to 10 per cent of what they used to be.

These are mainly due to over-harvesting and poor management of those fisheries. That is probably the biggest and most consistent threat to marine bio-diversity around the world. The main threats include over-fishing, degraded habitats, pollution and the arrival of invasive species.

But more problems arise from rising water temperatures and acidification due to climate change and the growth in areas of the ocean that are in low oxygen, and unable support life.

The most diverse region identified by the Census of Marine Life (COML) are around Australia and south east Asia. It is also a hot spot for terrestrial bio-diversity and has been known for about hundred years.

It looks like that region with the coral reefs has always had a very high rate of speciation. It also has very diverse range of habitats, from the deepest areas of the oceans to large areas of shallow seas, which can support coral reefs.

Australian and Japanese waters contain more than 30,000 species each. Next in line are the oceans off China, the Mediterranean Sea and the Gulf of Mexico.

An upwelling area is supposed to be a biological unit, as stocks of animals are maintained within it from year to year. The animals including fish might live in the upwelling area all through their lives.

Upwellings occurs in Andaman Sea between December and February.

The Andaman and Nicobar Islands comprise an arcuate chain of 348 islands of various sizes spread in a distance of about 1120 km and located between Lower Mayanmar and Upper Sumatra of Indonesia in the Bay of Bengal. The total land area of these islands is about 8,293 sq km. The islands have a mass of hills enclosing valleys and a large part of the territory is covered with dense evergreen tropical forests. The vegetation is mostly Mayanmar and Malay type. The climate is typically tropical, with heavy gales and rainfall. The soil varies from sandy to clay. Some of the shore support rich growth of mangrove vegetation and coral reefs of the fringing type.

The life in Andaman Sea has been much less affected by man. In the obscurity of its long distance from the mainland, its depths and the breadth, its ocean animals are difficult to hunt to extinction, while most of the marine vegetation and the bulk of the diverse invertebrate fauna are not useful to man. This is the reason for choosing the less disturbed Andaman Sea as a site for studying the fauna and flora and their biology in marine environment. In the sea, it has been said, that man reaps without sowing, and this is largely true. It follows that in this case the responsibility of ownership, which came with the domestication of animals and the care of crops has rarely displaced the unconsidered greed of the hunter. This is nowhere better shown than in man's persuit of marine fish, crustaceans and mammals.

The author during his tenure of directorship in Fisheries Department of Andaman and Nicobar Islands from 1981 to 1984, came in close association of Andaman Sea for developing fisheries wealth. He initiated catching more and more fishes and crustaceans from the sea to aid for nutritious diet to the people and to generate employment for the fishermen. Towards achieving this objective, the author, not only made indepth studies about the aquatic resources of Andaman Sea, estimated from the survey and explorations from time to time, but also personally took part in exploratory fishing to know about the fishing grounds, type of species available, their magnitude, area of concentration and suitable fishing gears for their exploitation.

The book, " Marine Biology " in twenty nine chapters, besides describing the marine environment and ecosystem has given an account of physiography and location of Andaman Sea, also gave an outline of different types of ecosystems (marine, coral reef, mangrove, sea grass and sea weeds, pelagic and benthic) occupying different niches of the sea associated with marine life.

The chemical features of the water and the sea bottom augmenting the primary production, through microbial realm, phytoplankton and algae forming the base of food pyramid was discussed in the book. The ecosystem associated flora and fauna, like algae, sea grass, zooplankton, sponges, coelentrates, chaetognaths, holothurians, echinoderms, mollusks, crustaceans, benthic macrofauna, and interstitial and benthic meiofauna have been highlighted in the appropriate chapters. Fishes and crustaceans, which form the main source of protein food for human population have been specially dealt in the book, as to the species of commercial importance, their availability, behaviour and migration pattern and their capture.

Once available, now endangered and almost extinct mammal, sea cow, *Dugong dugong* in the sea grass bed of Dugong Bay in Little Andaman finds a place in the book.

K.P. Biswas

Introduction

Marine ecosystem has varying diversity. The coast line comprises almost all types of intertidal habitat, from hypersaline and brackish lagoons, estuaries and coastal marsh and mudflats to sandy and rocky shores. The subtidal habitats are equally diverse. Each local habitat reflects prevailing environmental factors and is characterized biota.

Thus marine fauna itself demonstrates gradients of change throughout the Indian coasts and islands.

Out of the total 32 animal phyla, 15 are represented by the taxa in the marine ecosystem. They may constitute either migratory (pelagic crustaceans, coelenterates (medusae), cephalopods, fishes, reptiles, birds and mammals. The benthic macro-fauna comprises of resident species of polychaetes, bivalves, gastropods, sipunculus and mud burrowing fishes. Among invertebrates, the sponges, phoronids and echinoderms generally do not prefer an estuarine ecosystem.

Free swimmers or nekton are important components of marine biodiversity and constitute important fisheries of the world. The dominant taxa in the nekton are fish, others being crustaceans, mollusks, reptiles and mammals. Out of the total 22000 fin fish species, about 4000 species occur in the Indian Ocean, of which 1800 species were reported from the Indian Seas. A majority of nekton species is found in the coastal waters. It is estimated that 40 species of sharks and 250 species of bony fishes represented the oceanic varieties.

The global ocean balance between nitrogen fixation and the loss of fixed nitrogen through anaerobic ammonium oxidation (anammox) reaction and denitrification, due to water upwelling to the surface, are generally slightly depleted in nitrogen relative to phosphorus, where upwelling of nitrogen and phosphorus is substantial (eutrophic region), primary production is high, resulting in the sinking of large amount of organic matter. As this organic matter is broken down to its nitrogen and phosphorus are solubilized, a large fraction of available oxygen is consumed. In this suboxic waters, anammox and denitrification converts ammonia and nitrate to nitrogen resulting in a loss of biologically reactive nitrogen from the system and a marked decrease in the deep ocean N:P ratio. In Oligotrophic sea, on the other hand, the upwelling of nitrogen and phosphorus is low, primary production is reduced and often dominated by nitrogen-fixing cyanobacteria, which are favourd in low nutrient, nitrogen-depleted surface waters.

Andaman Sea, an oligotrophic one, was the least studied basin for its potential biological wealth and resources. The earliest sketchy account of fishery potential of Andaman Sea was reported from the results of a fishing trawler in 1908-1909. Thereafter, several cruises followed till 1979 with an ultimate objective of finding out the available marine biota and productivity.

The Andaman and the Nicobar groups of islands are located in the south-western part of the Bay of Bengal, between Latitude 6 and 14 degree North and Longitude 91 and 94 degree East. The two groups are comprised of 348 islands, of which only 40 are inhabited. They collectively have an area of 8293 Sq km, which does not include a number of exposed islets and rocks. All the islands are the exposed peaks of long ranges of submerged mountains, which extend from Mayanmar to Sumatra. A deep oceanic ridge, about 1500 meter in depth runs between the Nicobar and Sumatra and another oceanic ridge along 10 degree North separates the Andaman and Nicobar groups of islands. Most of these islands are surrounded by fringing reefs on their eastern side and barrier reef on their western side. They harbor a rich population of corals and mollusks, and most of them have luxuriant mangrove vegetation around them. Geographically, the sea on the east of both these groups of islands is termed as Andaman Sea. The sea on the north forms a part of the Bay of Bengal through the Preparis Channel and on the south it is connected with the South China Sea through the Malacca Strait. The Andaman Sea occupies an area of 6.02 X 100000 Sq km and has a volume of 6.6 X 100000 cubic km with an average depth of 1096 meters.

The Andaman and Nicobar Islands separate the Bay of Bengal from the Andaman Sea. The islands are fairly straight and gentle on the western side comprising coastal plains, whereas the eastern coasts are strongly indented and steep and in many places coral reefs and raised beaches, as high as, 20 meters above the sea level have been reported. Narrow channels separate the Andaman into North, Middle and South Andaman Islands, which are collectively known as the Great Andaman.

The Paleogene Andaman-Nicobar trough was folded and reverse faulted westward resulting in the emergence of the Andaman Islands during the Oligocene and late Miocene. The Andaman-Nicobar ridge is formed of serpentinite basement overlain by distorted Paleocene to Miocene and flat lying Pliocene to recent sediments. A sequence of Paleocene through Upper Miocene rocks including 3000 meter of Upper Eocene and Oligocene graywackes were deposited over the Andaman serpentinites under fluctuating shallow to deep water conditions.

These islands, which are a part of anticlinal belt passing from Arakan Yoma in Mayanmar through Andaman and Nicobar islands and Mentawai islands, west of

Sumatra, separate the Andaman Sea from Bay of Bengal, except the connections through Preparis Channel on the north with a depth of about 200 meters, the Great Channel on the south, 1800 meters deep and the ten degree channel in the middle with a depth of about 800 meters. A north-south arc of volcanic arcs and sea mounts, including Barren and Narcondum islands, in the Andaman Sea separates the large central basin from two smaller basins to the north and south. The fall of slope is also steep on the eastern side of the islands. The thermo-haline properties of the Andaman waters, within the surface mixed layer, thermocline and halocline downwards on the eastern and western sides of the island chains are the features special to the region.

The Andaman Sea is influenced by large quantities of freshwater runoff from the perennial rivers of Mayanmar, Thailand and Malaysia. This runoff largely influences the topmost layers. Below the surface layer, oceanic conditions prevail.

Oceanographic investigations in the Andaman Sea dates back to 1869, when Francis Day, a well known army officer and fishery biologist visited these islands. He recorded the occurrence of 136 species of fish in the Andaman waters.

Oceanic islands, or rather a group of islands, far away from continental margins, significantly influence, modify and change the general physico-chemical properties of the waters around them, mainly through changing the general circulation patterns and associated stratification-mixing processes. Especially, with a chain or cluster of such islands forming partial barriers and obstacles, the physical processes and hydrodynamic characteristic of the adjoining waters change in distinctive fashions more as a rule than as an exception. From the Hawaiian group of islands to the Andaman- Nicobar Island arc system, there are numerous examples to be found in all the major oceans and seas. The oceanic islands of the Pacific and Atlantic, at least some of them, have been quite intensively studied from physical, geological and marine biological points of view.

Keeping the Andaman group of islands in mind, more is known about the south-east Asian waters, from the Philippines to the Timur Sea and the strait of Malacca to the Arfuna Sea, as would be evident from the NAGA Report (Wyrtki, 1961). It is equally true and disappointing to mention that since the pioneering work of Sewell (1925) more than 94 years ago, precious little has been added to our knowledge regarding the Andaman Sea and the Andaman-Nicobar group of islands. Yet this is the region, where the complex air-sea interaction phenomenon release enormous amount of energy for the genesis of the devastating tropical cyclones, which hit the east coast of India and the north and north-eastern coast of the Bay of Bengal almost every year. What makes the seas around these islands even more interesting and a class apart from many other oceanic islands (barring a few) is the yearly cycle of the North-East and South-West monsoonal wind systems reversing the atmospheric circulation and the surface currents of the Bay of Bengal and the Andaman Sea from December to April and June to October with intervening transitional periods. Another feature which demarcates the Andaman Sea from the Bay of Bengal lies in the geological history of emergence and development of these two regions. The island arcs form a kind of barrier between the Bay of Bengal, which has more or less even bottom topography and the Andaman Sea with its three major basins.

Two cruises of RV Gaveshani in January and February, 1979, in the Andaman Sea revealed some detailed facts as to the habitat under sea, water qualities and faunal diversity in the region. The area studied was in between Latitude 6 to 14 degree North and east of Longitude 91 degree 30 minute East. The cruises were made from 29 January to 27 February, 1979, and during this period, the vessel covered a distance of 6300 line km and worked at 63 stations. Of these 50 were in the Andaman Sea and the rest in the Bay of Bengal.

At all the stations, water and plankton samples were collected using Nansen reversing bottles and large Van Dorn water samplers, Indian Ocean Standard Net, Heron-Tranter net and Neuston net. Bathythermograph, thermo-salinograph and a ship borne wave recorder were run at all stations. The conductivity-temperature-depth profiling system (CTD) with rosette samplers were operated at selected stations. In addition, intensive studies were conducted around the Little Andaman on the circulation pattern around the island and on its benthic communities. Off Hut Bay, in the Little Andaman Island, the ship took the measurement of currents and the flow pattern by launching a parachute drogue. Routine meteorological observations were also made at all the stations.

Some samples were extracted with hexane for the analysis of petroleum hydrocarbons, while many samples were filtered for the estimation of heavy metals, extra-cellular products, particulate organic carbon, dissolved glycolic acid and glycollates, chlorophyll and detritus. Samples were inoculated with carbon 14 and incubated on board for the study of photosynthetic productivity. Zooplankton samples, collected using three types of net were preserved for subsequent examination of biomass, secondary production, species composition etc. Samples collected with the Neuston net along the oil tanker route across the Bay of Bengal, were examined for floating tar balls and other particulate matter. Those which were found free from oil were preserved for the estimation of pesticides. Sediment samples collected, using a snapper and a grab, were processed on board for the examination of macro/micro/ and meio-benthos. Some of these samples were preserved for the analysis of radio-active isotopes. Fishes, caught by hook and line, were processed on board for the estimation of mercury and pesticides in their tissues. Experimental fishing with tuna long line was tried for three hours at a station, south of Great Nicobar Island.

Chapter 1
Marine Environment and Ecosystem

The oceans comprise, the largest of all biospheres or regions of life on the earth, being some 300 times greater by volume than the living space on and over the land. The environment of the sea is divided ecologically, according to depth, distance from the land and°of light penetration, in general, characteristic type of organisms occur in these various environmental divisions.

The primary division is the overlaying water mass, the pelagic zone and the land mass beneath it, the benthic zone. The pelagic realm is sub-divided horizontally into a neretic province and an oceanic province. The neretic province is composed of all the waters above the continental shelf extending offshore to a depth of 200 m. The oceanic province consists of all the waters beyond the continental shelf with a depth greater than 200 m.

Neretic Province

The neretic province, being contiguous with the continent and land mass, is well lighted, but a turbulent section of the ocean. Seasonal variation coupled with the influence moon characterizes the area by strong wave action, marked currents and broad changes in temperature, salinity and nutrients. Dissolved oxygen is high. Detritus in suspension plus extensive micro and macro-biological production imparts a characteristic green, grey, brown or red colour to these near shore waters. Representative waters include waters of shore line, kelp beds, coral or rock reefs etc. The number of potential habitats is thus large and varied.

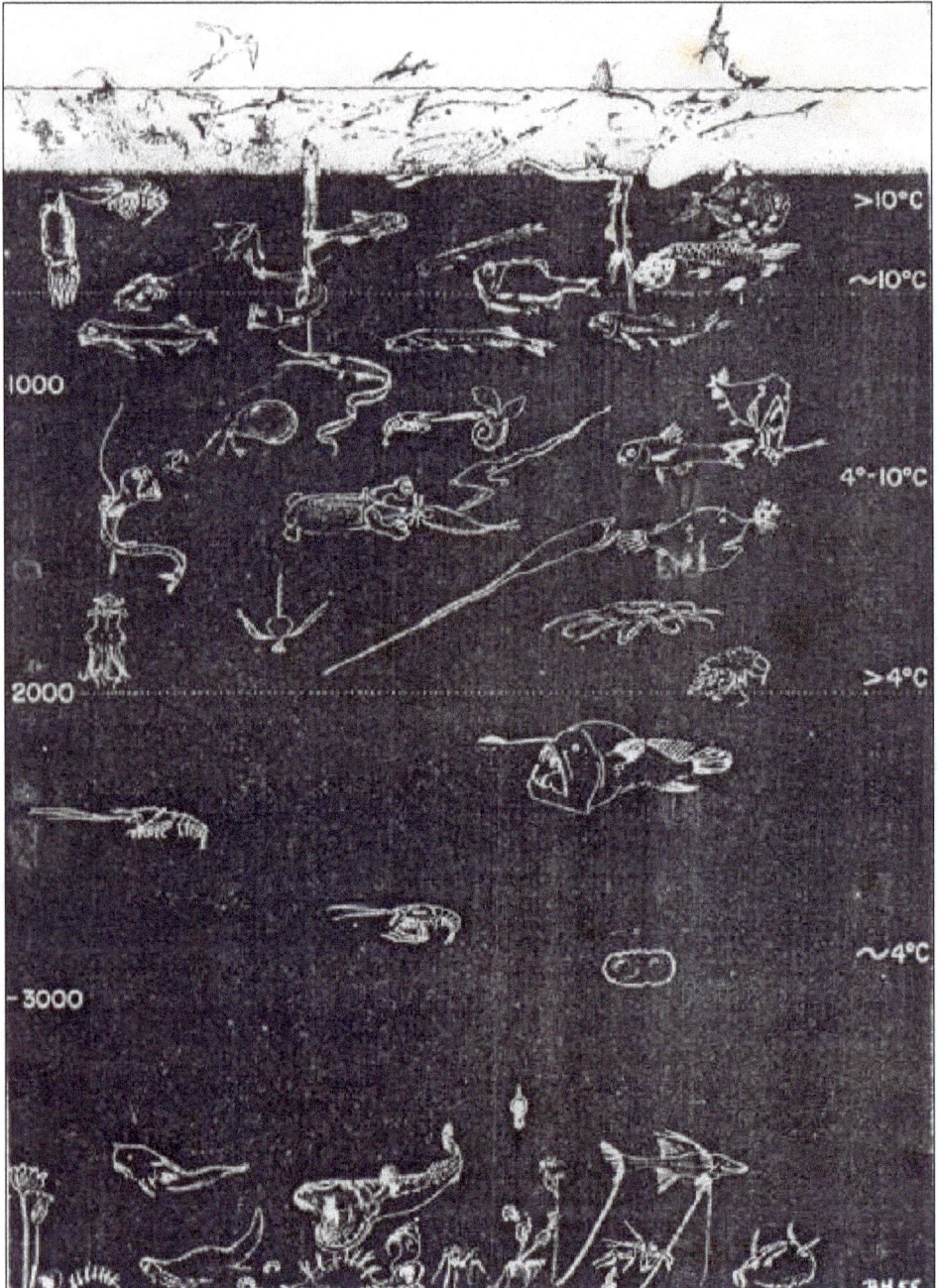

Figure 1.1: Vertical zonaation of sea

Oceanic Province

The oceanic province extends from pole to pole, from one continental shelf to another and from surface to the greatest depth. Seasonal fluctuations influence only

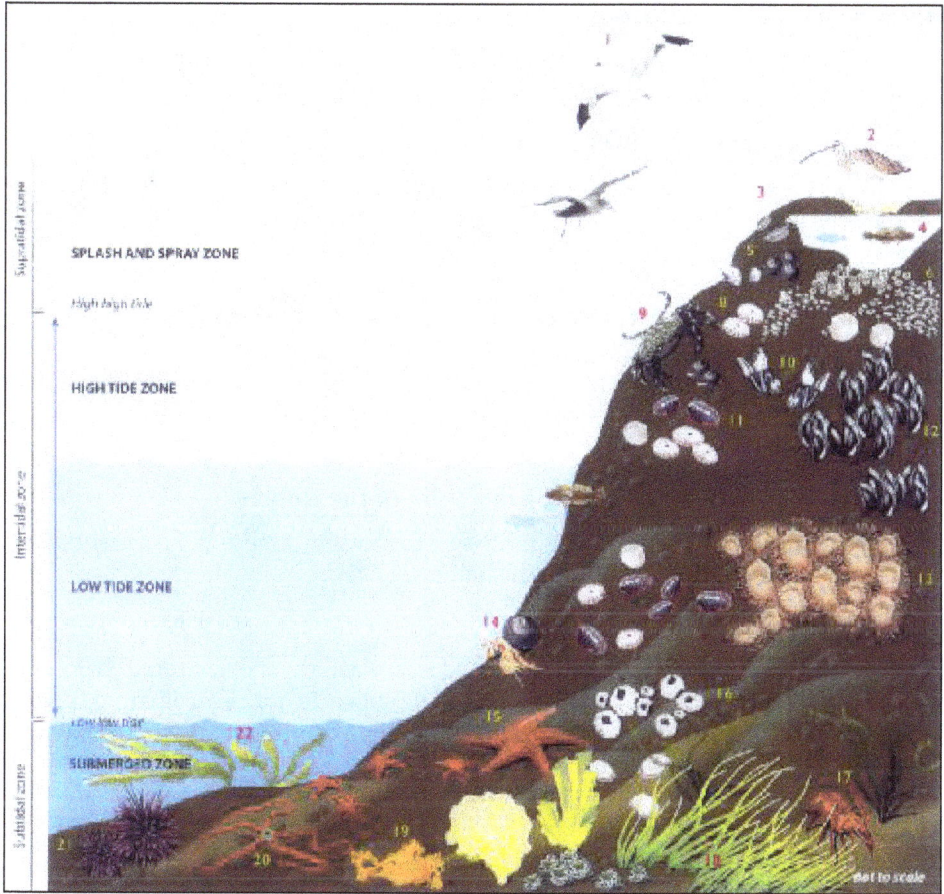

Figure 1.2: Vertical zonation of near shore sea

limited portions of this section, mainly the upper layers. The reminder of this province is fairly stable, uniform and with relatively few habitat types. Productivity is markedly less than the neritic province. Vertical divisions of the province are as under:

Epipelagic Zone (0-200 m)

The epipelagic zone of the oceanic province is characteristically blue in colour with good light penetration. The cool nutrient rich waters provide the basis for exceptional planktonic growth, which in turn, supports substantial population of larger animals.

Mesopelagic Zone (200-1000 m)

The mesopelagic zone is characterized by decreasing light intensity, predominantly in the blues, decreasing temperature and dissolved oxygen and steadily increasing water pressure. In general, conditions are relatively stable.

Bathypelagic Zone (1000-4000 m)

The bathypelagic zone is characterized by darkness, cold, high pressure and low biological activity. Seasonal variations are essentially nil.

Abyssopelagic Zone (4000 m + m)

The abyssal area of the ocean extends over half of the earth's outer perimeter. It is characterized by darkness, high pressure (200-1000 atmos.), cold (less than 4°Celsius) and low levels of dissolved oxygen. Physical change or variation appears slight, if any.

Benthic Environment

The benthic or land mass forming the bottom of the sea ranges from high tide to the maximum dept of 10000 m + m. Convenient sub-divisions are related to the units of pelagic environment.

Supralittoral (High Tide to the Spray Zone of the Beach)

A region of extreme variation, dry to wet to moist in a matter of moments. Few marine plants and animals occupy this habitat.

Littoral Zone (High to Low Tide)

This is essentially the intertidal area and is characterized by extreme changes in water depth (0 to several m) and turbulence. The substrate ranges from mud to sand to rock and is subjected to hourly, daily, monthly, and seasonal changes, which imposes harsh conditions for marine life.

Sub-Littoral Zone (0-200 m)

The sub-littoral zone is the underlying or bottom portion of the neretic province and, as such, is subjected to or influenced by the seasonal variations and conditions that impinge upon the later. The substrate consists of sand, mud, biological oozes, rocky outcropping, coral reefs, canyons, cliffs etc. These varying physical conditions act to create a great number of biotopes, habitats, and niches. Correspondingly, the area is rich in plants and animal life in number and species.

Bathyl Zone (200-4000 m)

The bathyl zone of the benthic environment corresponds to the mesopelagic and bathypelagic areas of the oceanic province. The substrate is composed of large areas of biological ooze and detritus. Darkness, cold, high pressure and low dissolved oxygen levels are the dominant features of the region of the bottom.

Abyssal Zone (4000-6000 m), Hadal Zone (Deeper than 6000 m)

The abyssal and hadal zones of the benthic environment form the bottom of the abyssopelagic portion of the oceanic province. Little is known of these regions. The substrate consists of biological ooze and detritus, while the waters are black and cold with high water pressure and contain low levels of dissolved oxygen.

Intertidal Area

In studying the distribution of plants and animals between tide marks in different parts of the world, it was found convenient to subdivide the tidal belt into strips or zones, each of them is characterized by distinctive features of its own. The three main zones of the shore are sublittoral fringe, midlittoral zone and supralittoral fringe. Fringe zones are boundary belts. Zones are related only approximately to the tidal levels. Generally the zones are, therefore, defined in terms of organisms.

Supralittoral Zone

The maritime belt lying near the sea, above the tide marks, but subject to some marine influence (*e.g.* to finely divided spray in rough weather). The lower limit of this zone is the same as the upper limit of the one below.

Supralittoral Fringe

From the upper limit of barnacles (in quantity of the nearest convenient land mark above this) *e.g.* the upper limit of littorinae or the lower limit of maritime land lichens of flowering plants. High water of spring tides invades atleast lower part of this zone.

Midlittoral Zone

From the upper limit of barnacles in quantity down to the upper limit of the zone below. This belt tends to be covered and uncovered every day, atleast in part.

Infralittoral Fringe

From the upper limit of any convenient dominant organisms (*Laminaria, Sargassum*) to extreme low water level of spring tides, or to the lowest level ever visible between waves. This zone uncovers only at the major tides and sometimes only in calm weather.

Infralittoral Zone

From extreme low water of springs to a depth, atleast to the edge of the continental shelf or to the lower limit of sea weed vegetation.

The Ecosystem of the Sea

Any area of nature in which materials are being exchanged between living organisms and their abiotic environment forms an ecological system or ecosystem. From the functional point of view, one can recognize the following steps in the operation of the ecosystem. (1) Reception of light energy, (2) Production of organic matter by autotrophic producers utilizing inorganic nutrients. (3) Consumption of the organic material by heterotrophic consumers. (4) Decomposition of organic material to inorganic nutrients, suitable for producers by saprophytic micro-consumers, chiefly bacteria and to a lesser extent, fungi. All of these steps and processes except (3) are essential to maintain the necessary cycling of the materials within the system. The operation of the system obviously involves production, growth and death of the living components, but the abiotic surroundings are always affected.

The effect of circulation on ocean temperatures based on the following conclusions, which are found on numerous and widely extended ocean observations.

1. The primary source of heat is the radient energy of the sun, both direct and diffuse, that penetrates the water (Murray, 1912; Gehrke, 1910; Helland-Hansen, 1911-12).

2. Absorption of this radiation directly heats the water in the upper layers (Nansen, 1913) and only a small fraction of this radiant energy penetrates below 25 meters (Krummel, 1907; Helland-Hansen, 1911-12 and Knott, 1903-05).

3. There is always a complex vertical circulation (Helland-Hansen, 1911-12; Gehrke, 1910; Nansen, 1913; Marray, 1912) due to lack of balance of the many forces acting on water particles. The resultant vertical flow through a finite section due to this motion may be very small and may be either upward or downward. That is, at the same time, some portions of the water are moving upward and others downward, thus tending to mix up the water at different levels. In this problem, the resultant of the upward and downward flow will be assumed to be zero.

4. The "mixing process" is most intense in the layers nearest the surface, owing to wave motion and other surface disturbances due to wind, but is present in some degree at all depths (Gehrke, 1909; Murray, 1898).

5. The amount of heat transferred from one level to another by conduction through the water is a negligible fraction of that carried by the water particles themselves as a result of the mixing process (Gehrke, 1909).

6. The mean annual rate of change of temperature with respect to latitude is practically independent of the depth within the upper hundred meters. This is revealed by a study of the average temperature of the North Pacific, tabulated with respect to latitude, longitude and depth (Schott, 1910).

7. At the time of year, when the surface temperature is a minimum, there is practically no variation of the temperature with respect to depth in the upper thirty meters (McEwen, 1916).

Chapter 2
Marine Biology

Marine Biology utilizes the basic knowledge obtained from biology of marine organisms and that from oceanography, so that relationship between the environment and aquatic resources could be explained and established. The links between the environmental and biological factors, as well as the effects of the changes on the potential fertility of the sea, and abundance, behaviour and distribution of the living resources of the sea are studied. Some of the gross objectives of marine biology are;

1. Study of physical and chemical factors and other processes related directly or indirectly with behaviour, distribution, abundance and availability of living resources with space and time;
2. Study of currents- upwellings and sinking (including mixing processes), water mass boundaries and their associated effects on marine animals and their biology;
3. Study of echosounding and location of animal concentrations;
4. Prediction of the major changes in the environment and animal resources.

Some Gross Features of the Oceans

Atlantic and Indian oceans are of same size and each is one third of the size of Pacific Ocean. However, the forms of each of them differs very markedly. Landmass boundary of Indian Ocean in its northern side, while there are openness of Atlantic towards both north and south, besides their dissimilar circulation pattern.

Ocean is related to each other dynamically. They interchange water from one to another. All the bottom and deep waters of the oceans was once formed at the surface in high latitude and then sinking and laterally spreading towards all parts of the oceans.

Effective separation of the animals and plants by the characteristics of the ocean water differ from one ocean to the another, or in the same ocean from one region to another. In this way distinctiveness between oceans or ocean regions could be observed. They are also distinctive in physico-chemical characteristics. The sea water is composed of dissolved and undissolved suspended matters. Dissolved matter includes, inorganic salts, dissolved gases and organic compounds, while the suspended matters include both inorganic and organic matters.

Elements Abstracted by Plants and Animals

Analysis of elements abstracted by phyto and zooplanktons as determined by Vinogradov and others are;

Phytoplankton – 100©. 17(N), 2.4 (P) and Zooplankton – 100©, 16(N), 2.4(P).

The relative proportions of these elements may not be constant for the same species and the proportions of nitrogen and phosphorus may vary, depending on the composition of sea water in which they occur. Some species have been reported to concentrate iron, manganese, nickel, cobalt, titanium, iodine in them. Calcium is absorbed in the formation of shells, skeleton, corals, pteropods and ooze globigein deposits. Silica is abstracted by sponges. Copper is found in the blood as haemocyanin in some species of crabs and oysters. Vanadium is found in tunicates and zinc is available in cyanea. The body fluids of many vertebrates resemble the composition of surrounding waters, although more concentration of potassium is found in highly developed species of cephalopods and less concentration of magnesium in crabs and lobsters.

Organic Productivity

The continuing interaction of components of the aquatic ecosystem is founded on energy transfer. Light is the source of the energy, only a fraction of which in aquatic ecosystem ultimately appears in fish flesh. Energy is first harnessed by water plants, some large, other small. The key to fish production is the harnessing that is done by microscopic algae, desmids and diatoms-like the phytoplankton. Phytoplankton has been termed, the real grass of the waters affording food stuffs, which are directly or indirectly converted into fishes. The harnessing of light energy for the manufacture of organic material is the process of photosynthesis. In this process, the green plants take carbon-dioxide from the water and build carbohydrates and other molecular components. Much more carbon is fixed in this process of primary production (first trophic level) each year in aquatic ecosystem than in terrestrial ones. Animal consumption and transformation of plant materials follows.

Herbivores, including some fishes are primary consumers. Subsequent grades of consumers are predatory. Since there is only fractional conversion from one trophic level to the next (about 10 per cent), it follows that the shorter the food chain, the greater the likelihood for survival abundance of fish species. Ecologists have likened this concept of producing to a pyramid either of biomass or numbers. The producers are at the base and exceed in mass and numbers; the primary; secondary, tertiary consumers are at various intermediate levels in the pyramid and the consumers with the longest chains, such as, highly predatory and deep sea fishes are at the top of the pyramid and are least in biomass and numbers. It also follows that in such non-parasitic food chains the predators not only decrease in number along the various links, they also grow larger than their prey in succeeding links.

Biochemical Cycles

Key components available for synthesis into protoplasm in the hydrosphere are circulated in biochemical cycles. Recgonized cycles include those for carbon, nitrogen and phosphorus. Initial sources of the key elements differ. Thus in the production cycles, the atmosphere is the source of carbon (from carbon-dioxide) and nitrogen in gases. Phosphorus originates from soil through erosion or bleaching. The productive phase of biochemical cycles is the fixation of elements into protoplasm by living things. Phytoplankton for example, is able to fix carbon, nitrogen and phosphorus. Fish that feed directly or indirectly on phytoplankton convert the plant fixed compounds of these three elements and others into flesh. The waste products and carcasses of both phytoplankters and consumers return the key elements directly or through bacterial decay to their biochemical cycles for reuse. The process of return may be very rapid or very slow. Phytoplankters, fish and other aquatic animals may promptly return carbon dioxide by respiration in the aquatic ecosystem. Phosphorus locked in the skeleton of fish may take years for its release after the fish has died and fallen to the bottom. The entire scene, however, is that of a continual, with the forces of production nourished from the land and the atmosphere consistently managing to exceed those of release. Death is required for balance, but it is the amount of the nutrient raw materials naturally available in an ecosystem that limits the size of the biomass, the system can support. The cycle of food in sea is that elements are abstracted by phytoplankton and then transferred to zooplankton, animals and bacteria.

Distribution of Nutrients

General distribution of nutrients are characterized by four different layers for nitrates and phosphorus. A surface layer in which concentration is low and relatively uniform with depth. A layer in which concentration increases rather rapidly with depth. A layer of maximum concentration, that is usually located somewhere between 500 to 1500 meter depth. A thick bottom layer in which there is relatively little change with depth.

Factors Influencing the Distribution of Nutrient Elements

Dynamic equilibrium exists in case of large scate distribution of phosphates, nitrates and silicates. As the local change is assumed to be zero, there should be a balance between the effects of diffusion, advection and the net effects of biological processes. In the surface layers, that is, in euphotic zone, the biological processes led to a net utilization of nutrient elements, and if the rate of utilization exceeds the rate of supply by diffusion and advection, their concentration decrease. This change take place during summer and spring, when plants production in euphotic zone is very low, diffusion and advection processes exceeds the biological utilization. A marked variation in the nutrient content seems to occur in the regions, where conditions (like temperature, light, biological and other agencies) are unfavourable for plant growth. Also fluctuations in biological utilization, and supply by diffusion and advection may vary during the course of the year, for example, variation in temperature distribution and wind effects, changes in current systems. The regions of convergences, where water with low nutrient concentration may occur below the euphotic layer, or the regions of divergences, that is, where due to upwelling nutrient rich subsurface waters are carried to the surface. Coastal upwelling caused due to prevailing winds is a seasonal and intermittent phenomenon, and such areas show a marked fluctuations in the nutrient content of surface waters and may give rise to an actual increase in nutrient content during high plant production.

Upwelling and Fish Production

The upwelling areas in the oceans are rich in fish populations. Much of the world's coast line in subtropical seas is in upwelling areas, where fish live in provision. Upwelling is found in the eastern boundary currents of the subtropical anticyclones. Also the biological effectiveness of upwelling is confined to the tropics and subtropics.

The coastal upwelling is bounded between 50-100 km offshore by a cell of convergence and divergence. Inshore of this boundary, there is a region of slow mixing of old and new upwelled waters, and offshore of it there is a region of divergence, secondary form of upwelling. The secondary upwelling, offshore of this boundary is associated with wind stress alone and is generated by the velocity of wind stress. There are wind driven upwelling and geostrophic upwelling, both in coastal areas and in oceanic areas. Coastal upwelling is generated by winds blowing parallel to the shore towards the equator, as in the major upwellings, or is generated geostrophically by tilt of a current. Some areas may include both wind generated upwellings and geostrophic upwellings, as for example, the Malabar coast.

The cause of high production in subtropical upwelling areas and in the Antarctic is often attributed to the associated presence of nutrients. The persistence of production in an upwelling area is due not to the presence of excess nutrients, but to the persistent addition of living material.

The production cycle in an upwelling area resembles that of temperate waters. The cool water originates from depths of less than 200 meters and it contains resident and sparse population of plants and animals, very like those of temperate waters in early spring. In the photic layer, the algae start to divide as in temperate waters, the increase in animal production is caused by the increase in plant population and must follow it in time. This delay may be as much as half a generation. It is this delay, which allow the production of large stocks of plants and later of animals as in temperate waters. If the mixing were complete and rapid, the new algal production would be eaten by older animals derived from earlier production as it upwelled. So for a productive upwelling, mixing should be rather a slow process.

The simplest way of assessing the production in the upwelling areas is to estimate gram carbon per square meter per day from radio-carbon measurements for a region and time period.

There are number of upwelling associated with south west monsoon in the Indian Ocean, which blows from April to September. They are in Somali current, off south west Arabia and off the Malabar coast of India.

During the south west monsoon, a southerly current is generated with a thermocline tilt along the Malabar coast of India. The upwelling is very shallow, originating from perhaps 20 meters, but the surface water is, as a consequence, 6°Celsius cooler. During July to October, low salinity water spreads over the surface from the rivers and it is at this period that the phase of algal production occurs. During September and October, the post monsoon period, there is a true wind induced upwelling between Allepey and Quilon, off Cochin, and a counter current is developed at 75-100 meters. The situation, off the Malabar coast, is complex and probably the production is quite high. It is clear that upwelling extends from Trivandrum as far north as Panjim, but because of its nature and shallow origin, reliable areas of upwelling can not be estimated with the methods used in other areas.

An oxygen deficiency can occur below the thermocline during the upwelling period (Bause, 1968). In the Bay of Bengal, upwelling has been detected, off Waltair in February-March. It probably extends up the coast to Sagar Island in June. Under the south west monsoon, upwelling probably occurs all along the coast of Orissa, parallel to the monsoon. Between December and February upwelling occurs in Andaman Sea in a rather limited area on the coast of Mayanmar, opposite to the Andaman Islands. Upwelling also occurs in the shallow Gulf of Thailand, on the west coast in August and on the north east coast in October and January. Slight upwelling occurs all the year round at 12°North and 101°West. Off South Vietnam, upwelling occurs during the summer monsoon between the central South Vietnamese coast at Nhatrang for some hundred of miles up the coast.

Fishes, characteristic of an upwelling area, are not caught outside it. As the stocks are maintained within it from year to year, it is reasonable to suppose that an upwelling area is a biological unit. The fish might live in the upwelling area all through their lives. The Californian sardines spawn right in the coastal upwelling zones. They live in or near the thermocline and so can not make use of the counter current in their migration. The vertical distribution of animals in the upwelling area has a special interest. Sardines, Sardinella and Anchovies live at or just above the depth of thermo-cline. Below the thermo-cline live horse mackerel probably in intensive and rather dispersed shoals. On the bottom at any depth of water, from 100 meters outward hake live close to the bottom, perhaps depending to some extent upon euphausids.

Chapter 3

Explorations

The earliest account of the fishery resources of Andaman Sea revealed from the results of the trawler Golden Crown (1908-1809) as a good potential without any specific data. Again towards the end of forties (1947), a private company, named Andamarine Development Corporation Ltd was set up for exploitation of fishery resources of Andaman Sea. The company started operation with an Australian built fishing boat, fitted with diesel engine, a flat bottom patrol launch and two dinghies. Though their Master Fisherman stated that the "potentialities are enormous", they did not succeed. In 1951, another fishing operation commenced in the Andaman Sea, but without success.

Very little work has been done on the oceanography and marine biology of Andaman Sea. The earlier explorations have been mostly on the fauna and some aspects of topography and hydrography of these waters. It is to the credit of Sewell (1925-1935) that the isolated and discontinuous observations carried out in the global expeditions of "Challenger", "Valdivia" and "Siboga" and the Bay of Bengal survey of "Investigator" were collected and some picture on the general pattern of the sea bed and deep sea deposits, marine metereology, temperature, salinity etc of the Andaman Sea could be compiled in a series of articles in the "Memoirs of the Asiatic Society of Bengal" during 1925-1929. The faunal features of the sea around Andaman were detailed in the various classic monographs of Alcock, Sewell, Annandale, Rao, Hors etc.

Subsequent work of any significance was the marine productivity work carried out by the Danish vessel, "Galathea" in 1954, which covered some parts of Andaman Sea. Some detailed investigations on the hydrographical and planktological features of the waters were made by the International Indian Ocean Expedition during the years 1959-1965, especially by the Russian vessel, "Vitiaz", the Indian vessel INS Krishna and to some extent by "Anton Brunn". Though the assessment of fisheries

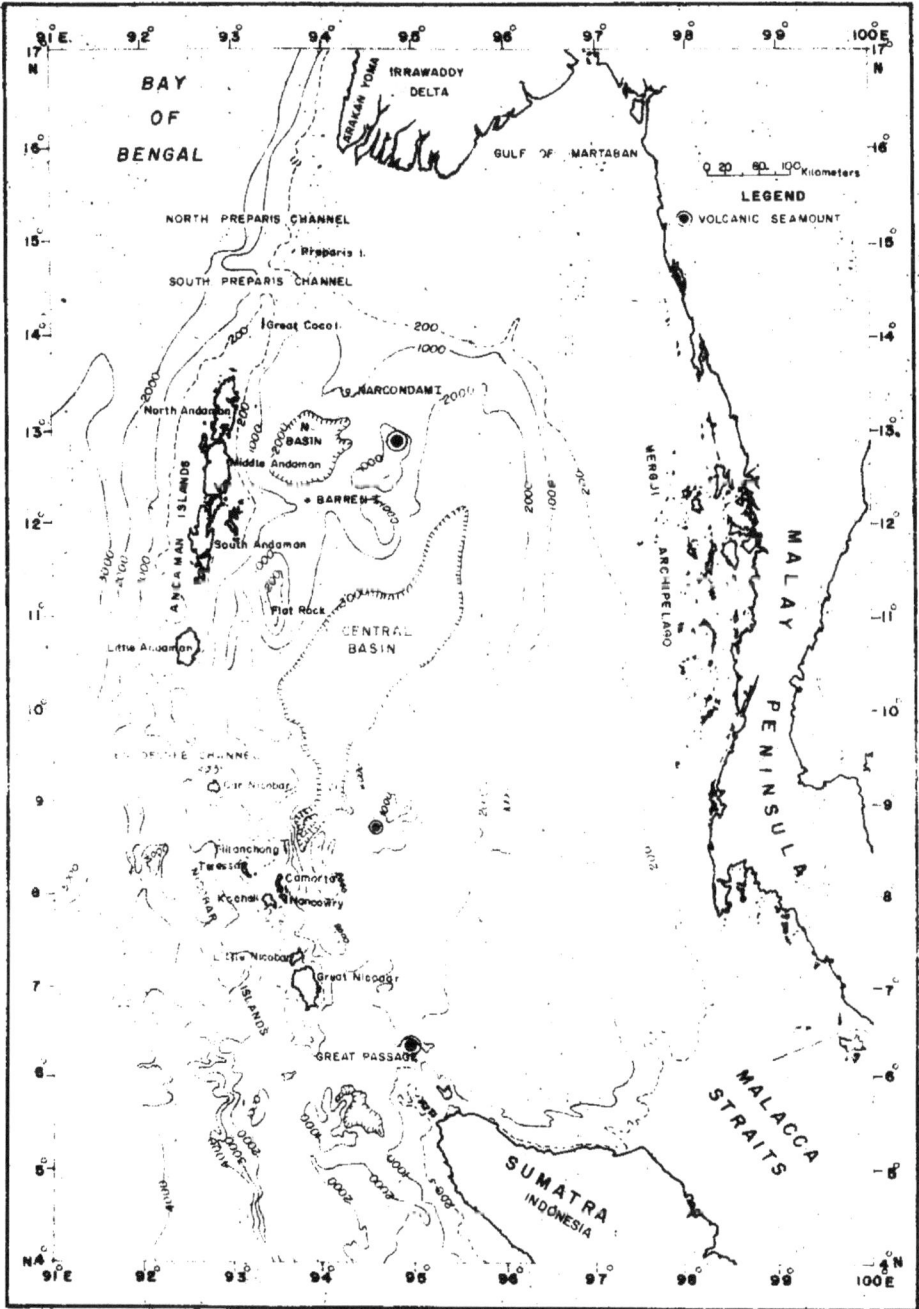

Figure 3.1: Andaman Sea from Encyclopaedia of Oceanography

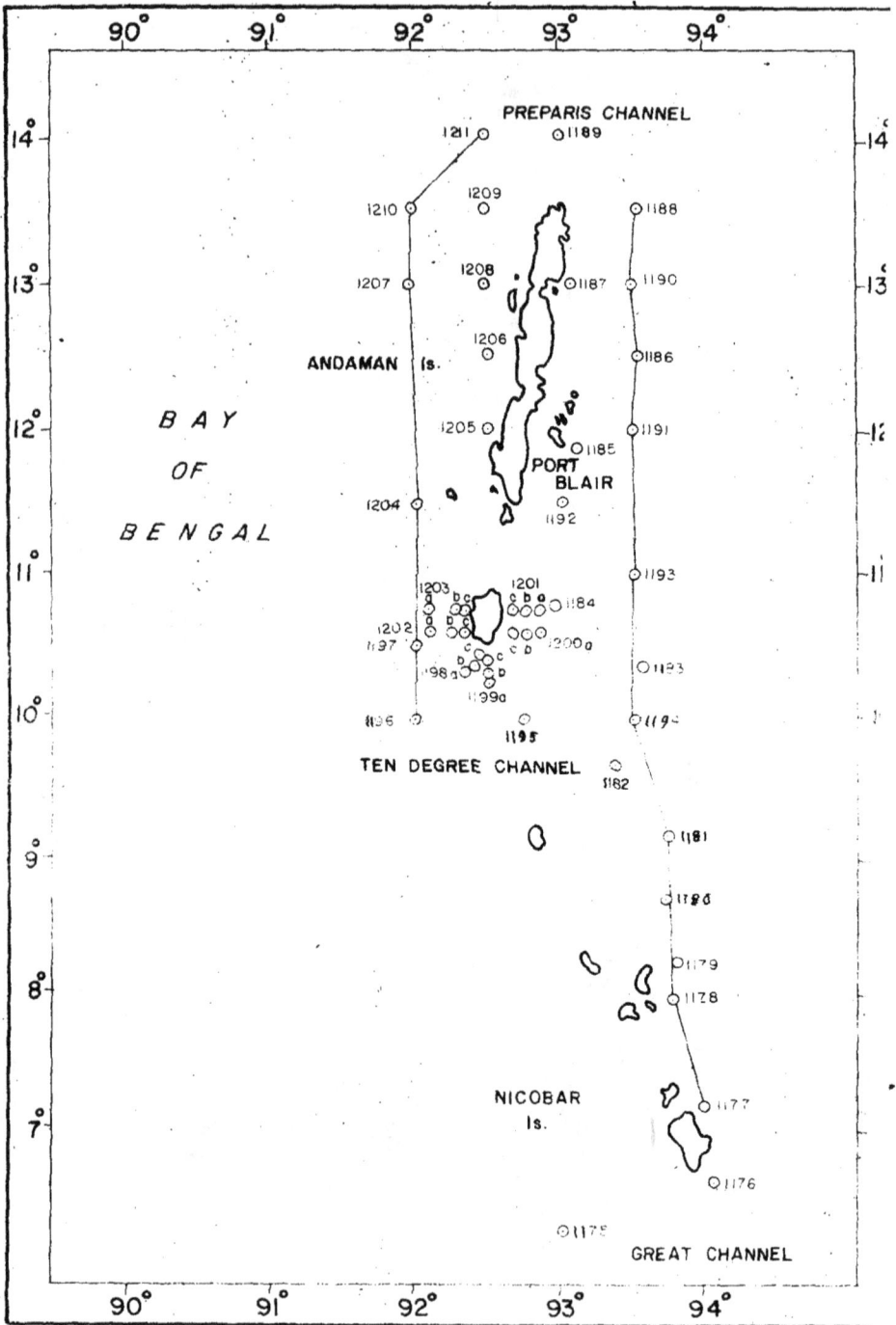

Figure 3.2: Location of Sampling Stations in Andaman Sea

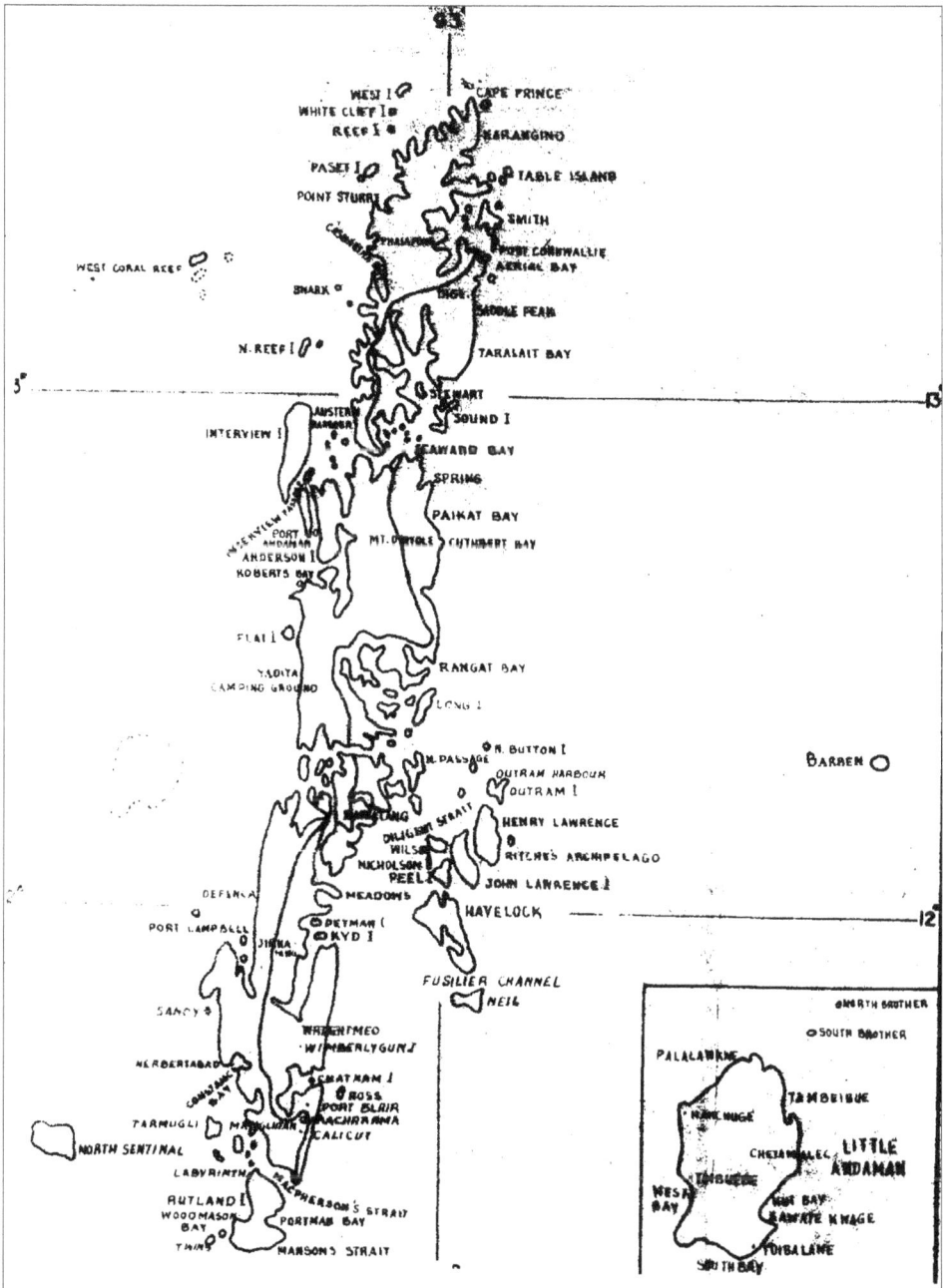

Figure 3.3: Andaman Islands

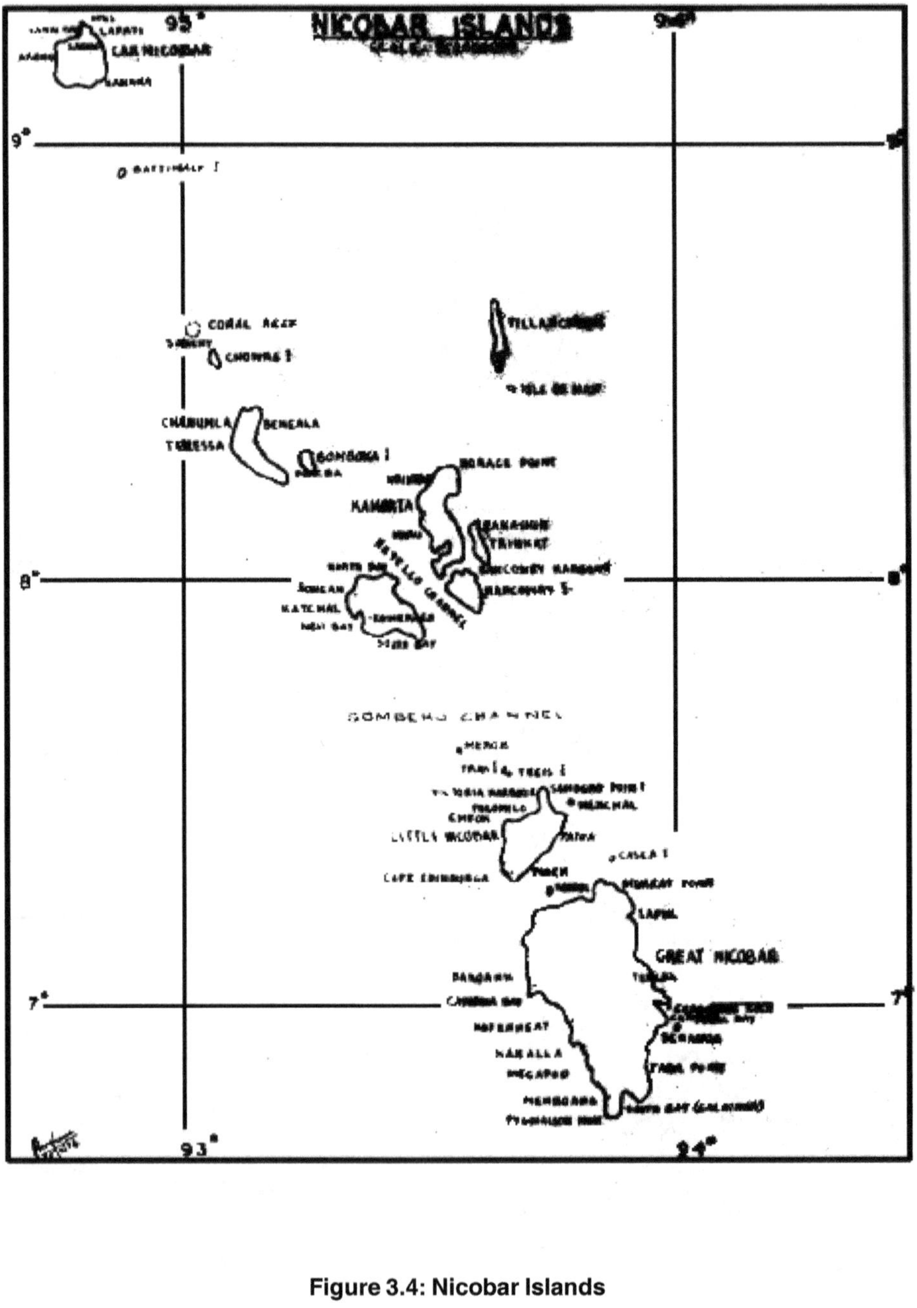

Figure 3.4: Nicobar Islands

potential was used as "an impressive argument to stimulate interest", when the programs of the International Ocean Expedition were formulated, "the actual fisheries work accomplished during the expedition itself has been disappointingly small" as succinctly put by Panikkar (1966).

Cruise I of RV Anton Bruun was carried out in the Bay of Bengal during the period of 12 March to 10 May, 1963. The cruise had two major objectives, (i) study of the hydrography, chemistry and plankton biology of the Bay of Bengal including Andaman Sea, particularly along its eastern boundaries in relation to possible upwelling areas produced by the north east monsoons and (ii) sampling of the benthic population on the Continental Shelf around the periphery of the Bay, and where possible, in deeper water to (a) assess the commercial fishery potential of the region and (b) obtain biological and geological specimens for systematics and distributional studies.

In the cruise track 40 stations in Andaman Sea from where measurements of temperature, salinity, dissolved oxygen, phosphate, nitrite, nitrate, silicate, primary productivity, phytoplankton and pigment measurements were taken. Hydrographic casts were made with Teflon-coated Nansen bottles to the bottom (in shallow water) or to 1000 meters. A few deeper stations were made in the central portions of the Bay. Samples for primary productivity and phytoplankton pigments were obtained with plastic water samplers from the depths to which 100, 50, 25, 10 and 1 per cent of the surface illumination penetrated.

Productivity measurements were made for 24 hours at natural, *in situ* light intensities, using an on-deck incubator with neutral density screens to stimulate the light at each depth sampled. Duplicate samples were exposed to fluorescent illumination at a constant intensity of 1000 foot candles for four hours.

Vertical net tows were made from 200 m to the surface with an IIOE standard net (one meter square mouth area, 0.33 mm mesh opening). The displacement volume of each sample was measured. Vertical plankton samples were also taken from 200 m to the surface with a half meter square mouth area, 0.064 mm mesh-opening truncated net for micro-plankton samples. In addition, oblique plankton hauls were taken with a half meter square area, 0.33 mm mesh opening net at all standard stations.

Bathythermograph observations were made at each station at an intervals of one hour or less, when running between stations. Bottom topography was recorded continuously with a precision echo-sounder. Meteorological observations were made and reported by radio to the International Meteorological Center at Bombay.

Bottom fauna were sampled with a "Gulf of Mexico" shrimp trawl and with a biological dredge. Bottom sediment samples were taken with a mud snapper and or with a Phleger corer.

In all, 92 standard samples, 33 trawls, 26 dredge, 30 sediments and 34 surface samples were taken during the cruise.

Preliminary Findings

The Andaman Sea is dominated during half of the year by the south west monsoon winds and during the other half by the north east monsoon winds. The latter develop as the result of a high-pressure area over the cold Tibetan Plateau in south China.

As a result of the northeast winds, the surface water in late March had been displaced offshore along the lee coasts of Thailand and Mayanmar and the sub-surface water had upwelled. This was evident by a gradual tilting of the upper isotherms toward shore on the eastern side of the Andaman Sea. Here the deeper isotherms displaced downward, which implied an accumulation of intermediate-depth water near the coast. This is a characteristic upwelling pattern.

Northward along the Mayanmar coast, the isotherms were tilted even more shore-ward and showed additional coastal upwelling. The strong tilting of the isotherms resulted from the effects of both the monsoon winds and the horizontal circulation. In late April, the monsoon winds had developed a strong upward displacement of isotherms off the east coast of India.

The direction of horizontal surface currents in the Andaman Sea was north west ward through the Strait of Malacca and south west ward in the northern part of the Andaman Sea. Both currents run to the west when they meet, and this results in a slow vertical motion in the eastern part of the Andaman Sea. These circulations also influence the vertical thermal structure.

Off the northern Mayanmar coast in early May, the flow was southerly down the coast. Since the ship was later set northward, all along the east coast of India from Hoogly Delta to Chennai and the isotherms tilted upward, it is fairly certain that a clock-wise surface circulation in the entire Bay of Bengal existed at that time.

Off the Andhra coast and the Hoogly River, the isotherms in the thermocline were tilted upward toward the shore, which implied upwelling and a northerly flow. However, the isotherms in the thermocline reached a maximum depth of about 100 miles offshore. They then rose to a crest at about 200 miles off the coast. This structure appeared to form a thermal dome, which might mean that a large clock-wise eddy occurs in the region.

Another explanation of the dome is that it may be the result of major internal waves, since large thermal oscillations were found in the southern part of the Bay of Bengal. The thermocline between Sri Lanka and Nicobar Islands contained vertical displacements of as much as 50 meters and with wave lengths of about 250 miles.

In the Andaman Sea, around Andaman and Nicobar Islands, the thermal structure was variable and patchy, a condition that may be due to large-scale turbulence caused by islands.

In the central Andaman Sea several rip tides were observed. These appeared as long lines of rough turbulent water that passed the ship, at speeds of about one knot, when it was halted on stations. These phenomena are believed to be instrumental in mixing the water column and aiding the vertical displacement of nutrients.

The primary organic production, which closely followed the pattern of upwelling was highest near the Thailand and Mayanmar coasts and around Andaman Islands. The central part of the Bay of Bengal showed extremely low production. In the latter part of April, the south west winds off the east coast of India started to blow and created a high primary production zone near shore that corresponded with the recent upwelling. However, the principal area of primary production was on the far eastern side of the Bay of Bengal and Andaman Sea.

The secondary production in the Andaman Sea closely followed the primary production, with high concentration of zooplankton found near the Thai and Mayanmar coasts. In the later observational period on the other side of the Bay of Bengal, the high zone appeared to extend all the way around the head of the Bay and partly down the coast of India. The central Bay ranks very low in production of the larger planktonic organisms, however, a strong echo was received from the deep scattering layer present throughout the Bay.

A Mexican shrimp trawl of 40 feet long and 42 feet wide was used to survey fish and shell fish resources from the R.V.Anton Bruun. Thirty one trawl hauls were made as detailed below; one haul in 280 fathoms and one in 1000 fathoms of water between Nicobar Islands and the coast of Thailand. Three on the continental shelf of Thailand. Twenty on the continental shelf and one in 200 fathoms and one in 1000 fathoms off Mayanmar and four near Port Blair in Andaman Islands.

The continental shelf generally was favorable for trawling, although outcroppings of coral were encountered in several regions. The precipitous nature of the continental slope at depths from 100 fathoms to about 400 or 500 fathoms usually prevented trawling. The bottom generally consisted of mud and was fairly level at depths greater than 500 fathoms.

None of the hauls yielded fish or shrimp in "commercial quantities". Catches of fish ranged from 2 kg to 161 kg per hour of trawling, and shrimp catches ranged from zero to 28 kg per hour of trawling. The overall average catch of fish and invertebrates amounted to less than 50 kg per hour of trawling.

The shrimp, mainly Penaeidae and Caridae, generally were small, ranging from 200 to 600 heads on count per kg. However, a few Penaeid shrimp weighing as much as 250 gram a piece were caught. Shrimp appeared to have a wide bathymetric range with as large catches being taken at depths from 150 to 200 fathoms, as were taken in shallow water of less than 30 fathoms.

Although the fish catches were small, they consisted of a great number of species. Best catches were made at depths from 8 to 30 fathoms, Fishes commonly caught included lizard fish (Synodontidae), goat fish (Mullidae), cat fish (Tachysuridae), queen fish (Carangidae), flute mouths (Fistularidae), slimys (Leiognathidae), threadfins (Polynemidae), silver bellies (Gerridae), and flat fishes (Pleuronectidae, Bothidae, Cynoglossidae, Soleidae and Psettodiae). The largest fishes taken were a 112 kg shovel nose shark and a 100 kg ray.

Much of the bottom trawled appeared relatively barren of life. Many shells of mollusks and other invertebrates were brought up in the trawl, but almost all of them were empty. Very few sea stars were encountered and hand lining yielded no fish.

Exploratory fishing in Andaman Sea has been carried out from October, 1971 to March, 1976. by adopting various fishing techniques, like, bottom trawling, purse seining, long lining, hand lining and trolling in order to assess the marine fishery resources in this region. Three fishing vessels, namely, M.T Matsya Vigyani, M.V

Meena Khojini and M.V Meena Prayas of 31.8 and 16.3 m length with a gross registered tonnage (GRT) 182.6 and 56.8 respectively were employed for the purpose.

In January and February, 1979 two cruises of R.V. Gaveshani were undertaken in the Andaman Sea. The area studied was in between Latitude 6 to 14°North and east of Longitude 91°30 minutes.The cruises lasted for 30 days (29 January to 27 February, 1979) and during this period, R.V.Gaveshani covered a distance of 6300 line km and worked at 63 stations. Of these 50 were in the Andaman Sea and rest in the Bay of Bengal. At all the stations, water and plankton samples were collected using Nansen reversing bottles and large Van Dorn water samplers, Indian Ocean Standard Net, Heron-Tranter net and Neuston net. Bathythermograph, Thermosalinograph and a ship borne wave recorder were run at all stations. The conductivity-temperature-depth profiling system (CTD) with rosette samplers were operated at selected stations. A similar profiling system, designed and fabricated at National Institute of Oceangraphy (NIO) was also run at a few stations to compare its accuracy with the other imported system. In addition, intensive studies were conducted around the Little Andaman on the circulation pattern around the island and on its benthic communities. Off Hut Bay, in the Little Andaman, the ship was anchored for 12 hours for the measurement of currents and the flow pattern by launching a parachute drogue. Routine meteorological observations were also made at all the stations.

In all, 503 water samples were analysed for salinity, dissolved oxygen, pH, phosphate-phosphorus, nitrate-nitrogen, nitrite-nitrogen, ammonia-nitrogen, silicate-silicon and fluoride. The total analysis carried out on board were 3126.

Some samples were extracted with hexane for the analysis of petroleum hydro-carbons, while many samples were filtered for the estimation of heavy metals, extra-cellular products, particulate organic carbon, dissolved glycolic acid and glycollates, chlorophyll and detritus. All these samples (1021 in all) were preserved on board for further analysis in the laboratory. Moreover, 104 samples were inoculated with carbon 14 and incubated on board for the study of photosynthetic productivity. In all 75 zooplankton samples collected using three types of net were preserved for subsequent examination of biomass, secondary production, species composition etc. Fourteen samples collected with the Neuston net along the oil tanker route across the Bay of Bengal, were examined for floating tar balls and other particulate matters. Those which were found free from oil were preserved for estimation of pesticides. Forty seven bottom samples collected using a snapper and a grab were processed on board for the examination of macro/micro/meio-benthos. Nine of these samples were preserved for the analysis of radio-active isotopes by Bhabha Atomic Research Center, Bombay.

About 50 fishes, caught by hook and line were processed on board for the estimation of mercury and pesticides in their tissues. Experimental fishing with tuna long line was tried for three hours at a station south of the Great Nicobar Island.

There after from December 1979 to February 1980, three more cruises of R.V. Gaveshani were conducted again in the same area. During these cruises, most of the stations around Andaman and Nicobar Islands and oil tanker route were repeated.

Chapter 4

Physiography and Location

The Union Territory of Andaman and Nicobar is a group of islands, both big and small numbering 321 and are located in the south eastern part of Bay of Bengal between Latitude 6 and 14°North and Longitude 91 and 94°East, of which 40 islands are inhabited at present. They collectively have an area of 8293 sq. km, which does not include a number of exposed islets and rocks making the total 556. All the islands are the exposed peaks of long ranges of submerged mountains, which extend from Mayanmar to Sumatra. A deep oceanic ridge about 1500 meters in depth, runs between Nicobar and Sumatra and another oceanic ridge along 10°North separates the Andaman and Nicobar groups of islands. Most of these islands are surrounded by fringing reefs on their eastern side and barrier reef on the western side. Geographically, the sea on the east coast of both these groups of islands are termed as the Andaman Sea. This sea on the north forms a part of the Bay of Bengal through the Preparis Channel and on the south it is connected with the South China Sea through Malacca Strait. These two separate groups of islands, namely, Andaman group and Nicobar group with different population and problems have been separated as two Administrative Districts of the Union Territory and are completely maritime in nature.

Having many unique characteristic of its own, the Andaman and Nicobar Islands, situated in the Bay of Bengal constitute the most isolated part of the Indian Union. The capital town, Port Blair is situated at a distance of 1255 km from Kolkata and 1190 km from Chennai, Prior to 1947, the island was known as penal settlement of the British Colonial administration. Only after independence, these islands were classified as a part "D" State and later constituting into a Union Territory in 1956 and were administered by the President of India through the Chief Commissioner till 12 December, 1982 and now through the Lieutenant Governor. The islands have a dominantly flat terrain, the highest point not exceeding 732 meters and are now endowed with forest plantation and marine resources because of its tropical climate and the seas around the islands.

The Andaman Sea is known to be rich in marine wealth. Its fishery potential has been estimated to be of the order of several lakh tones per year. Of this only about 16000 tonnes are exploited annually at present. The rich fishing ground of Andaman Sea attracted numerous ships, such as, "Challenger", "Valdivia", "Siboga", "Galathea", and "Vitiaz" to visit this region during their international expeditions. Even now, the rich marine wealth of Andaman Sea lure the foreign fishing vessels to intrude in the Exclusive Economic Zone (EEZ) of the Union Territory and are being captured some time.

All the islands of this area are of volcanic origin. They have steep slopes on all sides and hence within a short distance from the shore, open sea conditions are encountered. The coast line of all the islands grouped together is around nearly 2000 km, which constitute approximately 25 per cent of the country's total coast line. The area of the continental shelf is about 35000 sq km.

Oceanography of Andaman Sea

Andaman Islands, which are a part of anticlinal belt, passing from Arakan Yoma in Mayanmar through Andaman and Nicobar Islands and Mentawai Islands, west of Sumatra separate the Andaman Sea from Bay of Bengal except the connections through Preparis Channel on the north with a depth of about 200 m, the great channel on the south, 1800 m deep and the ten degree channel in the middle with a depth of about 800 m. A north-south arc of volcanic arcs and sea mounts, including Barren and Narcondum Islands in the Andaman Sea separates the large central basin from two smaller basins to the north and south. The fall of slope is also steep on the western sides of the islands.

The surface layer of the Andaman Sea is generally well-mixed to a depth of 100 meters, but a sharp decrease in temperature occur in the eastern side of the Andaman and Nicobar Islands at depths of 25 to 50 m. There is a wide fluctuation in surface salinity in the Andaman Sea, low salinity occurring during May to November. A recovery period in December and January and high salinity occur from February to April. Upwellings occur in Andaman Sea, with the aid of wind stress and Coriolis force. Tides in Andaman Sea are semi-diurnal. Spring tides range from 1 to 2 meters in the Andaman Islands.

Thermal Structure

The depths of the surface mixed layer varied from 44 to 58 m along the western section, with deeper layers towards south. The surface temperature varied from 27 to 28.5°Celsius, with lower temperatures towards north. Inversions at all places of the order of 0.2 to 0.75°Celsius were observed. At lat. 14°N and 93°E, on the western slope of the island arc, recorded the largest inversion of 1.4°Celsius in temperature from a depth of 31 to 60 m.

Along the eastern section, thickness of the isothermal layer varied from 26 to 50 m, shallower at lat.12°05 minute N and long. 93°30 minute E. It is observed that the inversions are stronger southwards starting from 13°30 minute N and 93°35 minute

E to 12°05 minute N and 93°30 minute E. The surface temperature ranged 28 to 28.5°Celsius.

In the western section, the main thermocline extended up to 200 to 350 m and along the eastern section 180 to 280 m, whereas, the thickness of the thermocline varied from 150 to 200 m and 140 to 250 m correspondingly. The isotherm in the thermocline, in general, are sloping down towards south in the eastern section. In both the sections, 12°Celsius isotherm, mostly demarcate the thermocline.

There existed a difference in the temperature profiles between the eastern and western sections beyond 1500 m. In the western section, the temperature decreased from 5°, down the deeper layers. At 2000 m depth, it is 2.84°Celsius at 14°N and 92°30 minute E near Preparis Channel (Lat. 13.5-14°N and Long. 92-93.5°E); whereas in the eastern section, the temperature beyond 1500 m depth were showing much higher values than the corresponding section west of the islands. At 1750 m, Lat. 13°N and Long. 93°25 minute E and Lat. 9°52 minute N and 93°30 minute E, the temperatures were 5.10° and 5.04°Celsius respectively. Still deeper at a depth of 2500 m, higher temperature was recorded at Lat. 9°12 minute N and 93°30 minute E. The higher temperatures recorded in these basins may be associated with heat flow from beneath or with volcanic activity associated with the islands and sea mounts. The islands Barren and Narcondam, are still active emitting fumes and lava occasionally. Higher temperatures in these basins may also be attributed to the enclosed nature of the basin with sill depths around 1300 m, above which the temperature distribution is similar on either side of the island chain (Ramesh Babu and Sastry, 1976).

Salinity

Lowest salinities have been observed at the surface varying between 31.2 to 32.6 ppt, along the eastern section and 31.87 to 32.15 ppt along the western section. Amongst the surface values, higher salinities were observed at Lat. 11°20 minute N and Long. 92°08 minute E and Lat. 9°52 minute N and Long. 93°30 minute E. The salinity profiles at most of the places of both the sides of the island chain are similar.

The major salinity features can be grouped into three divisions;

1. Surface isohaline layer, whose depth varied from 11 to 31 m along the western section and 19 to 48 m along the eastern section.

2. The halocline located in two parts; one the steep upper halocline, the other the less steep lower halocline. The depth of the sharp upper halocline beneath the isohaline layer, varied from 156 to 194 m along the western section and 133 to 194 m along the eastern section. The above figures indicate the depth where the feature ends. The salinity at the end of the halocline varied from 34.72 to 34.84 ppt along the eastern section and 34.76 to 34.80 ppt along the western section.

3. The value of the salinity along the salinity maximum varied from 34.85 to 34.87 ppt along both the sections. The layer of high salinity is lying at about 300 to 400 m depth. Beyond this layer, the salinity decreased with a very slow rate of 0.00016 ppt per 100 m on an average.

Wave Characteristics

The study of wave records obtained by the ship borne wave recorder of 15 minutes duration around Andaman and Nicobar Islands (45 records) revealed that in the sea around Andaman and Nicobar Islands, the significant wave heights and zero crossing periods are almost uniformly distributed from 0.6 to 1.4 m and 6 to 12 seconds respectively.

The wave records were analyzed following the method suggested by Draper (1966). For each record, the following parameters were evaluated. The height of the highest crest and depth of the lowest trough from the mean water line drawn visually, the number of upward zero-crossings and the number of crests. The height of the highest wave was then computed as the height of the highest crest plus depth of the lowest trough and corrected for depth attenuation effect due to the sensors being submerged below the sea surface. The significant wave height, the mean zero crossing perion and the spectral width parameter were also computed.

There is a fairly wide scatter of wave heights and periods. The character of the scatter diagram indicates a lack of dominance of local seas, which would have been the case if the highest waves were associated with shorter periods.

The spectral width parameter varies mainly from 6.0 to 7.75 with a dominance around 7.0. This distribution of spectral with parameter suggests that the wave component do not cover a very wide range of frequencies, that is, the long waves will carry few short waves on top of them. It can be concluded that in the area the wave characteristics are swell conditions with only slight influence of local sea. In cases, where wide range of wave frequencies exists, the spectral width parameter would be greater than about 0.85. In case of simple swell, the spectral width parameter would be less than about 0.40.

In contrast, along the 5°30 minute N latitude, the scatter diagram shows a slight dominance of local sea, since the higher wave heights are associated with lower periods. The spectral width parameter varies from 5.75 to 7.25 with a dominance between 6.25 and 7.0. This also suggests slight local sea conditions superimposed on swell conditions. The zero crossing period was mainly between 8 and 9 seconds and the significant wave height between 0.8 and 1.4.

During the same period the wave condition in the Bay of Bengal are best represented by significant wave heights between 0.5 m and 2.0 m and zero crossing periods between 5 and 9 seconds. In general, 80 per cent of the waves have heights less than 1.5 m.

It can be concluded that during the month of February, the wave characteristics around Andaman and Nicobar Islands are mainly swell conditions on which slight local sea characteristics are superimposed.

Surface Water around the Little Andaman Island

Little Andaman lies just north of the 10°channel, south of which lies the Nicobar group of islands. Its northern side is shallow with rocky bottom and is not fully charted for navigation. Oriented in a north-south direction, between latitude 10°55 minute and 10°30 minute N and longitude 92°22 minute and 92°36 minute E, this oval-shaped island is indented with several bays (Dugong, Hut, south and West bays and the Jackson creek). Though its western side has rather irregular bottom topography, compared to the eastern side, the shape of the island suits almost ideally to model studies of flow past a submerged obstacle with certain assumptions similar to such works as have been carried out by Hogg (1972) for Bermuda. However, such studies need a series of observational data on temperature, salinity, stratification, current measurements, wind stress etc, which are utterly lacking in the waters around Little Andaman.

The surface waters of Bay of Bengal (surface column of water from the island coast extending east, south and westward up to depths of about 300 m), which occupy depths of 100-150 m are easily distinguished by their contrasting properties. Of these, three water masses; the northern dilute water, a transition water and the Southern Bay of Bengal water, it is the latter which occupy the regions surrounding the Andaman during January-February, and by March, the surface water is largely southern Bay of Bengal water, (Gakagher, 1966). This water has a sigma-t value between 21 and 22. However, the north-eastern part of the Andaman Sea shows lower density (less than 20), which is corroborated by the surface salinity decreasing in a north-easterly direction, towards the Irrawaddy delta as shown in the atlas by Wyrtkl (1971). The same atlas also shows that the western part of the Andaman group of islands, during February, has higher salinity than the eastern side, which also holds true for surface temperature. This demarcation gets slightly intensified towards March-April.

In the equatorial region the water in the deeper levels, that is generally between 100-1000 m, it is quite different, when one compares the waters of the western side (south of Arabian Sea) with the eastern waters (south of Bay of Bengal). Though according to Sharma (1976), the flow here is mainly zonal and the equatorial Indian Ocean water acts as a barrier to trans-equatorial movement during the north east monsoon period, low salinity water flows into the western Indian Ocean from the Pacific and the incursion of Banda Sea intermediate waters gives rise to salinity maxima at different levels. Going along with such observations, it is expected that even north of 50°N, such salinity maxima could be expected during the north east monsoon, and hence the homogeneous (variation of salinity less than 0.30 ppt) water between 100 and 1000 m, which is found during south east monsoon period could not be expected to prevail in the eastern equatorial waters and the 10°channel, south of Little Andaman Island, where the flow (NE monsoon) is towards the west, would necessarily show regions of inhomogeneity (salinity maxima and the like) in the deeper parts.

After this brief account of the surface waters of the Bay of Bengal and the equatorial waters during the south west and north east monsoons, which might provide a backdrop, the findings and observations regarding the surface waters around the Little Andaman Island during February would perhaps become a little more clear if, at the same time, that the fact of bottom topography around such islands very considerably influence the currents flowing past the island and that the mixing processes are greatly enhanced by bottom friction is considered.

During the peak north east monsoon (January-February), the surface current in the Bay of Bengal flows in a large anti-cyclonic gyre. The eastern part of this gyre flows past the Andaman group of islands and joins the steady westward flow in 10°channel (which lies just south of Little Andaman). During this period of observation, a fairly steady wind of moderate strength (5-11.5 kts) was blowing from between north and north, north east. The air temperature and the surface water temperature were also nearly constant. The surface current was found to be of the order of about 2 per cent of the surface wind. This was substantiated by the results obtained from the parachute drogue and the current meter, when operated off the Hut Bay. The flow pattern around the island was made from the six sections giving the temperature structure around the island (barring its northern side). The isotherm on these sections sloped up towards the coast on all the three sides without exception. Directions of flow are thus easily obtained following the principle that (in the Northern Hemisphere) the isotherm slope up towards the coast on the right, when looking downstream.

Further away from the coast, however, the isotherms showed wavy natures, suggesting, at places, reversal of current directions (the possibility of the existence of internal waves). Such isotherms was identified in the three sections (II, III, IV), whereas they were identified in the upper layers only in the remaining two sections (I and V). It is interesting to note that though the flow around the island is southerly on the eastern side, east, south east to south east, on the western (lee) side and westerly on the southern side (as in the 10°channel), the flow pattern near the coast is in variance with this general pattern and forms zones of shear and eddies (near bays and shoals). These departures can not be attributed either to the tidal currents or the land-sea breeze system; the latter being almost non-existent and the former of very low magnitude. For the western side it could probably be said that the lee-effect produces the northerly flow on the south-eastern side of the island between the coast and the Dalrymple Bank. The same, however, may not be the case in the eastern side of the island, where it could be either due to the bottom topography (frictional influence) or a density variation caused by mixing as a combination of both.

In respect of the thermal and haline characteristics of the waters around this island, it should be mentioned at the outset that "Around the group of Andaman Islands, the continental shelf is very much wider on the western than on the eastern side and along the western edge of the shelf is a series of coral banks and islands"(Sewell, 1925). The above statement holds good for the Little Andaman Island very significantly. The effect of such bottom topography is clearly reflected, apart from the modifications that are superimposed on it by the lee-side effect of the flow around the island. The values of the salinity maxima are almost the same on all sides, whereas the depths at which they occur are not. On the western side, the salinity maxima occurs at 350 m, whereas on the eastern side, they are found at 330 m, which rises to 310 m on the southern side. The depth of halocline follows almost a similar pattern with slight variations. The depth of isohaline layer again shows the lee effect on the western side, which is as shallow as 15 m compared to 40 m on the southern and 25 m on the south-eastern side.

Regarding the nature of the physical properties of the surface waters around Little Andaman, the surface mixed layer decreases towards the coast. At the furthest off-shore, the mixed layer extends to 30 to 50 meters.

The thermocline has a moderately developed gradient and extends to the maximum depth of 125 m, where the temperature is of the order of 15 to 16° Celsius.

The surface waters appear to be fairly stratified. Salinity at the surface is around 33.5 ppt, with a sub-surface maxima of 34.3 ppt (where the pycnocline starts) at about 40 to 50 m. Below 150 m the salinity remains fairly stable, between 34.8 and 34.9 ppt. The surface temperature is fairly uniform and ranges between 28.25 and 28.5° Celsius. (2 m below the surface).

Temperature-Depth Profile

On the northern parts of the Little Andaman, the surface water column is warmer on the eastern compared to the western side. The mixed layer is sufficiently deep on the west as against the eastern side.

The isotherms show an upward tilt towards the near shore areas on both these sides. The temperature gradients within the thermocline are more or less uniform.

On the southern side, the picture is slightly different. The general pattern of the inclination of the isotherms is quite similar to that of the northern section. On the southern most side of the island, where the section radiate out towards the 10°channel, the sea surface temperatures are slightly higher on the eastward section. The temperature gradients are weaker on the southeasterly section below 150 m depth.

In the eastern side of the island, there is a marked temperature inversion of low magnitude at 20 m and 30 m with the depth of inversion increasing towards the south. A similar feature, with much less pronounced inverse gradient, occur around 50 m depth. Below 95 m depth the temperature gradient vary significantly though they are more or less uniform.

The 18° isotherm observed around 120 m depth on the north shows a dip at about 95 m on the south. Similarly the 15° isotherm moves up to 120 m (south) from 160 m (north). This feature coupled with a near isothermal layer with temperature 13.25° Celsius is clearly seen to move up to 150 m from 250 m depth from north to south on the eastern side of the island.

In the western side, the mixed layer column is very much shallow (12-15 m) and the thermocline begins at around 50 m depth. The temperature gradients are almost identical within the temperature inversion though noticeably present are of very low magnitude. An isothermal (26.7° Celsius) layer is present between 65 to 70 m.

A deepening in the depth of the isotherms with reference to the corresponding values on the eastern side is clearly discernible here. For example, 18° isotherm observed at 100 m on the west is located at 120 m, on the northern side and 95 m on the southern side. 15° isotherm is at about 140 m on the west, whereas on the eastern side they are found at 160 m and 120 m on north and south end of the eastern side of the island respectively.

Figure 4.1: Marine Micro-world

On the southern side, inversion are present at about 5 and 40 m, though they are much less conspicuous. The distribution of temperature with depth is relatively smooth except at 55-65 m level, 140 m and at 175 m when compared to other locations.

From salt and temperature distribution record (CSTD) many steps and inversions and differential lowering or rising of isotherms at different sections that both horizontal and vertical mixing are at play. However, the effect of wind which is only of moderate strength, and that of the tidal currents could only be of minor nature. It is also quite possible that turbulence and formation of eddies are due mainly to the shape of the island and the rough bottom topography play vital diffusion, that of salt and heat.

Chapter 5

Microbial Oceanography

Marine ecosystems are complex and dynamic. A mechanistic understanding of the susceptibility of marine ecosystems to global environmental variability and climate change driven by green house gases will require a comprehensive description of several factors. These include marine physical, chemical and biological interactions including thresholds, negative and positive feed back mechanisms and other non-linear interactions. The fluxes of matter and energy and the microbes that mediate them, are of central importance in the ocean, yet remain poorly understood. Detailed field studies over the past three decades have established the current "microbial loop" hypothesis, wherein microbes have a central position in the conversion of dissolved organic matter into higher trophic levels. An explicit and comprehensive test of the microbial loop hypothesis, however, has not yet been achieved. In addition, the central role of microbial activities in maintaining the oxidative state of our planet, and biogeochemical cycles other than the carbon cycle are not well captured in the current microbial loop hypothesis. Significant obstacles remain to be overcome in the measurement and modeling of ocean microbial processes.

The correlation between organism-and-habitat specific genomic features and other physical, chemical and biotic variables has the potential to refine our understanding of microbial and biogeochemical process in ocean systems.

In the past thirty five years, there has been a remarkable growth in understanding of marine micro-biota. During this time, researchers have recognized the crucial role that microbes play in ocean ecosystems. This stems in part from technical advances, such as, improved epifluorescence microscope techniques and ATP-based biomass metrics, that have revealed bacterio-plankton standing stocks that are several orders of magnitude greater than had been estimated by variable counting techniques. The use of radio tracers to estimate planktonic bacterial growth rates and turnover has also led to revised qualitative and quantitative models of microbial contributions to marine food webs. Research using quantitative auto-fluorescent cell counts and flow cytometry eventually led to the discovery of abundant photo-autotrophic picoplankton, including *Synechococcus* and *Prochlorococcus* species, that dominate photosynthetic activities in open ocean gyres. Around the same time, hydrothermal vents and their rich macrofauna and microflora were discovered, as well as the first bacterial isolates with an obligate growth requirement for elevated hydrostatic pressure. The development of cultivation-independent phylogenetic surveys using ribosomal RNA sequencing and fluorescence *in situ* hybridization set the stage for contemporary environmental genomic studies, soon after their development. RNA-based phylogenetic survey techniques using the technique polymerase chain reaction (PCR) revealed the widespread distribution and abundance of several previously unrecognized marine microbial groups, including *Pelagibacter* (SAR II) abundant new groups of planktonic marine Archaea and eukaryotic picoplankton.

More recently, the prevalence of bacterio-chlorophyll- containing and rhodopsin-containing bacterio-plankton was recognized, providing new perspectives on the nature of light-use strategies in ocean surface waters. Surprisingly, it has only very recently been realized that viral particles can exceed total microbial cell numbers by an order of magnitude in marine plankton and they represent potentially important vectors of bacterio-plankton mortality and lateral gene transfer.

The physical, chemical and biological changes that occur along a transect from the coast to the open sea are profound. It is too early to generalize extensively about specific biological properties of coastal bacterio-plankton differentiating them from their open ocean relatives. Some bacterial groups such as, Alpha-Proteobacteria in the *Roseobacter* clade seem more abundant in near shore waters than the open sea. The genome of *Silicibacter pomeroyi*, a member of *Roseabacter* clade, has some adaptations that may attune this bacterium to life in coastal environments. On the basis of its genome content, *S. pomeroyi* seems to consume not only dissolved organic carbon, but also reduced inorganic compounds, such as, carbon monoxide and sulphur, it harbors the genes required to generate energy from these compounds. Preliminary experiments confirmed that carbon monoxide and reduced sulphur compounds are important substrates for *S. pomeroyi* metabolism. In common with *T. pseudonana*, this coastal bacterium is equipped with a wide variety of nutrient transporters for uptake of different nitrogenous compounds, such as, amino-acids, ammonium and urea.

Microbiologists find it convenient to organize microbial groups into functional guilds involved in the specific processes, for instance, sulphate-reducers, methanogens and oxygenic photo-autotrophs. It is now well recognized that within any such group or indeed even within a species, tremendous biological diversity exists. Closely related physiological and genetic variants that seem to be "tuned" to specific ecological conditions have been referred to as "ecotypes".

One of the best documented examples of environmentally tuned ecotypes are photo-autotrophic *Prochlorococcus* strains isolated from different depths. The genome sequences of these ecotypes reveal some of the unique adaptive strategies employed in different parts of the depth continuum. Genes found in the low light strain, but not in high light adapted relative included those for nitrite transport and assimilation. This is consistent with the types of nutrient available at the base of the euphotic zone. High light adapted strain encoded many more genes for high light-inducible proteins than its low-light adapted cousin. By contrast, the low light adapted strain contained more genes associated with photosynthesis.

Biological processes occurring in natural microbial communities have diverse complex, interdependent intracellular and intercellular reactions. Much of this complexity is encoded in the structures, distribution and dynamics of interacting genomes in the environment. Recently, it has become possible to directly access the genomes of coexisting microbial species in natural communities en masse, without cultivation, using environmental genomic approaches. Variously termed environmental genomics, metagenomics or ecogenomics, cultivation independent genomic approaches can provide a new perspective on the naturally occurring microbial world.

Some of the first environmental genomic studies were focused on marine picoplankton. Initially large genomic fragments were cloned directly from marine microbial communities to survey genomic features of then un-cultivated phyto-genetic groups, such as, marine Archaea. Similar genomic surveys recently led to the discovery of rhodopsins in bacteria, a domain of life not anticipated to contain these photo-proteins. Subsequent field studies revealed that the bacterio-plankton rhodopsins were diverse, wide spread and abundant in the marine environment.

Ultimately, it may be possible to overlay microbial functional genes, metabolic pathways and taxonomic distributions upon other physical, chemical and biological oceanographic data to map microbial features onto oceanic provinces. Finally, the study of genome structure and dynamics *in situ* will provide the necessary information to interpret evolutionary processes, such as, genetic drift, recombination and lateral gene transfer, that drive microbial adaptation and divergence. These data in turn provide predictive tools to model the variable complexity of form, function and process that underpin the ocean's physical, chemical and biotic interactions.

Marine microbes are present at billions of cells per liter in sea water. They double every few days and are consumed at about the same rate by viral parasites and protistan predators. These activities capture and process energy and drive major elemental cycles. Hidden within these dynamic assemblages and diverse genomic structures are fundamental but, at present, incomplete lessons about environmental sensing, response and adaptation, gene regulation, species and community interactions and genomic plasticity and evolution. The genomic diversity, evolutionary dynamics and ecological processes, contained in these populations have global effects on the rates and flux of energy and matter in the sea, biogeochemical cycling, the Earth's atmospheric composition and global climate trends. Over the 3.8 billion years of life on Earth, microbes have been the stewards of geochemical balance and as biotic recorders of evolutionary history, their " biological memories" extend backwards further than other life forms. Additionally, they are Nature's biosensors *par excellence*, but we still need to decipher their code to interpret their outputs and understand the under-laying strategies and mechanisms of their survival in nature. The current convergence of microbiology, ecology, genomics and ocean science has the potential to be focused in unprecedented ways through the lens of microbial oceanography. By working together, molecular biologists, microbiologists and oceanographers have new opportunities to advance observation, method and theory, which together will better describe the living ocean system.

Chapter 6
Chemical Features

The Andaman Sea (partially isolated of the north-eastern Indian Ocean that lies enclosed between the coasts of Mayanmar, Thailand and Malaysia on the east and the chain of the Andaman and Nicobar Islands and Sumatra on the other) occupies an area of 6.02×100000 sq.km and a volume of 6.6×100000 cu km with a mean depth of 1096 m. It is connected to the Bay of Bengal by numerous channels, which are broadly, interruptions in the ridge that lies on the western boundary of the Andaman Sea. There are three main channels, whose depths increases as one proceeds from north to south. These channels are (i) The Preparis Channel, divided into north and south portions by the island of the same name; (ii) Ten degree channel, situated between the Andaman and the Nicobar groups of the islands and (iii) Great channel, lying between Great Nicobar Island and Sumatra. Towards the south, between Malaysia and Sumatra, the Strait of Malacca maintains the connection of Pacific Ocean water, flowing through the South China Sea, and Bay of Bengal. To get a picture of totality as the island arcs will influence the water characteristics, both on their east and west, the Andaman Sea have been defined as the part of the south-eastern Bay of Bengal lying between 6°and 14°North east of longitude 91°31 minute East.

Hydrographical Features

Very little is known about water masses of the Andaman Sea. Sewell (1932) suggested that the five different water masses, which were the characteristics in the south-western Indian Ocean from 55°S to 8°N should also be present in the Andaman Sea region. Ramesh Babu and Sastry (1976) suggested that water mass characteristics in the Andaman Sea and the Bay of Bengal are the same above 1300 m; while below this depth they may be different because of the presence of a sill at depth. They also suggested that the density distribution in the Andaman Sea is governed more by the distribution of salinity rather than temperature distribution.

The vertical variations on the eastern (13°N, 93°25 minute E) and western (13°N, 91°58 minute E) sides of the islands, indicate a well mixed surface layer to about 100 m. Mixing across this layer could be seen to about 150 m. Below this layer, sharp changes in values indicate the presence of a water mass extending to about 400 m in thickness. The sill depth between the Bay of Bengal and the Andaman Sea lie at about 1300 m, and poor water exchange below this depth makes water masses in the central region on the eastern side of the Andaman Islands relatively stagnant. Shallow sill depths at Duncan Passage, between South and Little Andaman and Car Nicobar, also contribute to this phenomenon.

In the vertical distribution of dissolved oxygen along both the sections, the magnitude and thickness of the oxygen minimum layer is almost of the same order in both the transects, excepting the central region of the eastern section. Oxygen values lower than 0.2 ml/l have been observed here, which is spread over a considerable depth. Even at the deeper layers, oxygen values are lower than those at the corresponding depths in other regions. This clearly indicates that the central basin of the Andaman Sea is cut off from its northern and southern regions by shallow sills, which result in poor water exchange and relatively stagnant water with low oxygen concentration.

It can therefore, be concluded that water masses on the western and the eastern margins of Andaman and Nicobar Islands have the same characteristics to a depth of about 1500 m. Water exchange between the Bay of Bengal and the Andaman Sea takes place through the Preparis Channel in the north-south direction and deeper water masses in the Central Andaman Sea are separated from the rest of the region by shallow sills and are relatively stagnant.

Water characteristics were further examined by silicate-density relationship at all stations. The water masses from 0 to 100 m have silicate concentrations ranging from about zero to 25 micro gram/l. In the Arabian Sea, this range of silicate values was observed at 200 m. Water masses from 101 to 500 m, 501 to 1000 m, 1001 to 1500 m have silicate values of 25 to 50 micro gram at/l, 30 to 75 micro gram at/l and 75 to 97.5 micro gram at/l respectively. The deeper layer in this region of Bay of Bengal has lower silicate concentrations as compared to similar depths in the Arabian Sea. This is probably, the result of advective mixing between several water masses in the north-eastern Indian Ocean. It has to be some kind of mixing as the source of origin of water below 1000 m is the same all over the northern Indian Ocean.

Observations in some parts of the Bay of Bengal indicated the presence of one thick oxygen minimum layer stretching from 100 to 1000 m. Two oxygen minima, separated by a layer of relatively high oxygenated water, were observed in the south-western Bay of Bengal, while only one oxygen minimum was reported in the north-western Bay. In the Andaman Sea, only one oxygen minimum layer, with the minimum concentration in the northern section has been reported by Ramesh Babu and Sastry, 1976.

The thickness of the oxygen minimum layer (1 ml/l) varies from 150 to 600 m at the stations on the north, east and south of the islands, with the lowest values recorded at the station on the north (0.1 ml/l). At the station on the western side, however, the oxygen minimum layer extends from 100 to 800 m, which is very similar to conditions in the open Bay of Bengal.

Ranges of values of nutrients are almost the same at all the stations. However, at stations on the east and the south, a maximum each for phosphate-phosphorus and nitrate-nitrogen at an intermediate depth with almost uniform concentrations extending to the bottom are noted. A similar feature can also be observed for pH values. These indicate the stagnant nature of the deeper waters at certain parts of Andaman Sea, due to poor renewal of water over the sill depth.

The magnitude of values for oxygen, phosphate-phosphorus and nitrate-nitrogen at these stations are relatively higher than those recorded for the corresponding depths in the Arabian Sea. The only exception is silicate-silicon, which is lower in the Bay of Bengal, as a whole, than the Arabian Sea. These indicate that the rate of oxidation of organic matter in depths of Bay of Bengal, including the Andaman Sea, proceeds at a faster rate resulting in quick depletion of dissolved oxygen. The faster rates of oxidation and oxygen depletion result in the sinking of organic matter in un-decomposed or partially decomposed state. In the upper 50 meters, nitrate-nitrogen is absent, while some phosphate-phosphorus is still present in water. At stations, nearer the coast, both nitrate-nitrogen and phosphate-phosphorus have been observed to be absent. This indicate that nitrate-nitrogen, in general, acts as a limiting factor for productivity in this region. Low availability and practically insignificant addition from land make the phosphate-phosphorus, at stations, nearer the coast, disappear also from the water column during periods of peak productivity.

On the western side (13°N, 91°58 minute E), oxygen saturation values were -3 to + 15 per cent at the surface, +1 to +14 per cent at 25 m and -5 to +10 per cent at 50 m. On the eastern side these values were -2 to +8 per cent at the surface, -8 to +5 per cent at 25 m and -63 to +5 per cent at 50 m. Conditions on the western side have been observed to be more or less normal for the open Bay of Bengal. On the eastern side, there is quite frequently, between 25 and 50 m, either a sharp fall in temperature associated with insignificant change in salinity or a temperature inversion associated with a sharp increase in salinity. In both the cases, the stability gradient in the water column will prevent molecular and advective diffusions of oxygen resulting in relatively higher values for under-saturation. This stability gradient might be also the result of river run-off.

The ratios of change in concentrations of nutrients (carbon, nitrogen and phosphate) in water are the same as their ratios of concentrations in plankton. In other words, calculation of the former will give an idea of the latter. An earlier study indicated AOU (Apparent Oxygen Utilization) : N : P in the waters of western Bay of Bengal as 260:12.4:1, which should result in C:N:P ratio of 105:12.4:1 for plankton.

The nitrate would act as a limiting factor for the production of phytoplankton in the Andaman Sea, which is in broad agreement with observations from other marine areas.

The upper 20 meter water mass gets the maximum influence of river run-off. Rivers draining into the region contain high amount of silicate (mean value of silicate in the Asian rivers is 11.7 mg/l). It is therefore, very probable that silicate seldom or never reaches zero in the surface layers of the Bay of Bengal. Silicate concentrations in the Bay of Bengal are less than those at the corresponding depths in the Arabian Sea.

Normally silicate-phosphate do not show linear relation in the ocean. However, in some areas, a linear relationship between these two variables has been approximated at selected depth intervals. The slope of regression line between 75 m and AOU maximum for silicate-phosphate relationship for the present set of values (obtained from Andaman Sea) gives a ratio of 29:1 by atoms.

In Arabian Sea waters, this ratio has been observed to vary from 36-40:1 by atoms. The lower value of the ratio is due to lower concentrations of silicate in the Bay of Bengal, as there is no appreciable difference between the concentrations of phosphate-phosphorus in the two areas.

Biochemical Relationships

Oxidative ratios have been applied to deduce the composition of plankton in Andaman Sea. In the Andaman Sea and in the Bay of Bengal in general, silicate and nitrate play a regulating role in making the composition of phytoplankton, somewhat different from the same in oceanic areas. This is the effect of river run-off.

The Arabian Sea and the Bay of Bengal have been observed to be areas, where the process of denitrification takes place at an intermediate depth. This process has been identified to be at the stage of ammonia, without reaching its completion in the Arabian Sea and at the concluding stage of gaseous nitrogen in the Bay of Bengal.

The average concentrations of ammonia between 75 m and AOU (Apparent Oxygen Utilization) maximum in the waters of Andaman Sea has been 2.05 micro gram-at/l. This indicates that the process of denitrification in the Andaman Sea exists at the intermediate stage of ammonia without proceeding to its completion.

The average concentration of nitrate from 75 m to AOU maximum layer is 20.2 micro gram-at/l with hardly any difference between its concentrations in the waters on the eastern and western sides of the Andaman and Nicobar Islands. This would mean that about 39 per cent of the available nitrate has been lost due to denitrification.

Figures of 40 per cent ranging from 30 to 40 per cent at the depth interval 100-800 m have been reported from observations in other areas of the Bay of Bengal (Naqvi, *et al*, 1978). In the Arabian Sea denitrification ranges from 40-45 per cent. Thus, the rate of denitrification in the Andaman Sea is in good agreement with similar rates for the different regions of the northern Indian Ocean.

Calcium, Magnesium and Fluoride Concentrations

The Andaman Sea is contained by a physiographic basin, which covers 798000 sq km. It is bordered by Mayanmar, Thailand and North-West coast of Sumatra. In addition, the Andaman and Nicobar ridge on its west separates it from the Bay of Bengal. In the Andaman Sea, there is a tremendous influx of fresh water from the Irrawaddy and Salween rivers, especially during south-west monsoon (Rodolfo,1966).

Similarly, the northern part of the Indian Ocean is also influenced by a large volume of river run-off. The waters of the Andaman Sea and the Bay of Bengal communicate through several channels, resulting in the possibility of the characteristics of the upper layers in the water in this area being modified by mixing and run-off.

Calcium

The average calcium concentration in the Andaman Sea was found to be 421±1.76 mg/kg with an average calcium/chlorine ratio of 0.02224±0.000045.

The values of calcium concentration, calcium/chlorinity and salinity in the eastern part (13°N, 93°25 minute E; 12°05 minute N, 93°30 minute E; 10°55 minute N, 93°30 minute E; 9°52 minute N, 93°30 minute E) of the Andaman Island, were pooled together and averaged accordingly, thereby making it possible to draw a generalized vertical profile, indicating the distribution of these parameters with depth.

Although calcium concentration increases steadily with depth in the first 150 m, the calcium/chlorine ratio decreases, owing to a more rapid increase in salinity, which reaches a maximum around 150-200 m. Thereafter, the salinity and the calcium concentration maintain a fairly constant trend, decreasing slightly, below 1000 m. The decrease in calcium/chlorine ratio in the upper water column, may also be due to the fact, that in the upper few hundred meters in the ocean conditions for lime-secreting organisms seem to exist.

The corresponding calcium/chlorine profile follow almost the same trend below 200 m, with increase and thereafter decrease below 1200 m. The contributing factors to such a pattern in the calcium/chlorine profile may be attributed to the fluctuating calcium concentrations, rather than the chlorinity, which remains uniformly constant below 200 m.

The present value of calcium/chlorine ratio (0.02224) is in close agreement with the mean values reported by Sen Gupta *et al* (1978) for the Bay of Bengal, Gulf of Mannar and Arabian Sea (0.02229). However, these values are higher than those reported by other workers for all oceans (Culkin and Cox, 1966; Riley and Tongudai, 1967), which are 0.02126 and 0.02128 respectively.

Naqvi and Reddy (1979) reported a value of 0.02168 for oceanic samples in the Arabian Sea. The high ratio may be the effect of dilution, due to river run-off and mixing between the two waters of Andaman Sea and Bay of Bengal.

Average of calcium ion concentration of the surface waters on the eastern side of Andaman Islands is 400±7 mg/kg with a average chlorinity of 17.80±0.19 per cent resulting in a calcium/chlorine ratio of 0.02245. Calcium ion concentration in the waters of river Ganges has been reported as 18.1 mg/l, while the world average is 15 mg/l (Livingstone, 1963). The lower concentration of calcium ion at the surface waters of the Andaman Sea, is therefore, the effect of run-off from Mayanmar and Thailand. The surface mean concentration of calcium ion in waters of off shore regions of Bay of Bengal is 414 mg/l at 18.4 ppt chlorine, resulting in a calcium/chlorine ratio of 0.02245. The similarity in the ratio between the surface waters of both the areas is due to the lower concentrations of both calcium and chlorine ppt in the Andaman Sea waters.

In coral reef areas, it is expected that enrichment of deeper waters with respect to calcium ion due to the dissolution from the bottom. The Andaman Sea is receiving a large volume of river run-off and having coralline rock on sand at most of the places. It is therefore, thought that it would be interesting to study enrichment of calcium ion in sub-surface waters with reference to the surface waters at places along 93°30 minute E meridian. The northern part of this section is separated from the southern part by a shallow bank (Invisible Bank). Waters starting from 100 m and up to about 200 m are enriched by 5-10 per cent with reference to the surface.

The enrichment is 10 per cent at 200 m on the Invisible Bank, south of which, this feature decreases. This enrichment is about an order higher than what it is in the Pacific, along 150°W and in the Sargasso Sea (Wilson, 1975). The difference is due to the coralline bottom in the Andaman Sea and large volume of river run-off at the surface.

The mean value of calcium/chlorinity ratio (0.02229±0.0009) in Andaman Sea (Lat. 12-13°N, Long. 92°30 minute E) agrees remarkably with that observed for the northern Indian Ocean. The high value is due mainly to the low chlorinity values as compared to calcium in the area. This element has a relatively high variability due to its involvement with the biosphere and carbonate system in the ocean.

Magnesium

Magnesium is the second most abundant cation in sea water. It is referred to as a "conservative" element. Its concentration in sea water at 35 ppt salinity is 1209 mg/kg bearing a magnesium/chlorinity ratio of 0.0665 (Wlison, 1975).

The principal source of dissolved magnesium in the oceans is sub-aerial weathering on the continents. Average river water contains 4.1 mg/l magnesium (Livingstone, 1963), and although it does not appear to take part in the biological cycle, geochemical processes do take place as a result of diagensis and weathering. However, due to its conservative nature in the sea water, its ratio to the chlorinity is more or less constant, except in near shore waters, where fresh water modifies the relationships between the ions.

Studies of water samples of Andaman Sea along the longitudinal transects, (92°30 minute E and 93°30 minute E) revealed that the vertical distribution of magnesium/chlorinity shows a pronounced maximum at depths of 75 m (0.06722). Below this depth, the magnesium/chlorinity curve decreases, and thereafter varied spatially with depth. A general decrease in Mg/Cl occurs below 1000 m. The maximum at 75 m may be due to water mass effect. In Andaman Sea, below 200 m, the Mg/Cl ratio curve fluctuates non-uniformly; 0.06621 at 200 m to 0.06658 at 600 m to 0.06642 at 2750 m.

The magnesium concentration gradually increases to a depth of 100 m in Andaman Sea. Thereafter below this, it remains uniform, decreasing very slightly below 1000 m.

The average value of magnesium in Andaman Sea was 1268±5.03 mg/kg, is lower than that reported by Sen Gupta *et al*, 1281±14 mg/kg and 1290 mg/kg for ocean waters (Wilson, 1975).

The average magnesium concentration at the surface stations on the eastern side of Andaman Islands was found to be 1185±11 mg/kg at 17.80±0.19 ppt chlorine resulting in Mg/Cl ratio of 0.06652. It is apparent that the low magnesium concentration is due to the influence of river water in Andaman Sea.

Fluoride

Fluoride is a semi-conservative constituent of sea water, yet its concentration lies more or less close to 1.4 mg/l at 35 ppt salinity. It bears a ratio of 7×10^{-5} to chlorinity. However, abnormalities have been reported from time to time.

In Andaman Sea, an overall average of 1.38±0.01 mg/l for fluoride concentration and a mean fluoride/chlorinity ratio of (7.15±0.05). 10^{-5} was obtained by analyzing 150 samples from different places.

The fluoride (mg/l) and F/Cl ratio, in the eastern region of Andaman Sea are higher than the fluoride (mg/l) and F/Cl values in western region, where the waters are influenced by the Bay of Bengal water masses. The high fluoride/chlorinity ratios in the rivers are caused by the fact that the river water holds fluoride in appreciable concentrations with very little or no chloride. Mixing of river water with sea water will therefore, increase the fluoride concentration and the F/Cl ratio. The F/Cl ratio in the Andaman Sea is regulated by river run-off.

In the eastern region of Andaman Sea, the vertical variation of fluoride with depth shows that the concentration of fluoride increases with depth, the first notable maximum occurring at a depth of 500 m. Above this depth between 150 to 400 m, the fluoride concentration remains fairly constant. This 150-400 m correspond to the zone of maximum oxygen concentration. Below 500 m, the fluoride concentration decreases slightly, and then increases marginally to the second maximum at 1000 m, thereafter it remains fairly constant, although a slight decrease is noticed.

The concentration of fluoride in the water column, below 600 m is higher than the concentration of fluoride in the water column at 75 m to 500 m. This increase in fluoride concentration is associated with the increase in oxygen concentration (600 m-2750 m) and corresponds to the enrichment of calcium. Fluoride can be lost to the bottom as its ion-pair CaF. Higher values of both indicate that fluoride is not precipitating in the Andaman Sea.

The overall concentration of fluoride in western region of Andaman Sea is lower. The vertical profile of fluoride and oxygen concentrations with depth follow almost the same trend as that of eastern region, although with exception it may be said that the trend in fluoride below 75 m is much more uniform. A maximum in fluoride value is observed at thermocline (75 m), while a slight minimum of fluoride concentration occurs at 1000 m, below which it increases gradually.

The upper layers (0-100 m) have a comparatively lower fluoride concentrations than the 100-1000 m and 1000-2000 m layers, in both eastern and western regions of Andaman Sea. The effect is due to the dilution of sea water. In the eastern region, the values of fluoride and fluoride/chlorinity are much higher than of western region of Andaman Sea. These fluoride concentrations throughout the whole eastern region may be due to different water mass characteristics in this area as a result of geographical variation.

However, high fluoride concentration at deeper waters may be due to the re-suspension of the colloidal fraction of the sediment (Brewer *et al*, 1969).

The depth-wise averages of fluoride, chlorine and F/Cl values from both eastern and western region of Andaman Sea reveal that, between 150-600 m, the zone of low oxygen concentrations in both the region, no depletion in fluoride concentration is noticed. On the contrary, the concentrations in both the regions remain fairly constant, increasing slightly in deeper water column. In deeper waters, below 600 m, fluoride concentrations tend to increase much more in eastern region than in western region.

Kullenberg and Sengupta (1973) reported that fluoride in the marine environment is regulated by evaporation, precipitation, settling particles and river run-off. The effect of dilution by river run-off and the resultant removal of fluoride from the water column should always be felt most in inshore waters.

Chapter 7

Ecosystems

Marine Ecosystems

Oceans cover over 70 per cent of the earth's surface area and account for 99 per cent of the volume that is known to sustain life. Coastal ecosystems, such as, estuaries, coral refs and mangrove forests, also contain significant diversity and are important to coastal communities.

In marine and coastal habitats, the biodiversity is highly pronounced. These habitats range from shallow coral reefs to the deep dark ocean floors, hard rock and soft sediments. Coral reefs are already known to be among the richest habitats in species diversity on the earth. The deep sea bottom, dark and subjected to very high pressure, is now known to be the dwelling place of millions of species of small invertebrates, including crustaceans, mollusks and worms changing their body forms and structures fitting to the habitat. Recently scientists have discovered numerous new species from apparently productive mid-twalight region.

Regardless of species count, marine animals are more diverse, than the land animals at the higher phyletic levels of evolutionary and taxonomic differentiation. Marine animals exhibit a correspondingly greater range of body forms and structures than are found among terrestrial species.

The biota of marine habitats also exhibit a diversity of survival strategies not found on land. The numerous plankton of the ocean drift passively in the water, depending on ocean currents to transport them to nutritional sources and new habitats. Filter feeders sieve plankton and other floating material for food. They range from microscopic zooplankton through barnacles and sea anemones to baleen whales.

Due to difficulties of explorations and collections, marine bio-diversity is less known. Very little is known about marine life, even in the most familiar seascapes.

For example, among twenty two phyla of marine meiofauna identified, only two were recognized in the past two decades. These animals, barely visible to the naked eye, live on grains of coastal and ocean bottom sands. Up to 10,000 such animals can be found in a single handful of wet sands. Yet only recently scientists begun to suspect the important role they play in the marine ecosystems. They are a major source of food for shrimp and bottom feeding fish, and they consume detritus and degrade pollutants in sea water that filter through coastal and near shore sands.

Marine environments encompass an impressive diversity of ecosystems and habitats. Coral reefs, among the earth's largest and oldest structures created by living organisms, are dwelling place to dense concentrations of species and complex webs of inter-species interactions. Some coastal habitats, such as, marshes, mangrove forests, and sea grass beds are characterized by high biological production. Though the species diversity in these ecosystems is relatively less, they are important in terms of fisheries and other ecosystem services they provide (for example, shore line stability in case of mangroves and sustenance of endangered species like, dugong in the sea grass beds).

Coastal ecosystems, such as estuaries, lagoons, mangroves, back waters, salt marshes, rocky coasts, sandy stretches and coral reefs, which are characterized by unique biotic and abiotic properties and processes, play a vital role in India's economy by virtue of their resources, productive habitats and rich biodiversity.

Andaman and Nicobar Islands have coastline of 1962 km. Realizing the importance of the coastal ecosystems and their multiple uses, the ever increasing human population exploit not only the biological resources, but also interfere with and modify the basic coastal processes.

Mangrove Ecosystems

Mangroves is a general term used to describe a variety of tropical inshore communities dominated by several species of trees or shrubs, that grow in salt water. It comprises of 12 genera of flowering plants in eight different families, the most dominant genera being *Rhizophora, Avicennia, Bruguiera,* and *Sonneratia.* Between latitudes 30°N and 30°S, the shoreline marsh vegetation is replaced by a community of mangroves. They form highly productive ecosystems as the inorganic nutrients brought in by the incoming fresh water from land run-off, are trapped to form the source of energy for many organisms. A mangrove ecosystem constitutes a reservoir, refuse, feeding ground and nursery for many useful and unique plants and animals confined to the region. Through the export of decomposable organic matter into adjacent coastal waters, the mangroves provide an important nutrient input and primary energy source of many tropical estuaries. The mangrove ecosystem also protects coastal areas from sea erosion and from the violent effects of cyclones and tropical storms. The warm calm water ways of mangroves provide shelter and rich food for many juveniles and larvae of fin fish and shell fish.

India has only 2.66 per cent of the world's mangrove covering an estimated area of 4827 sq km. Out of India's total area under the mangroves, 20 per cent are found on the Bay Islands of Andaman Sea (Andaman and Nicobar). The insular mangroves are present in the coastal areas of Andaman Sea (Andaman and Nicobar Islands), where many many tidal estuaries, small rivers, neritic islets and lagoons support a rich mangrove flora.

Mangroves is applied to the specific ecosystem of tidal forest in the inter-tidal zone in the tropics and subtropics. It is a peculiar habitat found at the interface between land and sea. Mangrove forests are among the world's most productive ecosystem, supporting genetically diverse groups of aquatic and terrestrial organisms. The system includes diversified habitats, such as, core forest, litter forest, mud flats, adjacent coral reefs and sea grass ecosystem. The highly variable habitat conditions make them profusely rich in biodiversity.

The richest mangrove communities occur in tropical and sub-tropical areas, between the 32°N and 33°S, latitudes, where the temperature ranges between 20 to 30° Celsius, having moderate to high monsoon precipitation (1000-3000 mm/annum), high humidity but with coastal aridity.

The total area of mangroves in India is about 6419 sq km, which is about 7 per cent of the world's total area of mangroves. The mangroves around Andaman Sea is about 1152 sq km.

The species diversity of mangrove flora is higher in the Indian mangrove ecosystem, compared to other mangrove ecosystems in the world. The Indian mangroves are represented by approximately 59 species (inclusive of some mangrove associates) from 29 families. Andaman and Nicobar Islands represent 43 species from 30 genera and 23 families.

Mangroves around Andaman Sea form dense forests on the shorelines, creating a secured habitat for a variety of fauna. Mangroves are transition ecosystems and give itself a richly diverse ecosystem. Mangroves may play a special role as a nursery habitat for juveniles of fish, whose adults occupy other habitats, such as, coral reefs and sea grass beds. They also serve as a breeding grounds for many fishes. The aerial roots, trunks, leaves, branches host other groups of organisms. The muddy or sandy sediments of the mangrove forest are home to a variety of epibenthic, infaunal and meiofaunal invertebrates. Because loose sediments surround them, the submerged mangroves roots, trunks and branches provide niche for epifaunal communities including bacteria, fungi, macro-algae and invertebrates. A number of crab species live among the roots, on the trunks or even forage in the canopy. Insects, reptiles, amphibians, birds and mammals thrive in the habitat and contribute to unique character of mangroves.

The total number of mangrove-inhabiting faunal species in India is 3111. This include 55 species of prawns and shrimps, 308 species of mollusks, 138 species of crabs, 745 species of other vertebrates, 546 species of fish, 7 species of fish parasites, 711 species of insects, 85 species of reptiles, 13 species of amphibians and 70 species of mammals.

Mangroves are the breeding and nursery grounds for several fish species. About 400 species of fish are reported to depend upon mangrove habitat. Besides, many other species visit the mangrove environment, some frequently and others occasionally. Some common species are scats, milk fish, mud skippers, mullets, cat fishes, perches etc. Mudskippers live on the mud flats associated with mangrove shores. These are fishes well adapted to alternating period of exposure to air and submersion and are seen frequently hopping along the mud at the water's edge. Thus respire under water like other fishes, but when out of water they gulp air. When submerged, they swim like other fish, but on land they move resorting to a series of skips. When a mudskipper is out of water, it carries in its expanded gill chamber a reserve of air, from which to extract oxygen. After a few minutes, when the reserve is exhausted, it is replenished from pool or from water in the burrows, which they dig.

Mangroves are the dwelling place for mollusks and crustaceans. About 20 species of shell fish and 229 species of crustaceans belonging to Indian mangroves have been recorded. High deposits of silt and detritus in the mangrove environment, a rich source of food is created, which is utilized by detritovorous organisms like crustaceans and mollusks. Many crustaceans in the mangroves make burrows, which are used for refuge, the feeding, as a source of water or for establishing a territory necessary for mating. Some may filter water through their burrows, feeding on suspended detritus and plankton, while others may breed there. These burrows play an important role in the mangroves, aerating, draining and turning the dense water-logged soil a direct benefit to the plants, which in turn, give them shelter.

Scylla serrata, the large edible swimming crab, inhabits the muddy bottom of mangrove estuaries, as well as coastal brackish water. It is a commercially important crab and is trapped in special nets throughout the country. Due to its habit of cutting stems of young plants, it is a pest for young plantations.

Fiddler crabs, in a mangrove area are characterized by the male, which are armored with a single huge pincer (claw), which is used as a courtship display tool rather than protection. The other pincer is small in the males and is used for feeding. Females have two small pincers of equal size. Fiddler crabs are semi-terrestrial and prefer to stay on protected sand and mud beaches of bays and estuaries. Their burrows are located in the inter-tidal zone, and at low tide the crabs come out for feeding and courting. Fish, large crabs, some mammals and birds prey upon fiddler crabs.

The mollusk, *Telescopium telescopium* is strongly associated with mangrove environment and is an indicator species of mangroves. The name is derived from the typical telescope like structure of the shell.

The zooplankton in the mangrove areas mostly include crustacean larvae. Larvae of several species are found in large quantities, since the mangroves are the breeding ground for a variety of organisms. Food in the form of suspended solids are plenty, while shelter is sought in the complex root systems of plants.

Olive Ridley (*Lepidochelys olivacea*) is the most common sea turtle in Indian waters. Other turtles in Indian waters are, the Green turtle (*Chelonia mydas*), Hawksbill sea turtle (*Eretmochelys imbricate*), Bibron's soft shell turtle (*Pelochelys bibrone*), Batagur turtle (*Batagur baska*) and Leather back turtle (*Dermochellys coriacea*). Sea turtles are valuable for meat, shell and plastron.

Dugong (*Dugong dugong*) or sea cow, though a frequent mangrove visitor of Andaman Sea, is not an exclusive mangrove dweller. It is found on the coastal shallow waters, where sea grass are abundant. It is a herbivorous sea mammal feeding alone or in a herd, in the day or night. It has a life expectancy of about 50 years. Dugongs are on the verge of extinction from Andaman Sea due to indiscriminate killing of the animals by Onge tribe for meat.

Otters are also visitors of mangrove in search of food and shelter. They are elegant and swift swimmer and constantly on the move. They feed on sea urchins, crabs, calms, mussels, squid, octopus and fish. Sea otter breed in the water itself.

Coral Reef Ecosystem

Coral reefs form the most dynamic ecosystem providing shelter and nourishment to many thousands marine flora and fauna. They are the protectors of the coast line of the maritime states. This unique ecosystem is most productive because of its ability to retain and recycle nutrient elements within the ecosystem as well as within animal-plant associations. Though they are the builders of the most massive structures ever created by living beings in the world, they are very fragile and vulnerable to natural disturbances and human activities.

In the Andaman Sea exists fringing reefs around many islands and a long barrier reef (329 km) on the west coast.

Reefs are home to more species than any other ecosystem in the sea. The total number of reef species in the world is still unknown, but up to 3000 species can be found together on a single reef in South East Asia and over 1000 on a single Caribbean reef. Reefs also contain many more major animal groups (Phyla) than any other ecosystem on land or in the sea.

In India, the reefs are distributed along the east and west coasts at restricted places. Fringing and barrier reefs are found in Andaman Sea.

A total of 199 species belonging to 71 genera were recorded from Indian reefs. Out of these, 155 species are hermatypes (reef building corals) and 44 are non-reef building corals. In twelve areas of Andaman Sea, alone reveals an addition of 111 species of scleractinian corals from the shallow regions up to 15 m depth. A total of 223 species of scleractinia from Andaman Sea have been recorded including the earlier report by Pillai (1983).

The coral reefs of Andaman Sea are of fringing type, and the coral fauna is diverse, when compared to other parts of Indian Ocean. The coral fauna in Andaman Sea is known to harbor 177 species among 57 genera and 15 families.

Andaman and Nicobar reef embrace 9 species of sea grass. The insular mangroves of Andaman and Nicobar Islands, where many tidal estuaries, small rivers, neritic islets and lagoons support a rich mangrove flora. The mangroves also serve as a wild life sanctuary for the islands. A total of 74 species of sponges have been described from Andaman Sea. The species concentration (35 species under 10 genera and 5 families) of Sipunculan fauna is dominant in Andaman Sea. Maximum abundance of Echiura has been reported from Andaman Sea.

The crustaceans rank second in the diversity of fauna in the coral reef ecosystem and many of them are exploited for commercial purposes. Marine crustaceans of India (94.85 per cent) constitute 121 species of stomatopods (4 families, 26 genera), 26 species of lobsters (4 families,11 genera), 162 species of hermit crabs (3 families, 40 genera); 705 brachyuran crabs (28 families, 270 genera); 84 species of shrimps and prawns (7 families, 19 genera); 159 species of caridea (15 families, 56 genera) have been recorded so far. Other than these 540 species of copepods, 104 species of cirripeds, 120 species of ostrocods have been recorded.

Andaman Sea have a rich molluscan diversity, which include over 1000 species from the marine region.

About 257 species of echinoderm are known to inhabit in Andaman Sea.

Genera of phylum Hemichordata, such as, *Ptychodera, Glossobalanus* and Glandiceps have been collected from the coral reef areas of Andaman Sea.

Over 1000 species of fish are found in the Andaman Sea. The category of fishes occurring in coral reef ecosystem includes groups, like, damsel fishes (52 species), butterfly fishes (32 species), sweet lips (16 species), angel fishes (16 species), parrot fishes (14 species), snappers (42 species), wrasses (53 species), groupers (43 species) and surgeon fishes (18 species). Another 20 per cent are cryptic and nocturnal species that are confined primarily to caverns and reef crevices during day light. The assemblage includes families, such as, the cusk eels, some groupers and their relatives, most of the moray eels and some scorpion fishes, wrasses and nocturnal families including the squirrel fishes, cardinal fishes and sweet lips. Another 10 per cent of fishes including snake eels, worm eels, various rays, lizard fishes, grab fishes, flat fishes, some wrasses and gobies dwell primarily on reefs covered with sand and rubble. A relatively small percentage (5 per cent) of the fauna is transient mid- water reef species that roam over large areas. This group includes most sharks, jacks, fusiliers, barracudas and scattering representatives of other families.

Four species of turtles in their marine environment are known from Andaman Sea. Andaman and Nicobar Islands have best nesting beaches for leather back, the hawksbill, green turtle and also the Olive Ridley. Adult sea turtles are completely aquatic and spend most of their lifetime in water, but they start their lives on land, that is, on sandy shores as hatchlings and then enter the aquatic environment. Turtles are often seen on reefs. The hawksbill is most closely associated with coral reefs, and is found around the reefs of Andaman Sea, nesting on small beaches and coral cays and feeding on sponges and invertebrates on nearby reefs. Green turtles, the loggerhead, Olive Ridley and leather back turtle are also seen swimming and feeding in coral reef environments.

Reef fisheries of Andaman Sea is important as a subsistence fishery for local people, harvesting snappers, groupers, emperors, breams, barracuda, jacks and other commercial inshore fisheries, such as, sprats, herrings, flying fish. Sea horse, sea cucumber and aquarium species were collected till 2002, but subsequently banned under Wild Life Act, 1972. There is no organized commercial reef fishery for food or ornamental fishes in Andaman and Nicobar Islands. Andaman and Nicobar Islands have 45 fishing villages and 57 fish landing centers. The marine resource potential

Figure 7.1: *Cheilosporum spectabile* 1, Red algae

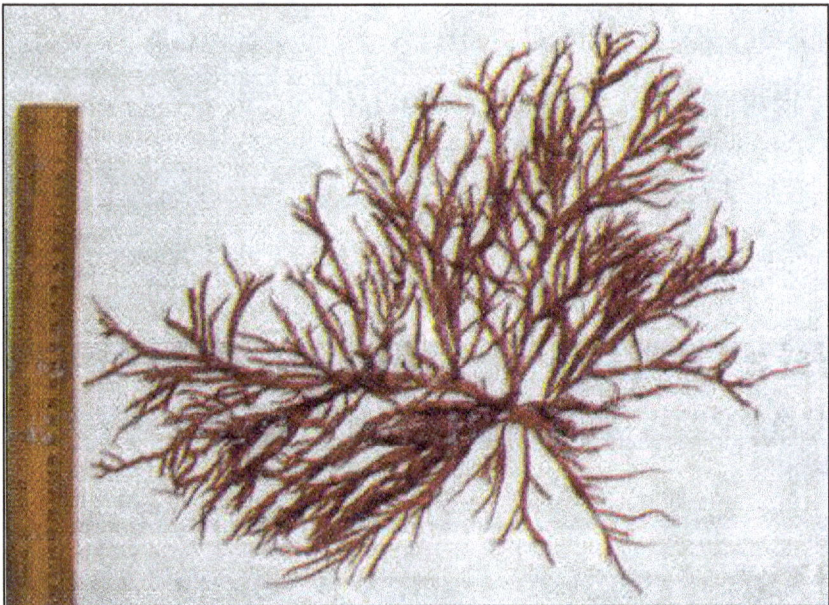

Figure 7.2: *Halymenia floresia*

has been estimated at 234500 tonnes of fish. A little over 11000 people depend upon the fishery for their income. The major type of fishing crafts of the islands are plank built country crafts, ranging in size from 5.4 to 7.5 m in length and motorized dug out canoe of 7.5 to 12 m fitted with inboard motors of 8 to 16 horse power. A few mechanized boats are also operated in Andaman Sea. Fish landing from Andaman Sea increased from 22674 tonnes in 1991 to 25477 tonnes in 1996. Most fishermen operate gill nets and hooks and line. Less important are cast net, shore seines, anchor seine. Seventy per cent of total landings are sold fresh, 21 per cent salted and dried and 9 per cent is frozen.

Sardines are most important in the catches (12 per cent), followed by perches (7 per cent). The catch consists of mainly groupers (Serranidae), snappers (Lutjinidae), jacks (carangidae), emperors (Lethrinidae), sweet lips (Haemlidae) and reef sharks.

Coastal communities including aboriginal tribes rely on reef fisheries. The fishery consists of primarily of fin fish, invertebrates (mollusks, crustaceans) and echinoderms. The fin fish catch is very diverse and includes both piscivores, such as, groupers, snappers, jacks and herbivores, like, rabbit fishes (Family Siganidae), parrot fish (Scaridae), surgeon fish (Acanthuridae). Reef fishery also includes of small pelagic (Scombridae, Clupeidae, Carangidae and Sharks), which move in and out of the reef areas in search of food and protection. Invertebrates consists mainly of giant clams (*Tridacna* sp.), top shells (*Trochus* sp.) and other bivalves. Sea cucumber (Holothurioidae) and spiny lobster (*Palenurus* sp.).

In Andaman Sea, bottom gill nets are set every evening and hauled in the morning to catch sharks. The major species caught include the Black-tip reef shark (*Carcharinus melanopterus*), Silver-tip shark (*C. albimarginatus*), grey reef shark (*C. amblyrhynchos*) spot tail shark (*C. sorrah*), white-tip reef shark (*Triaenodon obesus*). Of these the black-tip reef shark is more commonly caught. Groupers form on an average 5.8 per cent of reef fishes. The lobsters reported from Andaman Sea include three species belonging to family Nephropsidae, 7 species of Eryonidae and 2 species of Palinuridae. Stray numbers of *Panulirus homarus*, *P. ornatus* and *P. penicillatus* are caught with the fishes and prawns in the bottom set gill nets operated by the fishermen.

Andaman Sea, especially coral reefs are well known for their rich molluscan resources. The most important among them are the top shell (*Trochus niloticus*), Turban shell (*Turbo marmoratus*), pearl oyster (*Pinctada margaritifera*), giant clams (*Tridacna crocea, T. maxima, T. squamosa* and *Hippopus hippurus*), green mussel (*Perna viridis*), edible oyster (*Crassostrea madrasensis* and *Saccostrea cucullata*), scorpion shell (*Lambis chiragra*) and spider conch (*Lambis lambis*).

Sea Grass and Sea Weed Ecosystems

Sea grass are angiospermous plants adapted to grow in marine environment. Sea grass meadows are the nursery grounds for many commercially important shrimps, crabs and fishes. Its root mat adds stability to the sediments of coastal zone and the leaves help filter the water of suspended particles. There are 13 genera and about 52 species of sea grass distributed throughout the world.

The Gulf of Mannar, Marine Biosphere Reserve harbours 12 species of sea grass, namely, *Cymodocea rotundata, C. serrulata, Enhalus acoroides, Halodule pinifolia,*

H. uninervis (broad and narrow leaves form), *H. beccani, H. decipiens, H. ovalis, H. ovata, H. stipudacea, Thalassia hemprichii* and *Syringodium isoetifolium*.

Of these, *Enhalus acoroides* is the largest sea grass species in the Gulf of Mannar, reaching to a height of more than one meter. The sea grass prefer sheltered marine environment, where substratum is fine sand to clay. Normally it occurs in the tidal and sub-tidal zones. Some of the plants are seen established in Gulf of Mannar among *Cymodocea serrulata* and *Halodule pinifolia* sea grass species in depth of 40 cm during low tide. It is the only species, which shows aerial surface pollination. It has 3-4 linear leaf blades produced directly from the rhizome, although shoots with 7-9 leaves were also observed. Its leaf tips are rounded and with minute serrate projections when young. The underground rhizomes are stout (1-2 cm diameter) and branched, which can penetrate upto 30 cm deep. Roots are also well developed, both of which provide concrete anchoring system. The *E. acoroides* male flower is small, contains large pollen grains and has pendunculate spathe consisting of two connate blades, the inner and the outer. The female flower has ovary rostrate, six parietal placentas protruding far towards the center and with long pendunculate spathe consisting also of two free blades. The fruit of *E. acoroides* is ovoid or globular in shape with acuminate tips and the entire surface is covered with dense bifid projections.

During June 2008 (south west monsoon season), the Gulf of Mannar current was in the direction of south east and south west. The average wind velocity was 13.12 km per hour. This causes the fruits to dislodge from the plants in Pamban and Krusadai Island to reach shore of Rameswaram. The length of fruits varied between 5.7 to 6.5 cm and diameter 8.9 to 12.3 cm. There were 4 to 11 seeds per fruit and these seeds are edible and are eaten by the local people. The seeds of *E. acoroides* are more viable, compared to other sea grass species. Dispersal distance of sea grass seeds are limited to few meters and seeds settle rapidly into the sediment surface except during seasons of high wind velocity and tidal current. Seed deposition is more advantageous over vegetative propagation of new shoots for maintenance of sea grass meadows, frequently subjected to environmental stress.

Sea grasses of Andaman occur along the east and west coast of Andaman and Nicobar Islands in the infra-tidal and mid-tidal zones of sheltered localities. They are submerged monocotyledonous plants and adapted to the marine environment for completion of their life-cycle under water. They form a dense meadow on sandy and coral rubble bottoms and sometimes in the crevices under water. Among 14 species found along Indian coasts, the dominant species are, *Cymodium rotundata, Enhalus acorodies, Halodule pinifolia pinifolia, H. uninervis, H. wightii, Halophila beccarit, H. deeccipiens, H. ovalis, H. ovata, H. stipulacea, Syringodium isoetifolium,* and *Thalassia hemprichii*. About nine species are extensively found in Andaman Sea around Andaman and Nicobar Islands. The unique ecological importance of the sea grasses for the conservation of rare and endangered animals, like marine turtles, dugongs, some common echinoderms, juvenile prawns and fishes is very well known.

The sea weed communities prefer some what flat and rocky coastal wet lands that gradually slope towards the sea with a marked tidal effect of complete submergence during high tide and successive exposure during low tide. Their distribution extends from open shore formation to inter tidal lagoons, bays, rock pools, or in creeks and inlets beyond the low tide mark along the infra littoral region of the coast. Different species are abundant in the Andaman Sea.

Among 120 species of sea weeds so far recorded in Indian seas, the important sea weeds are, *Enteromorpha compressa, Ulva lactuca, Acetabularia crenulata, Dictyosphaeria cavernosa, Chaetomorpha media, Caulerpa corynephora, C. paltata, Odium iyengarii, C. tomentosum, Halimeda macroloba, Dictyota atomarica, Ectocarpus breviarticvulatus, Polysiphonia variegate, Grateloupia indica,* and *Sargassum duplicatum.* These plant communities serve as sustainable life support in the field of food, shelter, fertilizer, production of iodine, potash, glue, agar, algin, vitamins, antibiotic and others.

Pelagic and Benthic Ecosystems

The pelagic ecosystem is dominated by plankton, which is classified on the basis of size as picoplankton (0.2 to 2 mm), nanoplankton (2 to 20 mm), microplankton (20-200 mm) and meso plankton (more than 200 mm). In terms of trophic structure, the plankton can be grouped as phytoplankton and zooplankton. The former comprises of chlorophyll-bearing algal cells, coccoliths, diatoms and flagellates. The diatoms are dominant in the nutrient rich coastal waters, whereas the flagellates become important in offshore oceanic waters. The zooplankton includes such forms like the copepods, that spend their entire life in the planktonic phase and the larval forms of a range of invertebrates and vertebrates.

The second important group among pelagic organisms is the nekton. The fish constitute the dominant taxon in the nekton. About 4000 species of fish are known from the Indian Ocean, of which about 50 per cent occur in Indian seas. A majority of these species occur in coastal waters supporting valuable fisheries.

About 26 species of snakes belonging to one family, Hydrophiidae and five species of sea turtles were reported from seas around India.

The deep sea, a major part of the ocean, was considered as species-poor environment earlier. But now the estimates of the probable numbers of species in the deep seas range from 5 to 10 millions. Most of the new species would be microbes (bacteria and fungi). The discovery of deep sea hydrothermal vents with their characteristic animal groups and sulphur-based food chain in super-heated waters could not have thought twenty years earlier.

Chapter 8
Primary Production

Primary Production in Andaman Sea

Primary production in the surface waters ranged from 1.3 to 9.7 mg C/ cubic m /day with an average of 5.3. But in the euphotic column it was 120-615 mg C/sq m/ day (average 273). Surface chlorophyll *a* from whole phytoplankton ranged from 0.001 to 0.098 mg/ cubic m (average 0.033 mg/ cubic m). Phaeopigment values varied from 0 to 0.076 mg/cubic m (average 0.033). Nanoplaqnkton chlorophyll *a* ranged from 0 to 0.053 and phaeophytin 0 to 0.065 mg/cubic m. In other words, nano-

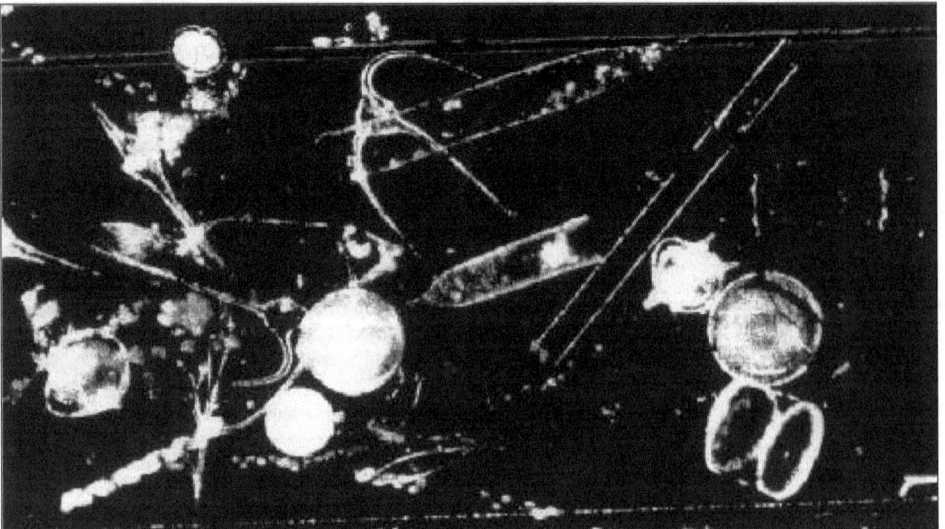

Figure 8.1: Plankton in the Deep Blue Sea

chlorophyll *a* varied from 27 to 100 per cent of the whole chlorophyll *a*. In euphotic column, chlorophyll *a* and phaeophytin of whole phytoplankton ranged from 0.53 to 11.0 and 0.17 to 15.23 mg/sq m respectively. The corresponding values of nano-chlorophyll *a* and phaeophytin ranged from 0.18 to 6.04 and 0 to 6.87 mg/sq m respectively. Particulate Organic Carbon (POC) at the surface ranged from 0 to 465 mg/ cubic m (average 132). The column values varied from 0 to 13.9 gC/sq m (average 4.59). Detrital carbon showed a range from 64 to 98 per cent of POC (average 87 per cent). The most abundant phytoplankters are *Navicula* spp., *Peridinium* spp., *Trichodesmium thiebautii* and flagellates. Dinoflagellates formed an important constituent of the population. Turn over rates of phytoplankton was estimated to be 0.1 to 1.9 days (average 1).

Table 8.1: Primary Production of Andaman Sea and other Regions

Area and Season	Primary Production mgC/sq m/day		
	Range	Average	
North Arabian Sea	109-2000	693	(Radhakrishna *et al.*, 1978)
Dec, 73 – May, 74			
North-east Arabian Sea	100-312	216	-do-
November, 1976			
Laccadive Sea			
October, 76	22-359	134	(Bhattathiri and Devassy, 1979)
December, 76	45-1167	358	
March/April, 78	254-829	372	
Bay of Bengal			
August-September, 76	49-606	275	(Radhakrishna *et al.*, 1978)
August, 1977	139-1188	641	(Devassy *et al.*, 1980)
August-September, 78	116-3408	983	(Bhattathiri *et al.*, 1980)
Andaman Sea			
February, 1979	120-615	273	(Bhattathiri and Devassy, 1980)

Ranges of particulate organic carbon (POC), phytoplankton carbon, zooplankton carbon and detrital carbon in the Andaman Seas and Laccadive Seas, as estimated are given below.

Area and season	POC mg/m sq	Phytoplankton Carbon mg/m sq	Zooplankton Carbon mg/m sq	Detrital
Andaman Sea				
February, 1979	975-17535 (7302)	30-432 (225)	108-648 (290)	761-17137 (6787)
Laccadive Sea				
October, 1976	2045-33200 (17425)	45-605 (308)	104-871 (365)	1900-32400 (16752)

Contd...

Contd...

Area and season	POC mg/m sq	Phytoplankton Carbon mg/m sq	Zooplankton Carbon mg/m sq	Detrital
December, 1976	1969-14754 (7194)	192-943 (530)	60-2040 (376)	1145-13945 (6288)
March, 1978	1546-6155 (3603)	83-770 (303)	75-672 (368)	984-5704 (2932)

Average values are given in parenthesis.

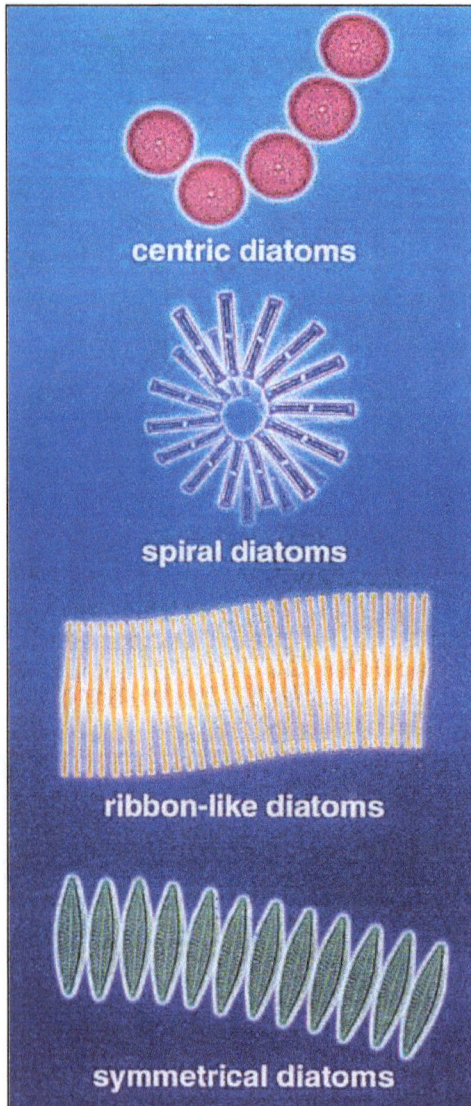

Figure 8.2: Phytoplankton Communities of the Sea, Diatoms

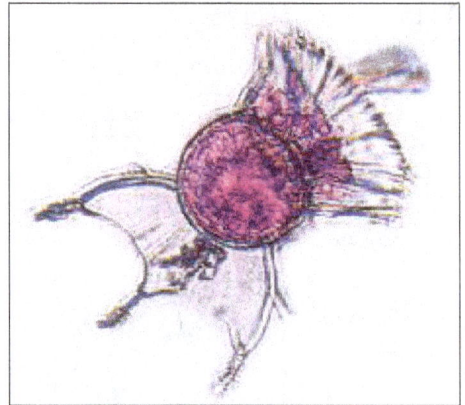

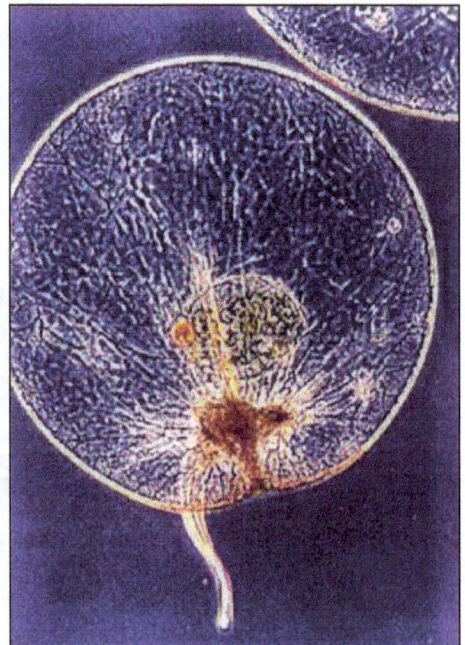

Figure 8.3: *Noctiluca* spp.

Andaman Sea extends southwards to a length of 1200 km from Irrawaddy Delta in Mayanmar and has a width of 650 km in an east-west direction and is bounded by the Malayan Peninsula in the east. The area of Andaman Sea is estimated to be 0.602 X 1000000 sq.km (Lyman, 1966). The two major rivers draining into this sea are Irrawaddy, which discharges at an average rate of 13560 cu.m/second, followed by river Salween (Coleman, 1968). The annual discharge of sediment by the Irrawaddy to Andaman Sea is estimated to be approximately 250 million tons. Some of the major rivers on the east coast of India namely, the Ganges, Brahammaputra, Mahanadi, Godavari and Krishna also discharge considerable volume of fresh water (39400 cubic m/sec, annual average) thereby reducing the salinity of Bay of Bengal surface layers (Rao,1975).

Production by Phytoplankton

Primary production at the surface waters of Andaman Sea, ranged from 1.3 to 9.7 mgC/cubic m/day (average 5.3). The standard deviation was 2.35. Column primary production varied between 120 and 615 mgC/sq m/day (average 273, SD 132).

Chlorophyll *a* and Phaeophytin

Whole chlorophyll *a* at the surface varied between 0.001 and 0.098 mg/cubic m. The average value was 0.033 mg/cubic m and standard deviation 0.019. Phaeophytin ranged from 0 to 0.076 mg/cubic m (average 0.033, SD 0.017). The corresponding values at the euphotic column were varying from 0.43 to 11.0 and 0.17 to 15.23 mg/sq m respectively. While the average value of chlorophyll *a* was 3.64 (SD 2.59), that of phaeophytin was 4.05 (SD 3.1). Nanoplankton chlorophyll *a* at the surface varied from 0 to 0.053 and phaeophytin from 0 to 0.065 mg/cubic m. Nano chlorophyll *a* contributed 27 to 100 per cent of the whole chlorophyll *a*. Nano chlorophyll *a* in the euphotic column ranged from 0.18 to 6.04 and phaeophytin 0 to 6.87 mg/sq m.

Phytoplankton Population

Quantitatively, phytoplankton population of the Andaman Sea is comparable to that of the Laccadive Sea for February/March period. *Navicula* spp,*Peridinium* spp., *Trichodesmium thiebautii* and flagellates dominated the phytoplankton population. Dinoflagellates formed an important constituent of phytoplankton in the Andaman Sea. It is known that dinoflagellates can thrive successfully in oligotrophic tropical waters. It has been observed that throughout the Andaman Sea, dinoflagellates are rich during and after north east monsoon.

Particulate Organic Carbon (POC)

Particulate organic carbon showed wide fluctuations. At the surface, its level increased from undetectable to 465 mg/cubic m. The average value was 132 mg/cubic m and standard deviation 121. Column POC varied from 0 to 13.19 gC/sq m with an average of 4.59 (SD 3.49).

Zooplankton

Zooplankton displacement volume was found to vary from 2.2 to 14.4 ml/100 cubic m or 3.6 to 21.6 ml/sq m. The sampling was done up to 200 m at places beyond shelf and almost to the bottom at shallower regions.

Environmental Features

Salinity at surface in the southern Andaman Sea ranged from 32 to 33 ppt (below 9°N). Along the 10°N Channel, salinity was about 32 ppt. On the eastern part of north Andaman, salinity ranged from 31.64 to 32 ppt at the surface. Salinity was higher on the western side of the Andaman Sea.

At most of the stations on the eastern Andaman Sea, oxygen was super-saturated at the surface and few meters below. But on the western side, super-saturation of oxygen was discernable even at 50 m.

Investigations made on the inorganic nutrients included phosphate-phosphorus, nitrate-nitrogen and ammonia-nitrogen. The integral mean concentration (IMC) of phosphate-phosphorus in the upper 50 m ranged from 0.17 to 0.54 micro g at/l except an isolated high value of 1.22 at 14°N and 93°E. However, the IMC values for 75 and 100 m column ranged from 0.20 to 1.45 and 0.38 to 1.63 micro g/l. At most of the stations, nitrate-nitrogen values (IMC) up to 50 m layer were zero, but the corresponding values for 75 and 100 m column varied between 0.41 and 6.51 and 2.32 and 9.53 micro g at/l. Nitrite-nitrogen was zero at surface as well as sub-surface levels at most places. IMC values of ammonia-nitrogen in the 50, 75 and 100 m column were between 0.39 and 3.82, 0.49 and 3.64 and 0.41 and 3.52 micro g at/l respectively.

The surface primary production of the Andaman Sea ranged from 1.3 to 9.7 mgC/cubic m/day. Measurements of primary production during March/April showed that surface production varied between 0 to 13.77 mgC/cubic m /day. The average production during February and March/April were 5.3 (SD 2.35) and 2.41 (SD 3.18) respectively. The range in column production (120 to 615 mgC/sq m/day) observed was well above the level found earlier (10 to 270 mgC/sq m/day). The average production was 273 and 135 mgC/sq m/day. The corresponding standard deviations were 132 and 75.

A comparison of the range in production and average of the Andaman Sea with those of the Arabian Sea and Bay of Bengal showed that the production in the Andaman Sea is comparable. The surface chlorophyll *a* reported earlier had a wider range (0-0.38 mg/cubic m) as compared to reported values in earlier paragraphs. Similarly the range in column chlorophyll *a* reported here was less than those found during March/April (2-45 mg/sq m). The average for the present and earlier values were 3.64 and 10.10 mg/sq m. Thus during February, although the chlorophyll *a* was less than that reported earlier, the production was high. Most of the earlier values of production were near Mayanmar coast, despite the production was less. In the Central Andaman Sea, the production was found to be 114 and177 mgC/sq m/day, when the phosphate-phosphorus was zero and nitrate-nitrogen was between 1.5 and 2.0 micro g at/l. The surface particulate organic carbon (POC) in the Andaman Sea ranged between 24 and 120 mg/cubic m, except for few high values near Myanmar coast (35 to 238 mg/cubic m). Zernova and Ivanov (1964) found that phytoplankton cells were higher in the Andaman Sea (more than 5000 cell/cubic m) compared to Bay of Bengal (500 cells/cubic m).

Zooplankton biomass (average 9.1 ml/sq m) observed during February, 1979 was mostly within the same range reported earlier. During March/April, 1963 the zooplankton displacement volume was found to range from 2 to 25 ml/sq m (with an average of 10.7 ml/sq m). The values reported during February/March, 1963 were between 0.5 and 4.0 ml/sq m (average 2.5 ml/sq m). But during April 1964, the zooplankton displacement volume was found to vary between 7.5 and 41.0 ml/sq m (average 19.1).

It was found that, in general the oxygen level on the eastern Andaman Sea was less than that of western side. On the eastern side oxygen concentration at 50 m was less than 4 ml near the islands, but was between 4.5 and 4.7 ml/l along 93°30 minute E. It was found in a section along 95°E and between lat. 9°30 minute and 15°N, the oxygen concentration in the upper layer during February/March, 1963 was 4.5 ml/l. It was also found that the thickness of the layer of oxygen concentration greater than 4.0 ml/l increased from 35 m in the northern and to 80 m at 11°N and then decreased southwards.

The surface temperature of the Andaman Sea remains near 28°Celsius throughout the year. Thermocline is shallowest during January/February (Colborn,1975). Since the Andaman Sea is a confined physiographic basin, flow to open ocean areas of Bay of Bengal occur through the channels around and between Andaman and Nicobar Islands. During January-April, the flow through the Strait of Malacca is strongest and northwestern. The surface circulation is controlled by the monsoon system of the northern Indian Ocean.

The ratio of detrital carbon to phytoplankton carbon varied from 4 to 148 (average 35) in the Andaman Sea, while the average value for different seasons from Laccadive Sea varied from 10 to 55.

The variation observed in zooplankton carbon was from 108 to 648 mg/sq m. This was 0.3 to 8.4 times (average 2.1), the phytoplankton carbon. During different seasons the average values ranged between 0.8 and 2.5 in the Laccadive Sea. These values are higher, when compared to oceanic regions.

The time required for complete turn over to phytoplankton ranged from 0.1 to 1.9 (average 1) and was in the same range found in the Laccadive Sea.

Though primary production in Andaman Sea is in the same range as found in the Laccadive Sea, it is much less than the values obtained for Bay of Bengal during August/September, 1978.

Extracellular Production in the Andaman Sea

Three depths were sampled within the euphotic zone for extracellular production by phytoplankton, which represented 100 per cent, 33 per cent and 1 per cent of surface irradiance. Particulate production in the Andaman Sea is low as compared to the Laccadive Sea and varies from 1.79 mgC/cubic m/hour to 0.18 mgC/cubic m/hour at the bottom of the euphotic zone at 9°50 minute N, 91°58 minute E; 10°40 minute N, 92°50 minute E and 11°20 minute N, 92°08 minute E. Column production values are between 80.3 mgC/sq m/day at 13°30 minute N, 93°35 minute E to 766 mgC/sq m/day at 9°50 minute N, 91°58 minute E.

The turn over rate of phytoplankton carbon, calculated by dividing the carbon equivalent weight of phytoplankton by the rate of addition of fresh organic matter (primary production) is approximately one day in this environment. Inspite of the fact that both phytoplankton carbon and primary production in the Andaman Sea are generally lower than in the Laccadive Sea, the relationship between the two variables is, on an average, similar in the two regions. This similarity is surprising because Nitrate-nitrogen in the Andaman Sea is almost absent from the surface waters (0 to 100 m) as compared to 0.29 micro g at/l in the Laccadive Sea. It is however, possible that the Ammonia-nitrogen is supplying the nitrogen for phytoplankton growth. Phosphate-phosphorus does not appear to be limiting in these waters.

Extra cellular production is an average of 54 per cent of total carbon dioxide fixation at the surface and decreases through 18 per cent at mid-depths, to 0 per cent at the bottom of the euphotic zone. This relationship of extra cellular products with depth is markedly different from that in the Laccadive Sea, where even at the bottom of the euphotic zone there was some extra cellular production. One factor influencing this pattern of extra cellular production could be the concentration of dissolved oxygen. In the Laccadive Sea, oxygen is super-saturated in the water column to the bottom of the euphotic zone, whereas, in the Andaman Sea even down to 1 per cent light penetration depth. It is established, at least in cultures and cell-free extracts, that higher partial pressure of oxygen substantially increase the rate of photo-respiration and, consequently, extra cellular production. This lower oxygen tensions at depth in the euphotic zone may result in decreased extra cellular production. The attenuation coefficient K does not vary significantly between the Laccadive and Andaman Seas.

Glycolate concentrations in the Andaman Sea range from 370 to 150 micro g/l and lie numerically between values from near shore areas (540 to 365 micro g/l) and the Laccadive Sea (79 to 92 micro g/l). The surface average in the region is approximately 280 micro g/l and decreases with depth, being about 270 micro g/l at depths of 70 m. Since there is no apparent extra cellular production at 70 m, the substance probably originates from path ways other than photosynthetic, such as, breakdown of large molecular weight compounds by bacterial action.

In the Andaman Sea, at this time of year (February), the well mixed layer extends some 25 to 30 m below the surface, which is well within the euphotic zone. Thus movement of phytoplankton cells across the boundary of the well mixed layer and gradations of both dissolved oxygen and nutrients must affect both primary and extra cellular production. The influences of bottom topography, particularly on the eastern side of the island chain, on water movement in the area and the consequent distributions of allochthonous material are not as yet fully understood. The fact that there is no discernable island mass effect in this data with regard to primary production suggests one effect of such distributions.

Detrital Content of Andaman Sea

Detrital carbon in the Andaman Sea constitutes about 92 per cent of the total particulate carbon, while phytoplankton and zooplankton constitute small fractions of total suspended matter. Latitudinal relationship between phytoplankton and zooplankton was found to be inverse. The Andaman Sea is oligotrophic in nature with low primary and secondary productivity and hence the production of large quantities of detritus appears to supplement the nutritionally inadequate food present in these waters.

Andaman Sea receives 250 million tons of sediment per year. The major rivers collectively discharge, on an average, 39400 cubic m/second/year of fresh water into the Andaman Sea.

Thus the large quantities of sediment and other materials brought to the Andaman Sea from outside make the study of detritus in this region of Bay of Bengal particularly important.

Suspended and Particulate Organic Matter

At deeper areas of Andaman Sea, within 0-200 m water column, the total suspended matter varied very little quantitatively. It ranged from 135 to 212 g dry weight/ sq m at deeper places. At the shallower places on the other hand, the range is somewhat large, from 55 to 146 g dry weight/sq m. This may be because of the island effect contributing to the suspended matter at shallow regions.

In contrast of the suspended matter, the particulate organic matter showed a wide range both at deep and shallow regions. The average value at deep regions was about 12 gC/sq m whereas, at shallow areas, it was about 6 gC/sq m.

Living Matter

There was a wide range in the values of phytoplankton carbon both at deep and shallow areas. The average of all the stations was 307 mg/sq m. Zooplankton carbon also showed a large variation with an average value of about 207 mg/sq m. This indicates that zooplankton carbon is almost similar to that of phytoplankton carbon.

From the latitudinal variation in phytoplankton and zooplankton carbon, the relation between the two appears to be inverse, that is, where the phytoplankton carbon was high, zooplankton carbon was low and vice versa. Maximum phytoplankton carbon was from latitudes 5 to 8°N, whereas zooplankton carbon was maximum at 11 to 14°N.

Detritus

Apart from one single value, where the detritus, like the POC value, was exceptionally high, at all other places, the range in value was small and did not vary significantly in relation to latitudes. Detrital carbon constituted about 92 per cent of the total particulate organic carbon in the suspended matter and the detrital energy varied from 25 to 479 kcal/sq m (average 96 kcal/sq m).

In the Laccadive Sea, the average contribution of detritus is of the order of 97 per cent. The average phytoplankton carbon 213 mg/sq m and zooplankton carbon 450 mg/sq m and the detrital carbon 2932 mg/sq m.

Both Laccadive and Andaman Seas are oligotrophic in nature with low primary and secondary productivity. The former is a coral sea with an archipelago of 24 atolls and many submerged banks of coral origin scattered in between. The Andaman Sea, on the other hand, has 15 islands of volcanic origin with luxuriant fringing coral reefs around many of these islands. Presumably the large quantities of organic detritus present in the suspended matter in the two seas is to supplement the nutritionally inadequate food in these waters. Some explanations as to the possible source of detritus is from the "organic pool" existing in the ocean. The contribution made by heterotrophic bacteria and other micro-flora to the quantity of organic matter present in the sea appears to be far greater than that contributed by phytoplankton and zooplankton. Detritus appears to originate from the dissolved organic matter as flakes or flocculent particles, which remain suspended enriching the food value of the water column.

Recent studies on the samples of "marine snow" or "macroscopic organic aggregates" from the shallow neretic waters have indicated that marine snow occurs in the sea in large quantities as amorphous and fragile particles. The organic matter contained in the marine snow forms a potential source of food for other animals. The snow is composed of passively sinking particles, such as, faecal pellets, thick-walled diatoms, frustules of diatoms, micro-flora etc, and motile organisms, such as, dinoflagellates and ciliates. The particles constituting marine snow range from 0.05 to 10 cm and these are of much ecological significance as micro-habitat and pelagic food sources. The snow contain nutrients in much greater concentrations than the surrounding waters. Marine snow is of common occurrence in the sea and has been reported from the atolls of Laccadives. Possibly the occurrence of this type of material in suspension increases the detritus content of the water column.

Primary Production

The average primary production of the Andaman Sea has been estimated to be 273 mgC/sq m/day. At this rate, the primary production of the entire Andaman Sea (area 0.602 X 10000000000000000 sq km) would be approximately 6 X 10000000 tons C/year.

Secondary Production

The average zooplankton biomass of the Andaman Sea estimated as zooplankton carbon is 288.8 mgC/sq m. Assuming that about 50 per cent of the biomass gives rise to daily rate of secondary production, the total production at the next trophic level for the entire area of the Andaman Sea would be of the order of 31.73 X 1000000 tons C/year. The 50 per cent figure is based on a series of conversions from biomass values to secondary production, which range from 3 to 98 per cent.

Tertiary Production

For the open ocean 0.1 per cent of primary production and 1 per cent of secondary production are to be taken for calculating tertiary production. The mean of the two gave a value of 18.70 X 10000 tons of carbon per year. In terms of live weight, the figure would be 18.70 X 100000 tons/year (using factor 10). In the Andaman Sea, like the Laccadive Sea, the major exploited stocks largely include long lived tuna, bill fish etc. Therefore 25 per cent of the potential yield could be taken as exploitable stock. This would amount to 470000 tons of fish per year.

Chapter 9

Wealth of Marine Life

Reef Associated Fauna

The Andaman and Nicobar group of Islands (530), located in the south-east of the Bay of Bengal (Andaman Sea), between latitudes 6-14°N and longitudes 91-94°E have got numerous coral reefs. Large number of individuals, called polyps in case of most of the corals, unite to form colonies (reefs). The first reef began to form in the world ocean with the advent of filamentary algae some two billion years ago.

Fringing reefs grow along the edges of continents and around islands, close to the shore, but sometimes separated from it by a shallow lagoon. Fringing reefs are common in Andaman sea around Andaman and Nicobar Islands. Where there is murky water, caused by soil run-off, fringing reefs rarely grow to any substantial depth.

Barrier reefs are found in Andaman Sea around Andaman and Nicobar Islands. Barrier reefs develop along the edges, around the islands that have become partially submerged, and separated from the island by a wide deep lagoon. Fragile corals grow on the lagoon side of the barrier than on the open side, where they would have to withstand the force of larger and more violent waves.

Andaman Sea has got a total coral reef area of 959.3 square km, of which reef flat, sand over reef, mud over reef coralline shelf, coral heads, reef vegetation and vegetation over sand represents an area of 795.7, 73.3, 8.4, 45.0, 17.5, 8.9 and 10.5 square km respectively.

The coral reefs of the Andaman Sea are of fringing type. A deep oceanic ridge along 10°N separates the Andaman and Nicobar groups of islands. The coral fauna of Andaman Sea is diverse, when compared to other parts of India.

Figure 9.1: Coral Polyps

Figure 9.2: Orange Cup Corals

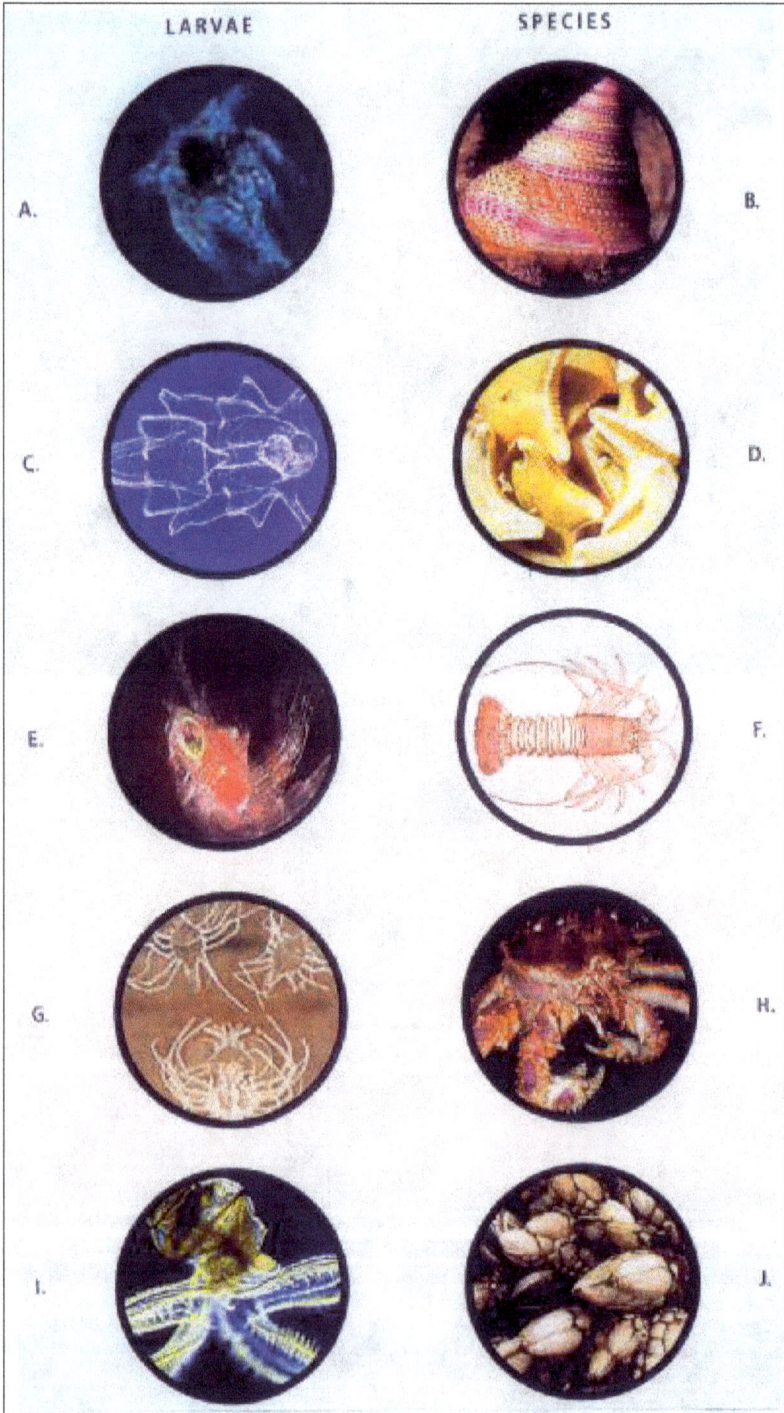

Figure 9.3: A: Nauplius larva of a barnacle; B: Blue-ring top snail; C: Starfish larva; D: Mud sea star; E: Blue king crab zoea larva; F: Spiny rock lobster; G: New Zealand red rock lobster larva; H: Blue king crab; I: Larva of the gastropod; J: Barnacle

Out of a total of 199 species under 71 genera, of whole of India, the Andaman Sea harbors 135 species in 59 genera. Among them 100 species are of Hermatypes and 35 species of Ahermatypes under 47 and 12 genera respectively. *Acropora* communities are reported to be the dominant.

Reef Associated Fauna

Sponges are asymmetrical benthic animals, and are strikingly coloured. They represent major component of reef communities. In Andaman Sea, 112 species of sponges have been found. Sponges contain several chemical compounds, which have therapeutic values, and not found in other animals. *Tethya crypta* have proven cancer inhibiting properties.

The crustaceans rank second in the diversity of fauna in the coral reef ecosystem and many of them are exploited for commercial purposes. The crustacean species in the Andaman Sea includes crabs, lobsters, prawns and shrimps, stomatopods, copepods, cirripeds and ostracods.

The Andaman Sea has a rich molluscan diversity, which include about 1000 species. Though molluscan diversity is the highest (3370 species) among all the invertebrates in the coral reef ecosystems, they are threatened by habitat alteration, and indiscriminate exploitation by man. Eight species of oysters, two species of mussels, 17 species of clams, six species of pearl oysters, four species of giant clams, one species of window pane oysters and other gastropods, like sacred chank, *Trochus, Turbo* as well as 15 species of cephalopods are exploited from coral reefs.

The echinoderms comprise species of starfish, brittle stars, sea cucumbers, sea urchins and feather stars. Among 765 recorded species in India, 257 species are known from Andaman Sea. Twelve species of Holothuroidea are exploited on a commercial scale for export. Sea urchins play an important role in creating reef sediment and keeping sea weed growth on the reef at bay, but large numbers can damage reefs. Crown-of-Thorns, star fish can devastate reefs in the course of its coral eating.

Fishes comprise about half of the total number of vertebrates. The exact number of fish species associated with coral reefs of India is yet to be determined, but the number of Fishes in the Indian Ocean is 1367, of which 750 species are found in the Andaman Sea. The fish groups occurring in coral reef ecosystem includes, damsel fishes, butterfly fishes, trigger fishes, file fishes, puffers, snappers, hawk fishes, triple fin wrasses, groupers, and gobies. Twenty per cent of coral associated fishes are composed of cryptic and nocturnal species that are confined primarily to caverns and reef crevices during day light periods. This assemblage includes such families as the cusk eels, some groupers and their relatives, most of the moray eels and some scorpion fishes, wrasses and nocturnal families including the squirrel fishes, cardinal fishes and sweet lips. Another 10 per cent of fishes dwell primarily on reefs covered with sand and rubble, including snake eels, worm eels, various rays, lizard fishes, grab fishes, flat fishes, some wrasses and gobies. Five per cent of the fauna is composed of transient mid-water reef species that roam over large areas, they are most sharks, jacks, fusiliers, barracudas and a scattering of representatives of other families.

Humphead wrasse, a rare coral fish, *Cheilinus undulates*, also known as Napoleon wrasse is the biggest member of the family, Labridae, and a giant among reef fish occasionally visit the coral reef of Andaman Sea. The fish sometimes measured 47.15 cm in total length and weigh 1.9 kg. In 1996, the species was listed as vulnerable on the IUCN Red list due to rapidly decreasing in numbers. Even though it is enlisted as endangered species, its exploitation is continued all over the world due to its high demand in Live Reef Food Fish Trade. There are number of reports about the substantial decline of the species due to large scale spear fishing, in which live *C. undulates* fetches top prices.

The dugong in Andaman Sea feed entirely on sea grass and other rich vegetation and are found near the reefs.

Turtles are often seen on reefs. The hawkbills are most closely associated with coral reefs, nesting on small beaches and coral cays and feeding on sponges and invertebrates on nearby reefs. Green turtles the Loggerhead, Olive Ridley and Leather back turtle are also seen swimming and feeding in coral reef environment.

The coral reef ecosystem supports a wide variety of flora and fauna and Andaman Sea is one of this kind. Coral reef fishes, generally recorded from Andaman Sea, consists of 183 species, belonging to 50 families as given below.

1. Family – Dasyatidae – *Neotrygon kuhlii, Pastinachus sephen, Himantura uarnak.*
2. Family – Myliobatidae – *Aetobatus narinari, Rhinoptera javanica.*
3. Family – Rhinobatidae – *Rhynchobatus djiddensis.*
4. Family - Acanthuridae – *Acanthurus gahhm, A. lineatus, A. mata, A. nigricauda, A. triostegus, A. xanthopterus, Zebrasoma velifer.*
5. Family – Antennariidae – *Antennarius pictus.*
6. Family - Apogonidae - *Apogon aure3us.*
7. Family - Aulostomidae – *Aulostomus chinensis.*
8. Family – Balistidae - *Balistoides viridescens, Odonus niger, Sufflamen fraenatum.*
9. Family - Belonidae – *Strongylura leiura, Tylosurus crocodiles crocodiles.*
10. Family - Caesionidae – *Caesio caerulaurea, C. cuning, C. xanthonota, Pterocaesio chrysozona.*
11. Family - Carangidae – *Alectis indica, A. ciliaris, Alepes djedaba, Atule mate, Carangoides armatus, C. chrysophrys, C. coeruleopinnatus, C. hediandensis, Caranx heberi, C. hippos, Gnathanodon speciosus, Megalaspis cordyla, Parastromateus niger, Scomberoides commersonnianus, S. lysan, S. tala, S. tol, Selar crumenophthalmus, Selaroides leptolepis.*
12. Family – Chaetodontodae – *Chaetodon auriga, C. collare, C. decussates, C. gardineri, C. lunula, C. octofasciatus, C. plebeius, C. trifasciatus, C. trifascialis, C. vagabundus, C. xanthocephalus, Heniochus acuminatus.*
13. Family - Chirocentridae – *Chirocentrus dorab.*
14. Family – Clupeidae – *Amblygaster sirm, Sardinella albella, S. gibbosa.*
15. Family – Dactylopteridae – *Dactyloptera orientalis.*

16. Family - Drepaneidae – *Drepane punctata.*

17. Family – Ephippidae – *Ephippus orbis, Platax teira*

18. Family - Fistulariidae – *Fistularia petimba.*

19. Family - Haemulidae – *Diagramma picta, Plectorhinchus albovittatus, P. diagrammus, P. gaterinus, P. pictus, P. schotaf, Pomadasys maculates.*

20. Family - Holocentridae – *Myripristis murdjan, Sargocentron rubrum.*

21. Family – Labridae - *Cheilimus chlorourus, C. undulates, Coris formosa, Halichoeres hortulanus, H. nigrescens, H. zeylonicus, Hemigymnus melapterus, Thalassoma lunare.*

22. Family - Latidae – *Psammoperca waigiensis, Gymnocranius griseus, Lethrinus harak, L. dentjan, L. mahsena, L. microdon, L. miniatus, L. nebulosus, L. ornatus, L. variegates.*

23. Family - Lutjanidae – *Aphareus rutilans, Aorion virescens, Lutjanus argentimaculatus, L. bengalensis, L. bohar, L. descussatus, L. fulviflamma, L. fulvus, L.gibbus. L. kasmira. L. lemniscatus. L. lutjanus. L. madras. L. quinquelineatus. L. rivulatus, L. russelli, L. vitta, Pinjalo pinjalo.*

24. Family - Malacanthidae – *Malacanthus latovittatus.*

25. Family - Menidae – *Mene maculate.*

26. Family - Monacanthidae – *Acreichthys tomentosus, Aluterus monoceros.*

27. Family - Mugilidae – *Valamugil seheli*

28. Family - Mullidae – *Mulloidichthys flavolineatus, Parupeneus barberinus, P. forsskali, P. heptacanthus, P. indicus, P. macronemus, Upeneus moluccensis, U. tragula, U. vittatus.*

29. Family - Muraenidae – *Gymnothorax favagincus, G. punctatus, Uropterygius concolour.*

30. Family - Nemipteridae – *Nemipterus furcosus, Scolopsis bimaculata, S. vosmeri.*

31. Family - Ostraciidae – *Ostracion cubicus, Tetrosomus gibbosus.*

32. Family - Pempheridae – *Pempheris molucca.*

33. Family - Platycephalidae – *Cociella crocodiles*

34. Family - Plotosidae - *Plotosus lineatus.*

35. Family - Pomacanthidae – *Apolemichthys xanthurus, Pomacanthus semicirculatus.*

36. Family - Pomacentridae – *Abudefduf bengalensis, A. saxatilis, A. septemfasciatus, A. vaigiensis.*

37. Family - Pricanthidae – *Priacanthus hamhur.*

38. Family - Rachycentridae – *Rachycentron canadum.*

39. Family - Scaridae – *Bolbometopon muricatum, Leptoscarus vaigiensis, Scarus ghobban, S. prittacus.*

40. Family - Scorpaenidae – *Pterois russelli, Rhinopias eschmeyeri, Scorpaenopsis gibbosa.*

41. Family - Serranidae – *Cephalopholis argus, C. boenak, C. Formosa, C. sonnerati, Epinephelus chlorosigma, E. coeruleopunctatus, E. corallicola, E. fasciatus, E. flavocaeruleus, E. longispinis, E. maculatus, E. malabaricus, E. merra, E. polyphekadion, E. tauvina, E. undolosus, Pogonoperca ocellata.*

42. Family - Siganidae – *Siganus canaliculatus, S. javas, S. lineatus, S. spinus.*

43. Family - Sillaginidae –*Sillago sihama.*

44. Family - Soleidae – *Pardachirus pavoninus.*

45. Family - Sphyraenidae – *Sphyraena barracuda, S. flavicauda, S. jello.*

46. Family - Synodontidae – *Saurida tumbil, Trachinocephalus myops.*

47. Family - Tetraodontidae – *Arothron hispidus, A. stellatus.*

48.   Family - Tetrarogidae – *Ablabys binotatus.*

49. Family - Terapontidae – *Terapon theraps.*

50. Family - Zanclidae – *Zanclus cornutus.*

Reef Associated Sponges

The part played by sponges belonging to the genera, *Cliona, Thoosa, Spirastrella, Halina* and *Samus* in the destruction of gregarious molluscan beds and coral reefs has been well documented in the past, and as understood at present, the sponges constitute a major group among twelve different taxa of marine plants and animals that cause considerable damage to the marine environment.

Damage caused to corals by boring sponges is rather widespread in all the morpho-zones of the reefs. Various borers resort to various methods in gaining entry into the hard calcareous substrate. In some, it may be by chemical means and in others by mechanical means. In yet others, it is effected by a combination of both the above means. In many cases, it cold be noticed that sponge infection may kill the colony either partly or fully. The sponge, as a group, dominated (87.2 per cent) among the various groups of boring organisms that destroy coral colonies. In an *Acropora* colony, weighing 600 g with a maximum diameter of 15 cm bearing 48 upright branches arising from a flat basal portion, when examined, it was found that branches were in a partly dead condition and when broken, were found infected with boring sponges. Other borers, mainly, polychaetes and sipunculids could also be noticed in some of these branches. While calculating the total number of boring organisms occupying the cut end of each branch, sponges belonging to five species was found to dominate over polychaete, mollusk and sipunculid.

Any localized death of a colony may not produce far reaching results, unless considerable damage occurs to the stalk portion. In branching corals, the larva soon after its settlement on the coral, flattens to form an encrusting mass. This is followed by the etching out of calcium carbonate particles to form an initial chamber. Further spreading of the sponge inside the coral is effected through chambers originating from the initial chamber and thus a mass of chambers is formed just inside the point of larval entry. These chambers may be seen filling the entire thickness of the branch and may open out at the other side (some times at several points) of the branch

Figure 9.4: Nudibranchs

through the excurrent and incurrent papillae. The number of papillae increases gradually on all sides of the branch. The death of branch actually occurs from this point onwards and as a result, the coral polyps above this point decayed gradually. Algae may get attached to such dead branches and grow. Filamentous algae of black colour usually colonize these dead branches, giving a black furry appearances to the branches under water.

In a branching colony, that too when it occupies the reef front zone, a partially dead and disintegrated stalk can result in the sliding away of the entire colony into deeper waters, where it will be buried by sediments. Branching colonies that occupy bottom level will never experience such a fate, since the interlocking of the branches of adjacent colonies will keep them in position even after the total disintegration of the stalk. And in such cases, colonies which do not have such interlocking, the tilting to one side will produce a result comparable to that seen in colonies growing along the reef front zone.

Too much accumulation of fouling organisms may cause considerable strain on the coral colony in which the stalk has already been weakened by borers. But such situation often turn out to be a blessing for corals, because massive sponges that grow across adjacent colonies will help to keep any coral colony, which is in distress in position. In the case of massive corals the question of getting themselves dislodged in such a manner, is quite remote. In such cases partial death is only possible and boring algae may flourish in the upper, well lighted side of massive corals, killing the polyps totally. But the coral colony compensates for the situation by accelerating its growth along the periphery producing a circular rim around the zone of dead polyps. In some cases, it is noticed that the overgrowing outer rim may curl in and completely cover the central area of dead polyps. If this is not possible, sedentary organisms attach themselves and grow in this area of dead polyps. *Tridacna* sp. may be seen attached to this area generally.

Mass transport of loosened corals and coarse detritus from the upper reef to the deep fore-reef slope is due to (i) total fall, (ii) slides and avalanches and (iii) reef subsidence and slump. The fore-reef slope, which is often covered with soft sediments becomes more stable and well drained, when coarse coral bits get deposited over it. This situation will indirectly encourage the colonization of corals. The minute calcium carbonate particles etched out from coral by activity of boring sponges constitute a major ingredient of the sediment fraction in any reef system. These particles have important lubricating and fluidizing properties which speed up the drainage of skeletal sediments from the reef.

Reef Associated Ornamental Fishes

The estimated area of coral reefs in Andaman Sea is 959.3 sq km. Reefs are home to a good number of aquatic species, far more than any other aquatic ecosystem. The category of ornamental fishes occurring in coral reef ecosystem includes the groups, such as, those of clowns, damsels, angels, lions, butterflies, triggers, puffers, snappers, hawk fishes, wrasses, groupers and gobies. Commonly available ornamental fishes in reef ecosystems are:

Sl.No.	Group of Fishes	Common Name	Species Name
1.	Pomacentridae	Sebae clown	*Amphiprion sebae*
		Blue damsel	*Pomacentrus caeruleus*
		Yellow tail damsel	*Neopomacentrus nemurus*
		Green chromis	*Pomacentrus pavo*
		Three spot damsel	*Dascyllus trimaculatus*
		Sergeant major	*Abudefduf saxatilis*
2.	Chaetodontidae	Pakistani butterfly	*Chaetodon collare*
		Vagabond butterfly	*C. vagabundus*
		Eight-banded butterfly	*C. octofasciatus*
		Racon butterfly	*C. lunula*
		Lined butterfly	*C. auriga*
		Banner fish	*Heniochus acuminatus*
3.	Pomacanthidae	Blue-ringed angel	*Pomacanthus annularis*
		Koran angel	*P. semicirculatus*
		Emperor angel	*P. imperator*
4.	Labridae	Cleaner wrasse	*Labroides dimidiatus*
		Moon wrasse	*Thalassoma lunare*
		Check board wrasse	*Halichoeres hortulanus*
5.	Acanthuridae	Surgeon fish	*Acanthurus mata*
		Powder blue surgeon	*A. leucosternon*
		Convict surgeon fish	*A. triostegus*
		Sail fin tang	*Zebrasoma veliferum*
6.	Scorpaenidae	Lion fish	*Pterois volitans*
7.	Pseudochromidae	Dotty back	*Pseudochromis fuscus*
8.	Balistidae	Orange-lined trigger	*Balistapus undulates*
		Red tooth trigger	*Odonus niger*
9.	Scaridae	Parrot fish	*Scarus ghobban*
10.	Tetradontidae	Puffer fish	*Canthigaster coronata*
		Guineafowl puffer	*Arothron meleagris*
11.	Haemulidae	Oriental sweet lips	*Plectorhinchus orientalis*
		Painted sweet lips	*Diagramma pictum*
12.	Zanclidae	Moorish idol	*Zanclus cornutus*
13.	Ephippididae	Bat fish	*Platax teira*
14.	Gobiidae	Gobi fish	*Istigobius ornatus*
15.	Holocentridae	Red coat	*Sargocentron rubrum*
16.	Muraenidae	Moray eel	*Gymnothorax meleagris*
17.	Lutjanidae	Mangrove jack	*Lutjanus argentimaculatus*
		Black spot snapper	*L. fulviflamma*

Contd...

Contd...

Sl.No.	Group of Fishes	Common Name	Species Name
18.	Holocentridae	Crown squirrel	*Sargocentron diadema*
		Sabre squirrel	*S. spiniferum*
19.	Serranidae	Square tail grouper	*Plectropomus areolatus*
		Camouflage grouper	*Epinephelus polyphekadion*
		Dwarf-spotted grouper	*E. merra*
		Coral grouper	*Cephalopholis miniata*
20.	Cirrhitidae	Dwarf hawk fish	*Cirrhitichthys falco*

Chapter 10
Microbial Flora and Fauna

Ecology of Microbial Plankton

In contrast to terrestrial ecosystems, micro-organisms are the main form of biomass in the oceans and form some of the largest populations on the planet.

The prevailing low-nutrient state of the ocean surface, in particular, characterize the ocean environment, which means, sometimes it is regarded as extreme ecosystem. Fixed form of nitrogen, phosphorus and iron are often at very low or undetectable levels in the ocean's circulatory gyres, which occur in about 70 per cent of the oceans. Photosynthesis is the main source of metabolic energy and the basis of the food chain. Ocean plankton account for nearly 50 per cent of global carbon fixation, and half of the carbon fixed in the organic matter is rapidly respired by heterotrophic micro-organisms. Most cells are freely suspended in the mainly oxic water column, but some attach to aggregates. In general these cells survive either by photosynthesizing or by oxidizing dissolved organic matter or inorganic compounds, using oxygen as electron acceptor.

Microbial cell concentrations are typically about 100000 cells per ml in the ocean surface layer (0-300 m), thymidine uptake into microbial DNA indicates average growth rates of about 0.15 divisions per day. Efficient nutrient recycling, in which there is intense competition for scarce resources, sustains this growth, with predation by viruses and protozoa keeping populations in check and driving high turn over rates. Despite this competition, steady-state dissolved organic carbon concentrations are many times higher than carbon sequestered in living microbial biomass. However, the average age of the dissolved organic carbon pool in the deep ocean, of about 5000 years suggests that much of the dissolved organic matter is refractory to degradation. Although dissolved organic matter is a huge resource, reviling atmospheric carbon dioxide as a carbon pool, chemists have been thwarted by the complexity of dissolved organic matter and have characterized it only in broad terms. Although most of the major microbial plankton clades have cosmopolitan distributions, new marine microbial plankton clades continue to emerge from studies that focus on unique hydrographic features. For example, the new Archaea, Crenarchaeota and Euryarchaetota were discovered at the brine-sea water interface of the Shaban Deep in the Red Sea.

There are also significant differences between coastal and ocean gyre microbial plankton populations. Typically, continental shelves are far more productive than ocean gyres, because physical processes, such as, upwelling and mixing bring nutrients to the surface. As a result, eukaryotic phytoplankton make up a larger fraction of the biomass in coastal seas, and species differ between coastal and ocean populations. Most of the bacterial groups found in gyres, also occur in large numbers in coastal seas, but a number of microbial plankton clades, particular members of the Beta-Proteobacteria, have coastal ectotypes or appear to be predominantly confined to coastal seas.

One of the most enigmatic microbial groups in the ocean is the marine group I Archaea. Geochemical evidence suggests that these organisms are chemoautotrophs. In the 1990s, Delong and Fuhrman established that archaea are widely distributed and numerically significant in the marine water column. The marine group I Archaea are crenarchaeotes. They predominantly occur in the mesopelagic, but are found at the surface in the cold waters of the southern ocean during the winter. Fluorescence *in situ* hybridization technology was used to demonstrate that marine group I Archaea populations comprise about 40 per cent of the mesopelagic microbial community over vast expanses of the ocean, making them one of the most abundant organisms on the planet. All of the marine Archaea remain uncultured.

The above concepts are general in nature and do not take into account specialized adaptations that allow organisms to exploit specific features of the ecosystem. For example, most marine bacterio-plankton seem to be lone cells that drift in the water column, but there is a subset of marine microbial groups that specialize in the colonization of particles. These cells have a special geochemical significance because particles sink, carrying carbon and other nutrients from the surface into the mesopelagic. This is the essence of "biological pump", which effectively reduces the partial pressure of carbon dioxide in the ocean surface, thereby increasing the rate at which atmospheric carbon dioxide dissolves in sea water. A particularly important class of particles are macroscopic aggregates, known as marine snow, which tend to form in the lower photic zone and disappear as the particles sink into the upper mesopelagic. Delong *et al* investigated organisms associated with marine snow and found little overlap between these organisms and the species freely suspended in the water column. Members of the Planctobacteria and Bacteroidetes phyla are particularly abundant on marine snow particles. Most Planctobacteria have specialized holdfast structures for attachment to surfaces. Bacteroidetes members frequently use gliding motility to traverse surfaces, and have been implicated in degradation of biopolymers associated with detritus. The recently discovered phylum Lentisphaerae has an unusual phenotype that suggests it might participate in particle formation. Lentisphaerae form a three dimensional polysaccharide net that makes water viscous.

The function for this net has not been proven, but the hypothesis that it serves to trap sinking particles fits the location of this organism, which resides in the upper mesopelagic, just beneath the photic zone.

Viral mortality and predator-prey relationship dominate current thinking about interactions between microbial plankton species, but it is difficult to believe that the water column is not spawned more insidious biological devices for gaining on the competition. A number of recent reports have focused on interactions between bacterioplankton and phytoplankton. Several strains of algicidal bacteria that produce algicidal peptides, antifouling agents or antimicrobial peptides have been described. The common genus *Pseudoalteromonas* crops up repeatedly in this context. But for now, the passive "microbial loop" model in which heterotrophic microbial plankton acquire organic carbon that is lost from phytoplankton by leakage or lysis, reflects the prevailing view.

Ocean's Microbial Realm

Smallest organisms on the earth running its largest ecosystem, the ocean. Planktonic microbial communities, comprising bacteria, viruses and the smallest protests (eukaryotes less than 5 micron in size), which inhabit the surface sunlit layers of all water bodies, run their food webs in a strikingly similar manner. The bulk of the cyanobacteria and algae and protozoa are represented by relatively few genera from several ancient widely separated races, suggesting that existent species within the food web have coevolved. Although the species composition varies, the structure of the food web in terms of the relative proportions and the maximum abundances of the plants, animals and bacteria is remarkably conserved, contrasting strikingly, with the huge variation in biomass and abundance of planktonic organisms larger than about 10 micron.

At the micrometer-sized scale, water is so thick and viscous that attempting to swim through it would be futile. Yet some peculiar looking bacteria and other strange organisms would be zooming through the honey thick medium with effortless ease.

Viscous forces rule the micrometer-sized organisms just as the force of gravity rules those large organisms. The microbial realm in viscous media is full of activity. The medium is continuously and thoroughly mixed by molecular diffusion, supplying dissolved nutrients to cell surfaces with such rentless efficiency, that it matters little, whether a cell stays stationary or moves around at a few body lengths per second. Yet a bacterial cell has to race against its medium at a speed of about 20 body lengths per second to experience a change of surroundings. This speed could be attained by *Escherichia coli*, our gut bacteria, though some planktonic bacteria can move five times faster.

These motile bacteria are propelled by rigidly rotating flagella, which are driven by reversible rotary motors. It was calculated that rotating this propeller costs very little energy and, in a viscous world, bumping into an object causes no damage. So bacteria can afford to move around at their maximum speed, restlessly searching for nutritive hot spots, such as, leakage of large organisms or flocks of detritus. But many bacteria are non-motile and thrive on nutrients supplied by molecular diffusion.

When resources are plentiful, microbial growth rates increase significantly, but cell numbers do not exceed a threshold value. The stable bacterial numbers, ranging around a million cells per milliliter, are explained by the heavy predation pressure of a few genera of flagellated protozoa that hunt, capture and ingest bacterial cells individually. If a bacterial population manages to grow through this predator gauntlet, it is decimated by viruses. So the numbers of bacteria and also of other microbes, are kept within boundaries by the double lock of predators and pathogens. As in an overgrazed pasture, potential resources are under exploited, when the predation pressure is heavy, avoidance and defense mechanisms will be selected for, planktonic bacteria hide forming spores, grow large by forming filaments, produce tough cell walls, secrete toxins and can also flee at high speed. Successful escape reactions can be triggered at encounter, during handling and even after ingestion. So not only do motile bacteria race to find hot spots, but they are also difficult to catch. As their wake is obliterated almost instantaneously by diffusion, they can not be tracked by predators. But such large particles (the nutritive hot spots) can be tracked, so bacterial populations growing on or inside them are vulnerable to attack. Many bacteria will be eaten, others will escape from the hot spot. Thus, the rate at which bacteria break down large particles will be slowed by the presence of predators.

Apparently, the stability of bacterial numbers is maintained by a set of interacting biophysical rules that balance the growth and loss rates of cells within the microbial size range. As one ascends the food chain, these rules weaken with increasing organism size, the ciliate predators of microbial organisms can increase their populations by orders of magnitude. So microbial production is channeled up a food chain that is tethered at its base. The excess nutrients are taken up by large algae that end up in large zooplankton. In all cases, the end result is large particles that sink from the surface layer faster than they can be broken down by bacteria. This rain of large particles impoverishes the productive layer and provides food for the deeper living organisms below, with immense consequences for aquatic ecology, ocean biogeochemistry and global climate.

The microbial network that stretches like veil across the ocean surface functions as a self-restrained, semi-permeable membrane that retains small particles, but lets bigger ones pass through. A closer look at the rules of the game in the fast lane of the microbial realm would require an *in situ* computerized tele-microscope. There are in orders of magnitude more bacteria in the ocean than there are stars in the universe.

Microbial Fauna on the Sediments of Andaman Sea

Marine bacteria (heterotrophic) on the sediments of Andaman Sea showed wide variations in total bacterial counts from 2.1X1000 to 3.7 X 1000000. No significant correlation has been observed between organic carbon values and the activity of heterotrophs. The ratio of gram positive to gram negative bacteria was 1:15.4 and the majority of isolates (55.11 per cent) were coccobacilli followed by pleomorphic forms (26.26 per cent).

The total maximum counts of heterotrophic bacteria in the sediments from the Andaman Sea were recorded at 10°38 minute N and 92°45 minute E (3.7 X1000000), where organic carbon was also maximum (14.8 mg/g), and minimum counts at 8°12 minute N and 93°36 minute E (2.1 X1000), where carbohydrate-C concentrations were high (12.62 mg/g) in the sediment. Generally, where the sediment consisted of yellow sand, organic carbon values are high. But bacterial total counts show no significant correlation with these type of sediments. It is possible that other factors may be limiting the growth of organisms. Low values of organic carbon (4.5 mg/g) were also found in latitude 11°50 minute N and longitude 93°05 minute E. Total nitrogen results showed a maximum at !2°05 minute N and 93°30 minute E (1.25 mg/g) and minimum at 12°32 minute N and 92°30 minute E and 13°N and 92°20 minute E (0.35mg/g). Low carbohydrate was found at 10°40 minute N and 92°20 minute E (0.45mg/g).

Among a total of 213 bacterial colonies isolated, 180 colonies remained viable during purification. Majority of colonies are circular in shape (66.66 per cent), approximately 5 mm in size (69.44 per cent), opaque in appearance (58.33 per cent), white in colour (61.11 per cent) and smooth in margin (62.62 per cent). In cell morphology studies gram positive : gram negative ratio is 1:15.4. The different groups found are Cocci (4.44 per cent), Coccobacilli (51.1 per cent); Rods (17.77 per cent) and pleomorphic bacteria (26.66 per cent). !3.33 per cent of the bacteria were of the sporulating variety, the majority of bacterial isolates were non-motile.

As regards physiological characteristics of the organisms, the majority isolated were nitrate reducers (95.55 per cent), glucose fermentators (86.11 per cent), catalase producers (100 per cent), starch hydrolyser (84.44 per cent) and cellulose decomposers (80.55 per cent). Isolates were not selective towards pectin as a substrate.

Bacterial isolates showed that although 45.5 per cent of the isolates were capable of growing at 10 to 15°Celsius, they could not tolerate low pH. Effect of salinity was moderate. Interestingly all isolates tested for their oxygen requirements seem to be facultative anaerobes in the laboratory. More than 50 per cent of the isolates showed an ability to utilize sugars, alcohols and carboxylic acids. Phenol is used by very few isolates (14 per cent) at certain stations in the Andaman Sea.

The majority of organisms were gram-negative and non-motile. Bacteria culture from these sediment samples exhibit a wide range of physiological activities. More than 50 per cent of the organisms are nitrate reducers, gelatin liquifiers, catalase, amylase, and cellulose producers. Similar types of characteristics were reported from different marine environments. The nature of the sediment in Andaman Sea varies between coralline sand and clay. At 11°20 minute N, 92°08 minute E; 12°32 minute N, 92°30 minute E and 13°N, 92°20 minute E, the bottom was sandy clay and this appears to favor the distribution of heterotrophic bacteria as far as total counts are concerned.

Although these organisms have not been identified to the species level, the cultures show great deal of overlap in physiological characteristics. For example, at 10°30 minute N and 92°15 minute E and at 10°38 minute N and 92°45 minute E, nitrate reducing activity is shown by 100 per cent and gelatin liquefaction activity by 100 per cent of the total organisms cultured. Morphological identifications show that similar types of colonies were involved in both cases. This suggests that the organisms may be able to synthesize necessary enzymes depending on the nature of their substrates. Such behaviour is typical of facultative chemoorganotrophs.

Three stations at 10°28 minute N, 93°E; 10°40 minute N, 92°20 minute E and 13°N, 92°20 minute E showed the highest value for total bacterial counts and meiofauna density. The constituents of organic matter, such as, carbohydrate carbon and sediment nitrogen do not show any correlation with total counts either. Since the organisms isolated appear to be facultative chemo-organatrophs, the absence of such correlation is expected.

Interestingly, phenolic degraders are present at only 8 stations out of the 19 samples tested from the Andaman Sea. These 8 stations are situated mostly in the southern part of the Andaman Sea and 35 per cent of the organisms isolated from the stations are found to be phenol degraders.

It is interesting that higher concentrations of such bacteria occur to the north-west of the 10°channel and south of Little Andaman Island. The reasons for such a distribution is not clear. With the available data, it is difficult to assess whether precursors of phenolic compounds are produced autochthonously by anaerobic degradation of organic matter or whether such compounds allochthonously produced are transported to this area by physical processes of current combined with bottom topography of the area. It may be significant that, oxidisable organic carbon at these stations is generally low, as is the dissolved oxygen content in the overlying water column. The sediment is composed of sand, sand and clay or clay alone and is not significantly different from that of other places.

Physical and chemical data from this region indicate oxygen depletion with depth in the water column and it is possible that the sediment, at least in some areas, is oxygen deficient in the month of February.

Table 10.1: Colony and Cell Morphology of Bacterial Isolates Obtained from Sediments of Andaman Sea

	Per cent of Isolates			*Per cent of Isolates*
1. Shape Circular	66.66	5. Margin Smooth		62.62
Irregular	19.44	Entire		25.0
Spreading type	8.33	Rough		12.77
Branching type	5.55	6. Gram (+)		6.1
2. Size Less than 5 mm	25.0	7. Gram (-)		93.88
Equal to 5 mm	69.44	8. Cocci		4.44
More than 5mm	5.55	9. Cocco bacilli		51.11
3. Appearance Opaque	58.33	10. Rods		17.77
Transparent	8.33	11. Sporulating		13.33
Translucent	17.77	12. Pleomorphic		26.66
Granular	15.55	13. Motility		32.22
4. Colour Colourless	22.22			
White	61.11			
Pink	8.33			
Creamish (yellow)	8.33			

Table 10.2: Biochemical and Physiological Characteristics of Bacterial Isolates from Andaman Sea

Sl.No.	Groups	Per cent of Isolates
1.	Oxidase (+)	66.22
2.	Hydrogen sulphide formers	34.44
3.	Nitrate reducers	95.55
4.	Gelatin liquifiers	73.88
5.	Glucose fermentators	86.11
6.	Catalase producers	100.00
7.	Cellulase producers	80.55
8.	Amylase producers	84.44
9.	Pectinase producers	40.00

Table 10.3: Growth Requirements of Bacterial Isolates: Physical

Sl.No.	Growth at	Per cent of Isolates
1.	Temperature 10-15°Celsius	45.5
2.	pH 4	19.44
3.	pH 6	23.33
4.	pH 8	100.00
5.	pH 10	61.11
6.	Salinity 7-10 ppt	82.22
7.	Salinity 17-20 ppt	100.00
8.	Oxygen depleted medium	100.00

Table 10.4: Nutritional Requirements of the Bacterial Isolates

Sl.No.	Organic Compounds Tested	Per cent of Isolates
1.	Glucose	94.44
2.	Fructose	98.33
3.	Ethanol	66.66
4.	Butanol	61.11
5.	Glycerol	42.22
6.	Acetic acid	51.11
7.	Citric acid	54.44
8.	Phenol	13.88

Chapter 11
Nanoplankton

Nanoplankton from Recent Sediment off Andaman Islands

Andaman and Nicobar islands separate the Bay of Bengal from the Andaman Sea. The islands are fairly straight and gentle on the western side comprising coastal plains, whereas, the eastern coasts are strongly indented and steep and in many places coral reefs and raised beaches, as high as 20 meters above sea level are reported. Narrow channels separate the Andaman into North, Middle and South Andaman Islands, which are collectively known as the Great Andaman.

The Paleocene Andaman-Nicobar trough was folded and reverse faulted westward resulting in the emergence of the Andaman Islands during the Oligocene and late Miocene. The Andaman and Nicobar ridge is formed of serpentinite basement overlain by distorted Paleocene to Miocene and flat lying Pliocene to Recent sediments. A sequence of Paleocene through Upper Miocene rocks including over 3000 meters of Upper Eocene and Oligocene graywacks were deposited over the Andaman serpentinites under fluctuating shallow to deep water conditions.

The samples collected in the vicinity of Andaman Islands are largely clays except four, which are terrigenous sands. Smear slides of the clays were prepared and scanned under Reichert polarizing microscope and relative abundance of nanoplankton was estimated following Bounreaux.

The microfossil assemblage in the sediment is a complex mixture of calcareous nanoplankton, foraminifera, diatoms, silicoflagellates, radiolarians, ascidians, holothurians, chrysomonads etc. A total of 48 species of coccoliths and discoasters in the sediment, which is an admixture of modern and fossil flora of 14 and 24 species respectively, was recorded.

Of the 14 modern flora encountered (*Braarudosphaera bigelowi, Ceratolithus cristatus, Coccolithus pelagicus, Cyclococcolithus leptopora, Emiliania annula, E. hguxleyi, Discolithina japonica, Gephyrocapsa oceanica, Helicosphaere carteri, Rhabdosphaera* sp., *Scyphosphaera apsteini, Syracosphaera* sp., *Thoracosphaera* sp., *Umbilicosphaera sibogae*), *Gephyrocapsa oceanica* is the most abundant found in all the stations. Owing to the limited magnifications of the light microscope and the close affinity of various species with each other, it was not possible to identify the different species of *Gephyrocapsa* and therefore, all the gephyrocapsiids are grouped together. Following *Gephyrocapsa oceanica*, other species, such as, *Cyclococcolithus leptopora, Helicosphaere carteri* and *Ceratolithus cristatus* are also present in the sediment of all the stations in varying proportions. The relative abundance of *H. carteri* increases with depth. *Thoracosphaera* sp., *Umbilicosphaera sibogae* and *Discolithina* sp. are rare. The remaining seven species, *Emiliania huxleyi, E. annula, Coccolithus pelagicus, Rhabdosphaera* sp., *Syracosphaera* sp., and *Braarudosphaera bigelowi* occur in small quantities in a very few samples.

The majority of modern flora in the assemblage indicate their tropical affinity, among which, *E. huxleyi* and *C. leptopora* are euty-thermal, and the overall assemblage indicates a temperature range from 18 to 28°Celsius.

In addition to the 14 modern species of nannoplankton, 24 reworked species (*Discoaster aulakos, D. barbadiensis, D. herggrenii, D. bollii, D. brouweri, D. calcaris, D. challengeri, D. deflandrei, D. exilis, D. intercalaris, D. levinii, D. lodoensis, D. pentaradiatus, D. quinqueramus, D. saipanensis, D. surculus, D. tamales, D. variabilis, Coccolithus miopelagicus, Reticulofenestra pseudoumbilica, Sphenolithus obies, S. dissimilis, S. neoabies, S. obtusus*) were also encountered, which range in age from Eocene through Pliocene, though the majority belong to Miocene. It is possible that these might have been derived from the erosion of the Miocene formations in the region.

It is a very well established phenomenon that, owing to their minute size, the nannoplankton is more susceptible for reworking into younger sediment without sustaining any sign of mechanical damage. This poses a problem in differentiating and delineating the biostratigraphic zonation and in assigning a precise age of the sediment. Similar phenomenon of redistribution of nannoplankton has been encountered in the Arabian Sea sediment. It has been reported from the slope region off Bombay that Cretaceous and Tertiary fossils are being mixed up with the Recent sediments deposited by the Indus River into the Arabian Sea. Besides, it has also been reported in the Gulf of Kutch and South East Arabian Sea. The Gulf of Kutch is yet another example of the presence of reworked fossils. They range in age from the Cretaceous to Quarternary and the mixing is mainly caused by the very strong tidal currents prevailing in the Gulf of Kutch. Furthermore, it is inferred that the Gulf is one of the contributors of reworked fossils to the western slope and also to the deeper regions.

Frerichs (1967) reported that the Pioneer dredge eight sampled radiolarite shale with Upper Miocene fauna. In addition, Pioneer 12 and 13 have recovered Miocene calcarenite and calcilutite from two sites at a depth of 1000 m on the eastern slope of Invisible Bank to the east coast of Andaman. As has been suggested earlier, the source of reworked fossils is mainly from the late Neogene submarine outcrops in the

Bay of Bengal and Andaman Sea regions, which might contribute the older floral components and get mixed up with the Recent sediments. Also there is every possibility that the Andaman Islands, which are of Tertiary formations would have for some extent act a provenance in supplying the older fossils through degradational processes. However, the supply may not be regular/continuous for want of any drainage system properly on the islands. In addition, the coral reefs present along the coasts are to some extent act as barriers to the transport of sediments to the deeper regions. The Andaman and Nicobar Islands are insignificant provenances for the basin. Therefore, as an alternative, it is felt that the turbidity currents/ or submarine slumping must have played a major role in bringing about the processes of mixing. The processes of slumping, transports accumulates often strata on top of younger formations causing stratigraphic reversals. In case of Andaman Sea, it appears that the turbidity currents to be the prominent one. Because of the major rivers, such as, Irrawaddy to the north of Andaman Islands flush out enormous amount of terrigenous material into the sea, which generally generate turbidity conditions. The increased detrital sediments that might have resulted by the turbidity currents is the dominant effect on the *G. oceanica* abundance. It is felt that the abundance of *G. oceanica* to the increased sedimentation rate in the sediments of North Florida itself. The LANDSAT imageries of this region also indicate that the area is influenced by turbidity. This is also substantiated by the presence of ascidians found very commonly in all the stations. Occurrence of fossil spicules of genus Micrascidites indicates the turbidite deposits. The turbidites of Miocene to Pliocene are good examples of a well preserved redeposited nanno-fossils in Coral Sea assemblage.

Summerising, the source of the reworked fossils may mainly be outcrops of Miocene occurring on the Andaman Sea floor and in part from the turbidites resulting from deposition of sediment by the Irrawaddy River.

Chapter 12

Phytoplankton and Algae of Tropical Seas

The dense thallus of *Nostoc* under the family Nostocaceae of Cyanophyta division is composed of numerous intertwined, contorted filaments with thin walls and a thick gelatinous sheath. Gelatirous thallus is extremely variable as to size and shape, solid or hollow, floating free or attached. Growing in sea or on surface of damp soils, *Nostoc* contains a fair amount of protein and the Chinese use sea weed jelly made from species of *Nostoc*. Cosmopolitan in distribution, *Nostoc commune occur* in Andaman Sea.

The colourless rhizomes of *Caulerpa* under the family Caulerpaceae of Chlorophyta division are attached to sand or to dead pieces of coral by means of colourless rhizoids. The rhizomes give rise to a great number of green shoots, which are simple and phylloid, or exhibit axis bearing ramuli of various shapes. A large widely distributed genus in the tropics. Most of the species occur in shallow water of the littoral region, below low tide level, preferably in lagoons protected by coral reefs.

The species, *Caulerpa racemosa* with varieties like, *clavifera, laetevirens, uvifera* and *Caulerpa sertularioides* are important in diets as salad in the Philippines, Island of Guam, and South Bali and are also available in Andaman Sea.

The plants under the genus *Chaetomorha* of family Cladophoraceae are composed of unbranched filaments, which sometimes fasten themselves by means of more or less branched rhizoids. In other species, however, they are not attached at all, the filaments being twisted together forming intricate clumps resting on stones or other algae. The genus is found both in the sea and in brackish water all over the world. They are mainly attached to the rocks in the littoral zone (ultra-littoral belt).

Chaetomorpha crassa occur in Indian Ocean and Andaman Sea. The Chinese at Ambon collect these weeds in the months of August and September (dry months). After being soaked a night in fresh water, they are dried in the sun. The dried weeds are boiled and eaten with bacon.

The thallus of the species under the genus *Codium* of family Codiaceae, distributed widely is spongy. The erect, richly dichotomously branched threads of the plants are anchored by a small cushion like disc or by tufts of rhizoids on coral reefs or on exposed rocks in the surf (littoral zone).

Codium tenue and *C. tomentosa*, available in Indian Ocean and Andaman Sea are eaten raw as salad in Philippines, eaten as food in several parts of Malaysia. In the Hawaii Islands, the species is eaten as raw, usually with tomatoes, after being thoroughly washed in fresh water. It is never cooked or blanched, for it becomes soft and disintegrates quickly with heat. The ascorbic acid content of the species, when kept in sea water in a refrigerator for two days after gathering was found to be low, only 5 mg of ascorbic acid per 100 grams.

The thallus of *Enteromorpha* under the family Ulvaceae is tubular, simple or with more or less numerous branches. The species occur in the upper parts of the littoral zone, near the shore in quiet waters, attached to small stones and rocks by means of rhizoids. *Enteromorpha* species are capable of enduring a wide range of salinity, therefore, often occurring in brackish waters of estuaries.

Enteromorpha intestinalis and *Enteromorpha prolifera*, inhabitants of Indian Ocean and Andaman Sea, is a good source of vitamin A and less good source of vitamin B. The value of *Enteromorpha* in diet is its flavor and its tendency to prevent constipation. The chemical analysis of the plant reveal; water-90.34 per cent, protein (N X 6.25) – 2.82 per cent, fat – 0.0485 per cent, ash – 1.59 per cent, calcium – 0.17 per cent and phosphorus – 0.0336 per cent.

Eaten in the Philippines and in the Malay Peninsula. It is eaten by the Japanese as a supplementary article of the diet with boiled rice only after being sun dried. The species also serve as food for milk fish (*Chanos chanos*) in the Philippines, where fish feed on algal growth in the pond other than this green algae and *Chaetomorpha* area, they develop an unpleasantly strong fishy taste. It would therefore, appear that the food reserve in the green algae has a direct bearing on the palatability of the milk fish.

The flat broad and membranaceous thallus of *Ulva lactuca* of family Ulvaceae consists of two layers. It is attached by a small disc to rocks, stones, valves etc. On the upper part of the littoral zone (ultra and supra-littoral belt). These sea weeds prefer spots, where the water is polluted by organic matter and where supplies of fresh water mingle with sea water.

Being cosmopolitan in distribution, the species occur in Andaman Sea. The plant (sea lettuce) is eaten widely in the East in soup, as salad and that it is employed for garnishing. The plant contain, 50 per cent of carbohydrate and 15 per cent protein if water is reduced to 19 per cent.

The thallus of the gnus *Hydroclathrus* under the family Encoeliaceae of Division Phaeophyta is, in the full grown state, hemispherical and regularly pierced with orbicular holes. Growing older these holes gradually dialate more and more, thus forming an open meshwork up to 25 cm in diameter. The young thalli, in the earliest stages are hollow and not perforated. On account of this, the mature specimens have the aspect of a sponge, which is loosely attached by the lower surface to the substratum.

Hundreds of these yellowish olive, net like expansions are sometimes heaped together by wind and waves in the lagoons of the coral reefs. The colour, however, rapidly changes and the sea weeds become dark brown in a dried state, whereas the plants lose rigidity at the same time. Distributed in Indian Oceans, the species is eaten raw as salad in the Philippines.

The thallus of *Padina* of family Dictyotaceae, is flat leafy, fan-shaped, fringed with articulated hairs, and marked at regular distances with concentric lines. The fronds are entire or sometimes split into sections. The tissue is differentiated into two kinds, a single upper and lower surface layer of assimilating cells, and a central one consisting of two or more layers of parallel epipedal cells.

Male and female gametangia on the same plant, occurring in concentric zones between the fringed bands. The tetraspores are developed on separate plants of the same appearance and are scattered along the margins of the zones.

The property of fronds of reflecting prismatic colours, whilst growing under water, together with the fan-like shape, have given the sea weeds the popular name of "Peacock's tail".

The algae grow on somewhat sandy bottom in lagoons etc, attached to little coral pieces, shells and stones by organs, which are densely coated with woolly hairs. They occur in the lower part of the littoral and in the highest part of sub-littoral region. The species is sometimes eaten as a salad, but frequently it is collected for preparing a gelatin-like sweet meat. *Padina australis* occurs in Indian Ocean and Andaman Sea.

The plants belonging to the genus *Sargassum* of family Sargassaceae have the appearance of land plants showing a differentiation into "roots", "stems" and "leaves". The vertical dichotomously branched axis is attached to the substratum by a distinct, often massive, lobed holdfast (the "roots"). The axis has a short stipe-like portion (the "stem") and bears in the upper parts the repeatedly alternate branches on all sides with broad to filiform, entire or serrate, leaf like organs (the "leaves") showing a mid-rib. Moreover, it is provided with stalked air-bladder (vesicles) and more or less forked, cylindrical receptacles, developing in the axis of the so-called leaves.

The stipes are attached to rocks, dead coral pieces etc, in the sub-littoral region, sometimes forming a dense zone around a coral island being settled upon the abrasion lay of the reef. On the other hand, masses of *Sargassum* plants are often seen floating in the sea having been torn off the shore by storms and carried in the open water by currents. The plants regenerate from broken thalli.

The ash of burnt *Sargassum* species contains a large quantity of potash and carbonate of soda. Therefore, the plants are cultivated in the coastal regions of some countries and the ash sold under the name of "kelp". Iodine is another valuable constituent found in the tissues, especially in those weeds collected in the open sea. Chemical analysis of Indian *Sargassum* species are found to contain, moisture (air dry sample) – 11.72 per cent, loss on ignition – 61.63 per cent, insolubles – 0.17 per cent, solubles – 26.48 per cent, nitrogen – 1.02 per cent, potash – 5.99 per cent and phosphoric acid – 0.21 per cent.

Sargassum aquifolium, S. granuliferum and S. polycystum occurs in Indian Ocean and Andaman Sea. The species is used as fertilizers in Philippines and India. The tips of the weed are eaten raw or cooked with coconut milk.

Turbinaria belonging to family Sargassaceae are attached to the substratum by short branched holdfasts, sending up one or a few simple or branched axis. The stiff cylindrical main axis or stipe bears leaf-like organs. These have the appearance of a triangular shield or disc with a serrate or nearly entire margin, tapering downward usually in a pyramidal fashion to the petiole. The air-bladders when present, are immersed in the leaf-like organs near the end. The receptacles grow into dense clusters from the base of the petioles. The plants of this genus occur in the sub-littoral region of the warm waters of the tropics and sub-tropics, growing on rocks and pieces of dead corals. *Turbinaria* species contain, moisture (air dry sample) – 6.13 per cent, loss on ignition – 63.07 per cent, insolubles – 0.50 per cent, solubles – 30.30 per cent, nitrogen – 0.96 per cent, potash – 9.11 per cent and phosphoric acid – 0.27 per cent.

Turbinaria ornate, T. conoides occur in Indian Ocean and Andaman Sea. The species are eaten raw and cooked with coconut milk and also used as manure in coconut plantation.

The dull purplish thallus of the plants belonging to the genus *Porphyra* is flat and thin, foliaceous, one or two cells thick. The plants are very short stalked and attached by a very small holdfast, expanding above into a soft slippery blade with an entire or lobed margin. The species of this genus grow in the littoral zone on stones or rocky parts, and show marked resistance against desiccation.

The species of *Porphyra* furnish a plentiful supply of food to the people in quite distant regions of the earth. In Scotland, Wales and Ireland, *Porphyra* species are used as laver-bread. It is usually served with bacon for breakfast, made into small flat cakes and fried crisp in the bacon fat or heated with butter, lemon-juice and pepper, and served with roast mutton.

Porphyra tenera is cultivated on a large scale in several parts of Japan. In Tokyo Bay, at low tide, men go out in small boats with bundles of bamboo prepared for the purpose. In order to make holes, up to 10 or 15 feet deep at high tide, conical wooden boxes or frames are pushed down into the muddy bottom. Then the bundles of bamboo are planted or set out in these holes.

The spores of *Porphyra* float in the water before dropping down and becoming attached to some object. In the culture area, they attach themselves to the twigs of bamboo bushes and develop so rapidly that by January they have grown into full-sized plants. The women and girls then finish the business of harvesting the crop.

The fresh red fronds of *P. tenera* picked from the twigs are thrown into tubs of fresh water and stirred with long sticks in order to remove the sand and other foreign substances. The plants are then sorted and chopped with sharp knives into fine particles and spread out in sheets of a uniform size on bamboo mats placed on inclined frames in the open air. In a short time the *Porphyra* sheets are dry. To get them ready for the market, they are stripped from mats, pressed and then made into bundles of ten sheets each, the dimensions being approximately, 25 to 35 cm the sheets have the appearance of dark brown mottled paper with a glossy surface.

Before being eaten it is crisped over a fire, which changes the colour to green. Then it is rubbed between the hands and the small pieces are dropped into soup or sauces to give a pleasant flavor. Now-a-days it is often preserved in tins, after having been boiled with soy bean sauce. For preparation of "sushi", first cold boiled rice is spread over a sheet of *Porphyra*. On the rice are laid pieces of meat or fish. The entire sheet or layer is then rolled, like an American "Jelly roll" and cut into slices.

The dried product of *Porphyra*, called "gamet" used by the Filipinos in Hawaii. Gamet may be served in several ways. It is added to hot soups containing vegetables, fish, shrimp or chicken. It may be cut into small pieces or strips, softened in a small amount of water and served with salad vegetables especially, sliced or mashed tomatoes. It is sometimes fried dry and crisp in a very small amount of fat and served with rice or vegetables.

Acanthophora spicifera, belonging to the family Condriceae, an erect plant and form bushy with a base of fibrous holdfast strands. They are up to 2 meter high with dull grayish purple colour. The branches are long, cylindrical and stiff, bearing short spur-branchlets, especially developed on the smooth main axis. These spur-branchlets are rather crowded with short spines, spirally disposed. The seaweeds occur in shallow water, in lagoons, on rocks and dead pieces of corals near the low tide line. In Kangean Island, the sea weeds are first immersed in hot water and after that eaten with red pepper.

Caloglossa adnata, belonging to the family Delesscriaceae, has flat thallus and dichotomously branched with a pronounced midrib. The secondary branches arise from this midrib. The terasporangia are distributed near the tips of the fronds in parallel transverse lines. The cystocarps are on the upper branches on the under side of the midribs. The male plants are smaller than the females, The colour is violet.

The sea weeds are attached to the breathing roots of the mangrove trees or to the rocks by means of holdfasts creeping beneath the dichotomies. They are distributed in the upper part of the ittoral region and even above that level where the sea weeds are wetted by the spray. It is eaten raw or boiled.

The plants of the genus *Eucheuma* have cylindrical to flattened main axis and branches of a pale reddish colour drying to yellow brown. Branching irregularly alternate to some what di-or trichotomous in appearance. Main axis and branches roughened by blunt nodules or spines, which harbor the gemetangia. The central tissue is composed of colourless filaments and surrounded by large cells grading to a small-celled epidermis. The general substance of the plants is firmly gelatinous to somewhat cartilaginous. The sea weeds occur in the sub-littoral region attached to a firm substratum. *Eucheuma* species is used along with other species of this genus (*E. gelatinae, E. horridum, E. muricatum, E. serra*) in the manufacture of agar-agar.

Gelidium amansii, belonging to the family Gelidiaceae, have flattened or cylindrical, with many usual lateral branches. The substance is cartilaginous. The main axis is erect, arising from a somewhat fibrous base. The axis is formed of long cylindrical cells (medulla) and a compact cortex of cells becoming small at the surface. Very slender refractive filaments (rhizines) are extend in this cortex. The organs of reproduction are immersed in swollen branches.

The plants occur on a firm substratum, especially coral fragments in the sub-littoral region in exposed places. The gelatin-like substance derived from *Gelidium* species and certain other red sea weeds is named agar-agar. The agarophytes are collected by divers and simply dried and bleached in the sun on the beach. Then the sea weeds are boiled with several times their weight of water. Sometimes some vinegar or coconut milk is added. After straining, the extract is ready to eat.

In the case of sea weeds sold to Japanese factories, situated in mountainous areas, the dried sea weeds are successively cleaned, washed in fresh water, drained and bleached on bamboo frames. Subsequently the agarophytes are fused into sheets and the jelly is extracted by boiling. After straining, the jelly is poured into wooden trays and allowed to cool. At a certain stage of this process, the hardening jelly is cut into bars of certain size. The most important use for agar-agar is in micro-biological culture media but it is extensively used for human and animal food as roughage, as a stabilizer in bakery and diary products and in making jelly desserts, confectionaries, icings and salad dressings. Agar-agar is employed in canning meat and poultry, in laxative preparations, as a constituent of medicinal tablets, pills and capsules, in numerous pharmaceutical emulsions and ointments, in cosmetic creams and jellies, in paints and varnishes, in wire drawing lubricants, in liquor clarification and in creaming of latex and as dental impression mould base etc.

Gelidium contains, ash- 4.23 per cent, lime – 0.28 per cent, magnesium – 0.52 per cent, nitrogen – 2.01 per cent, crude protein – 12.56 per cent, fiber – 17.89 per cent, galactan – 23.70 per cent, pentosan – 2.30 per cent, methyl pentosan – 0.93 per cent, reducing sugars after hydrolysis with dilute acid – 23.20 per cent.

The plants of widely distributed genus, *Gracilaria* belonging to the family Gracilariaceae are bushy and irregularly or dichotomously branched. The branches are compressed or flattened. The whole plant is of a firm cartilaginous substance, horny when dry. The thallus consists of two layers of cells, a central one (the medulla) of large angular, colourless cells, and a cortical one of small assimilative cells.

Sessile on the branches are cystocarps, containing the sexually produced spores in the form of prominent swellings. Tetrasporangia are scattered among the cortical cells. Plants are light pink to purplish coloured, occurring on rocks and broken corals in the littoral region.

Gracilaria confervoides as a commercial source of agar-agar has been duly recognized and well established. *Gracilaria* contains moisture – 10.71 per cent, insolubles – 2.41 per cent, solubles – 15.29 per cent, nitrogen – 1.48 per cent, potash – 0.66 per cent, phosphoric acid – 0.07 per cent. Loss on ignition has been found to be 71.59 per cent.

The species like, *Gracilaria confervoides, G. crassa, G. eucheumoides, G. lichenoides, and G. tacnioides* have been found to occur in Indian Ocean and Andaman Sea.

Grateloupia filicina, belonging to the family Grateloupiaceae, is bushy plants and of moderate size, up to 60 cm high. The branching is generally pinnate and complanate, the branches being flattened in the middle part and tubular at the base and the tips. The texture of the thallus is fleshy to membranous. A medulla is present, consisting of slender anastomosing filaments surrounded by a jelly and a cortex of moniliform filaments, loosely or solidly united with mucus. Gametangia are scattered over the surface. Colour reddish-brown to violet. Occurring in shallow water of the littoral region on a firm substratum.

The species has a pleasant flavor even to one unaccustomed to eating sea weed. In Hawaiian Islands, the sea weeds belonging to this species are eaten with limpets.

Halymenia durvilliae, belonging to the family Grateloupiaceae consist of a compressed, flat thallus of moderate to considerable size (up to 50 cm long) has a soft fleshy consistency. The branching is dichotomous to pinnatifid. Structurally the thallus shows a few, slender filaments embedded in a soft jelly scattered through the surface cellulose. Occurring on coral reefs in the littoral region. Extremely variable. The species is eaten in Philippines and in the Hawaii Islands.

The bushy plants of the genus, *Hypnea*, belonging to the family Hypneaceae, occurring in most warmer seas, consists of a fleshy, cylindrical thallus, which is alternately branched. The longer branches are frequently curled at the tips, the shorter ones are often spine-like acute. The central axis (medulla) is segmented and surrounded by a cortex with larger cells toward the center and minute coloured ones toward the surface.

Gametangia is situated in swollen branches. Colour brown to yellowish-red. Substance sub-cartilaginous. Occurring in the lower part of the littoral region, where water is quite and in rock pools. The species has an agar-agar content of 30 per cent. It has a very low viscosity and a setting point just above 30°Celsius. *Hypnea musciformis* gives an excellent bacteriological agar.

Hypnea cervicornis after being brought ashore are dried on the white coral beach for three days. Subsequently they are washed in fresh water and again exposed to the open for three days. The colour of the sea weeds then turns yellow-white. After boiling, filtering and cooling the jelly is eaten with palm sugar and rasped coconut.

Liagora farinose belonging the family Helminthocladiaceae, occurring in all warmer seas form rather compact tufts. It is mostly dichotomously, but also laterally branched. Surface farinaceous, however, this being covered by calcareous deposit. When decalcified the axial, longitudinal branching can be easily seen. The corticating filaments hardly extend beyond the calcification, hence the absence of a distinct colour. The assimilating filaments are moniliform.

Gametangia in node-like discs of the upper part of the frond. Older plants look like coralline algae, but they are not so stiff and brittle. The species show much variation in habitat. Occurring on rocks and broken pieces of coral in the littoral region.

Nearly cosmopolitan, *Rhodymenia palmata*, belonging to the family Rhodymeniaceae has a flat frond, which is dichotomously or palmately divided at wide angles. The interior cells are large and uncoloured; the cortex being composed of minute coloured small cells. Gametangia are scattered over the thallus. Occurring on rocks and dead pieces pf coral in the littoral region. The sea weed is used as vermifuge.

Sarcodia montagneana, belonging to the family Sarcodiaceae consists of a flat, flabellately divided thallus, frequently shortly stipitate and provided with marginal prolifications. The prolifications are often again dichotomously divided and are gradually attenuated toward the base. Margins acute or dentated. Structurally the thallus shows number of slender, long filaments and a cortex of cells gradually diminishing in size toward the surface.

Gametangia are scattered over the surface of the thallus. Colour pink when fresh, but dark red when dried. Occurring in the lower parts of the littoral region attached to a firm substratum, but also in the sub-littoral region. Much used as food in the Moluccas.

Sea Grass

Sea grass are angiospermous plants adapted to grow in marine environment. Sea grass meadows are the nursery ground for many commercially important shrimps, crabs and fishes. Its root mat adds stability to the sediments of the coastal zone and the leaves help filter the water of suspended particles.

The Andaman Sea harbors 12 species of sea grass, namely, *Cymodocea rotunda, C. serrulata, Enhalus acoroides, Halodule pinifolia, H. uninervis* (broad and narrow leaved forms), *H. beccari, H. decipiens, H. ovata, H. ovalis, H. stipulacea, Thalassia hemprichii* and *Syringodium isoctifolium*.

Of these, *Enhalus acoroides* is the largest sea grass species, reaching to a height of more than one meter. This sea grass prefers sheltered marine environment, where substratum is fine sand to clay. Normally it occurs in the tidal and sub-tidal zones. Some of the plants are seen to establish among *Cymodocea serrulata* and *Halodule pinifolia* sea grass species in depth of 40 cm during low tide. It is the only species, which shows aerial surface pollination. It has 3 to 4 linear leaf blades produced directly from the rhizome, although shoots with 7 to 9 leaves were also observed. Its leaf tips are rounded and with minute serrated projections when young. The underground rhizomes are stout (1-2 cm diameter) and branched, which can penetrate up to 50 cm deep. The roots are also well developed, both of which provide concrete anchoring system. The *E. acoroides* male flower is small, contains large pollen grains and has pendunculate spathe consisting of two connate blades, the inner and outer. The female flower has ovary rostrate, six parietal placentas protruding far towards the center and with long pendunculate spathe consisting also of two free blades. The fruit of *E. acoroides* is ovoid or globular in shape with acuminate tips and the entire surface is covered by dense bifid projections.

Large number of fruits of this plant, during south-west monsoon, with an average wind velocity of 12-13 km/hour causes the fruits to dislodge from the plants to reach the shores. The length of the fruits varied between 5.7 to 6.5 cm and diameter 8.9 to 12.3 cm. The number of seeds per fruit varied between 4 to 11. Seeds were in the length range of 0.6 to 1.3 cm and diameter 0.6 to 1.5 cm. The seeds are edible and are eaten by the local people. Some of the seeds were observed to be germinating within the fruit.

The seeds of *E. acoroides* are more viable, when compared to other sea grass species, seed production, seed dispersal and seedling recruitment serve as important mechanism in maintaining the genetic diversity of plant. Dispersal distances of sea grass seeds are usually limited to a few meters and seeds settle rapidly into the sediment surface except during seasons of high wind velocity and consequent tidal current. Seed deposition is more advantageous over vegetative propagation of new shoots for maintenance of sea grass meadows frequently subjected to environmental stress.

Dinoflagellates

The dinoflagellates are part of the marine plankton population and are important for their share in the primary productivity of the sea. The marine phytoplankton (minute chlorophyll-containing organisms) populations are often mainly composed of diatoms.

Dinoflagellates are placed in a phylum designated for them alone, Pyrrophyta. Like diatoms and coccolithophorids, certain dinoflagellates produce a wall material, partly cellulose, that is preserved by the fixatives that are used in microscopy and that can be recognized in non-living plankton samples. These dinoflagellates are called the armored forms and they produce a series of plates that have distinctive shapes, sizes and alignment for each of the various species. There are other dinoflagellates that produce no plates of armor and are quite devoid of wall materials. These unarmored cells are usually unrecognizable in preserved plankton samples, because the cell membrane does not retain its shape in commonly used preservatives like formaldehyde.

Dinoflagellate cells are very complex, exhibiting a wide variety of structure and form, containing many different organelles, some of which are unique to this group of organisms. The cells of most dinoflagellates are oriented from top to bottom, and two flagella producing quite different movements are attached on the vertical (lower) side of the cell, and located within two distinct grooves of the cell surface membrane. In the girdle groove, which encircles the cell, lies a flattened ribbon like flagellum that beats with a rapid helical action in the groove and this motion seems to cause most of the forward thrust of the cell. The cell swims rapidly in a spiral path often rapidly rotating in either direction of its own body axis. The second flagellum lies in a longitudinal groove, the sulcus, that extends toward the posterior on the ventral side of the cell. The flagellum in this groove occasionally seems less active than the girdle flagellum and beats in a planar wave, probably contributing to the spiral path of the cell as well as to the forward propulsion. The force causing cell rotation was attributed to the shape of the cell in *Ceratium* (an armored genus with long horns), but remains unexplained in less unusual species. It seems possible that the angle of the groove may play a major role in the rotation of dinoflagellate cells.

Unarmored dinoflagellate cells include a wide range of shapes, from almost to spheres to long, attenuated cylinders and ellipses. The ventral face is usually flattened or concave, the dorsal (upper) side is convex. The shape of many species suggests that they have become more streamlined during their evolution. A streamlined body of a size comparable to a dinoflagellate cell would have approximately 20 per cent less drag than a large spherical-shaped body with the same surface area. This is a significant reduction in drag, and presumably confers an advantage in natural selection.

Dinoflagellate cells range in size from extremely minute forms, with a diameter approximately 5 micron to very large species with cells visible to the naked eye, namely *Pyrocystis, Noctiluca*. The very small species are usually highly motile, and do not sink as rapidly as the larger cells because of the increased effect of friction in relation to volume in smaller bodies. Larger forms are often spherical and usually contain an enormous vacuolar system, which contributes to cell buoyancy. In *Noctiluca*, vacuoles contain an acidic solution, indicating that sodium, potassium and other positive ions, resulting the change in density of the vacuolar solution that cause the organism to float.

The nucleus is a very conspicuous organelle in dinoflagellate cells and is often the largest. Dinoflagellate nuclei are quite distinct from those of most other type of cells, in that the chromosomes retain their condensed appearance throughout interphase, when they can be clearly recognized as thickened, thread like structures. There are often photosynthetic pigments in dinoflagellates, and these are carried in special bodies, the chromatophores. Chromatophores are sometimes are green, more commonly, they are some shade of yellow or brown and in some cases, they are blue or red. Colourless, spherical or ellipsoidal bodies, attached to chromatophores, called pyrenoids, where reserve food is stored.

The "nanoplankton" that often contributes a major fraction of primary productivity in certain parts of the ocean consists to some extent of unarmored dinoflagellates and very small armored species. Most photosynthetic unarmored dinoflagellates fall within the size of 5-90 micron to which 55 to 72 per cent of the photosynthetic activity in the Indian Ocean has been attributed. A small photosynthetic *Gyrodinium* occurred at all of the stations (including Andaman Sea) studying productivity in the Indian Ocean. On the second cruise of the *Anton Bruun* and at many places, it was the most abundant of the photosynthetic phytoplankton observed. The species, *Gymnodinium* was also common in Indian Ocean samples along with several small species of armored dinoflagellates of genus *Oxytoxum*, responsible for most of the primary productivity in some areas of Indian Ocean.

Coccospheres

With the discovery of enormous concentrations of these primitive plants live near the surface of the ocean, their importance as primary producers in the food chain leading through the zooplankton to the fishes became apparent and a study of their migration and seasonal fluctuations is now accepted as an essential part of fisheries research.

The living coccospheres, an unicellular algae, belong to the Chrysophyceae are very primitive plants related to the sea weeds. Nearly all the species live in the open ocean. They are too small to be collected in a plankton net, and are usually obtained by centrifuging samples of water taken at measured distances below the surface. Like all other plants, they use the energy of sun light to synthesize food stuffs from carbon dioxide and water and possess chromatophores for this purpose. A few species have been observed to ingest solid food, but this method of nutrition is exceptional and always subordinate to photosynthesis.

Most coccolith-bearing algae live in the open ocean for their dependence on light for photosynthesis, and are mainly concentrated in the top 100 meters of water. Many species can not tolerate the fierce sunlight that enters the water near the equator, and prefer to live at a depth with subdued light. In the tropics (Andaman Sea), the maximum concentration is at about 50 meters. In the tropical waters, there is a much greater variety of species, inspite of the smaller total number of cells per litre. Not many species tolerate a salt concentration much different from that of the open ocean at salinities higher than 36 ppt or lower than 25 ppt.

Coccospheres are capable of reaching enormous concentrations, when conditions are particularly favorable. Spectacular proliferation (13 million cells per litre) takes place, the water becomes discoloured, when unusually liberal supply of nutrients were provided at the water of the Oslo Fjord by the drainage from the harbor and the city.

Distribution of Phytoplankton and Chlorophyll *a* around Little Andaman

The distribution of surface chlorophyll *a* and phaeopigments from whole and nannoplankton was studied around Little Andaman at locations between 10°15 minute N, 10°46 minute N, 92°7.5 minute E and 92°53.5 minute E. In the region, temperature ranged from 28.1 to 28.67°Celsius, salinity from 31.97 to 32.6 ppt in the upper 85 meter water column. While chlorophyll *a* values of the former ranged from 0.02 to 0.085 mg/cubic m (average 0.043, SD 0.015), phaeopigments ranged from 0.017 to 0.068 mg/cubic m (average 0.039, SD 0.015), Nannoplankton chlorophyll *a* concentrations varied between 0.013 and 0.085 mg/sq m (average 0.034, SD 0.016) and phaeopigments from 0.014 to 0.064 (average 0.029, SD 0.016). The highest chlorophyll *a* concentration was obtained at 10°42 minute N and 92°40 minute E. Nannoplankton chlorophyll *a* also showed equally high concentration at this location. This was probably due to the abundance of flagellates at this place. Though, in general, nanno chlorophyll *a* was less than the whole chlorophyll *a* values, at five locations, the former equaled that of the later. At most of the places phaeopigment values were lower than those of the corresponding chlorophyll *a* and nanno-chlorophyll *a* concentrations. However, at few locations phaeopigments were higher than those of the chlorophyll *a* concentrations.

Total phytoplankton ranged 1200 to 4000 cells/litre. Flagellates occurred at all places in appreciable quantities, followed by diatoms, dinoflagellates and blue green algae. *Navicula* spp. was the most abundant, diatom followed by *Coscinodiscus* spp. and *Denticula* sp. Other diatoms (*Amphora* sp., *Diploneis* sp., *Eucampia* sp., *Fragilaria oceanica*, *Licmophora* sp., *Nitzschia closterium*, *Pleurosigma* sp., *Rhizosolenia* sp.) occurred in very negligible quantities. Bulk of blue-green algal population was composed of *Trichodesmium thiebautii, T. erythracum*, the former being present at all the places.

At 10°28 minute N and 92°35 minute E and 10°38 minute N and 92°35 minute E, parameters, such as, whole chlorophyll *a*, nanno-chlorophyll *a*, phaeopigments, primary productivity and particulate organic carbon were studied from five depths. These are shallow banks with live corals and calcareous algae. At the first place (10°42 minute N and 92°40 minute E), whole chlorophyll *a* showed a progressive increase from surface to 50 m (1 per cent light depth). Phaeopigments showed the same trend. Nanno-chlorophyll *a* and phaeopigments also followed an increasing trend up to 50 m layer. However, the highest values of primary productivity and POC were observed at 20 m and 0 m respectively. Conditions were very similar at 10°38 minute N, 92°35 minute E also. Highest values of whole chlorophyll *a*, nanno-chlorophyll *a*, pheopigments and POC were observed at 1 per cent light level (30 m), though production was highest at 60 per cent light level (5 m).

Depth wise distribution of phytoplankton at above two locations ranged from 1400 to 5000 cells per litre. Flagellates occurred at various depths of both the locations, maximum being at 1 per cent light level (30 m).

Similarly diatoms and dinoflagellates were also maximum at this depth. Distributional trend of diatom population was similar to that of surface. The quantitative distribution of phytoplankton was not very different from surface to 1 per cent light depth. More abundant diatoms were *Navicula* spp., followed by *Denticula* sp. and *Coscinodiscus* spp. Other diatoms, such as, *Nitzschia* spp., *Amphora* sp. and *Cyclotella* sp. were also present in negligible quantities. The blue-green algae, *Trichodesmium thiebautii* occurred at various depths, followed by *T. erythracum*, which was less abundant and found mainly at the surface. Qasim (1979) recorded a *Trichodesmium* bloom from the Malacca strait during April, 1979. During the pre-monsoon, the occurrence of these blue-green algae is a wide spread phenomenon in the west coast of India. At the beginning of a *Trichodesmium* bloom, *T. thiebautii* also occurs in abundance along with *T. erythraeum* though the later takes over completely. Moreover, *T. thiebautii* found to be more distributed in the open ocean as evidenced from Laccadive Sea and Little Andaman Sea. Dinoflagellates were as abundant as diatom at 10°45 minute N, 92°40 minute E. *Peridinium* spp. was the most predominant form at various depths of both the above locations around Little Andaman. This was followed by *Ororocentrum* spp., *Ceratium* spp. and *Gonyaulax* sp. in the order of abundance. Movachan (1973) has recorded that throughout the entire Andaman Sea, dinoflagellates were rich during and after northeast monsoon. Many of the photosynthetic dinoflagellates are able to tolerate low inorganic nutrient levels and this may be the reason why dinoflagellates successfully thrive in the oligotrophic tropical waters. The photosynthetic rate of *Ceratium furca* was found to be depressed by an increase of nitrate in culture relative to the diatom *Biddulphia sinensis*. In general, the phytoplankton population of Little Andaman Sea is quantitatively comparable with that of the Laccadive Sea.

Most of the features (chlorophyll *a*, phaeopigments and primary productivity) from Little Andaman coral banks were comparable with those of Sesostris Coral

Bank, Bassas de Pedro and Gaveshani Bank, and the highest concentrations of chlorophyll *a*, phaeopigments and primary productivity were observed at 1 per cent light level (50 m).

An assessment of the phytoplankton population of the Little Andaman Sea during the month of February shows the importance of dinoflagellates as constituents of primary production, since they are able to thrive successfully in oligotrophic tropical waters unlike the diatoms. The coastal waters, where the nutrient levels are naturally higher can sustain a richer diatom population.

Table 12.1: Distribution of Surface Chlorophyll *a*, Phaeophytin, Nanno-chlorophyll a, and Phaeophytin (mg/cubic m) around Little Andaman Island during February, 1979

Location	Chlorophyll a	Phaeophytin	Nannochlorophyll a	Phaeophytin
10°15 minute-10° 28 minute N;				
92°21-25 Minute W 10°12-24 minute N;	0.041-0.052	0.034-0.064	0.013-0.042	0.019-0.064
92°31 minute E 10°37-38 minute N,	0.045-0.068	0.043-0.068	0.031-0.045	0.020-0.060
92°39-52 minute E 10°46-47 minute N;	0.029-0.040	0.021-0.053	0.024-0.040	0.014-0.029
92°39-52 minute E 10°36-37 minute N;	0.020-0.047	0.020-0.052	0.014-0.047	0.015-0.052
92°7-22 minute E 10°43-45 minute N;	0.028-0.085	0.025-0.053	0.020-0.085	0.020-0.053
92°6-18 minute E	0.033-0.037	0.017-0.044	0.019-0.035	0.015-0.029

Table 12.2: Distribution of Surface Plankton (Cell count or filament/litre) around Little Andaman Island during February, 1979

Phytoplankters	Locations					
	1198	1199	1200	1201	1202	1203
Diatoms						
Amphora sp.		100			200	
Denticula sp.	200	100	400		100	100
Coscinodiscus spp.	400	500	500	200		
Diploneis sp.	100	100				
Fragilaria oceanica					400	400
Eucampia sp.					200	
Licmophora sp.					100	
Navicula sp.	1500	1400	700	900	1300	800
Nitzschia closterium	100					
Pleurosigma sp.	100			100		
Rhizosolenia sp.				100		

Contd...

Table 12.2–*Contd...*

Phytoplankters	Locations					
	1198	*1199*	*1200*	*1201*	*1202*	*1203*
Green algae						
Flagellates	3400	4400	4300	3300	5100	3600
Blue green algae						
Merismopedia sp.		400	400			
Synechocystis sp.			200	200		200
Trichodesmium erythraeum	400	200		500	400	
T. thiebautii	1700	1300	600	1100	1800	1000
Dinoflagellates						
Ceratium spp.	100	100	100	100		
Peridinium spp.	1100	400	700	800	400	300
Prorocentrum sp.	200		200	400		100

1198: 10°15-28 minute N and 92°21-25 minute E; 1199: 10°12-24 minute N and 92°31 minute E; 1200: 10°37-38 minute N and 92°39-52 minute E; 1201: 10°46-47 minute N and 92°39-52 minute E; 1202: 10°36-37 minute N and 92°7-22 minute E; 1203: 10°43-45 minute N and 92°6-18 minute E.

Table 12.3: Depthwise Distribution of Phytoplankton at 10°35 minute N and 92°40 minute E (A) and 10°45 minute N and 92°40 minute E (B) around Little Andaman Island during February

Phytoplanktons	(A)					(B)				
	0m	*5m*	*10m*	*20m*	*50m*	*0m*	*5m*	*10m*	*20m*	*50m*
Diatoms										
Amphora sp.	300	100	100		200					
Coscinodiscus sp.	300				100	100		200		100
Cyclotella sp.			100		100					
Denticula sp.	100	300					100	400	200	100
Diploneis sp.										100
Grammatophora sp.										200
Licmophora sp.										100
Navicula sp.		200	300	400	400		200	400	100	700
Nitzschia sp.								200		300
Pleurosigma sp.										200
Rhizosolenia sp.								200		
Surirella fluminensis										100
Thalassiothrix frauenfeldii										100
Triceratium favus					100					

Contd...

Table 12.3–*Contd...*

Phytoplanktons	(A)					(B)				
	0m	*5m*	*10m*	*20m*	*50m*	*0m*	*5m*	*10m*	*20m*	*50m*
Green algae										
Flasellates	1200	700	600	600	1600	700	900	1600	800	1400
Blue green algae										
Synechocystis sp.						200				
T. thiebautii	300	400	300	400	100	200	100	600	100	100
Dinoflagellates										
Ceratium sp.	100				100		300			
Peridinium sp.	300	100	100	200	400	200	1000	200	600	900
Prorocentrum sp.	100	100		200	300	400				300
Gonyaulax sp.					200					200
Silicaflagellate										
Dictyocha sp.	100	100					100	100		

Table 12.4: Depthwise Distribution of Chlorophyll *a* (chl *a*), Phaeopigment (Phaeo), Primary Productivity (pp) and Particulate Organic Carbon (Poc) in Little Andaman Sea

	!0°37 minute N *92°42 minute E* *(mg/cubic m)*					*!0°47 minute N* *92°42 minute E* *(mg/cubic m)*				
Depth (m)	*Chl a*	*Phaeo*	*Nanno*	*pp*	*Poc*	*Chl a*	*Phaeo*	*Nanno*	*pp*	*Poc*
0	0.034	0.034	0.024	3.6	120	0.047	0.052	0.047	5.4	120
5	0.036	0.031	0.020	5.7	60	0.039	0.026	0.032	9.6	120
10	0.050	0.023	0.025	4.5	60	0.040	0.017	0.040	4.1	30
20	0.089	0.058	0.029	8.0	30	0.139	0.067	0.110	4.2	130
50	0.189	0.174	0.111	5.2	60	0.140	0.098	0.113	1.2	180

pp: in mgC/cubic m/day; Poc: in mgC/cubic m.

Chapter 13
Zooplankton

Zooplankton Abundance in Andaman Sea

The zooplankton biomass and abundance of major groups from the Andaman Sea consists of Copepoda, Chaetognatha and Tunicata. The average zooplankton biomass in February, 1979 was estimated as 5.6 ml/100 cubic m and is generally poor compared to reported values from east and west coasts of India. Copepoda formed the dominant component followed by Chaetognatha and Tunicata. The average standing crop of zooplankton was estimated to be in the order of 288.8 mgC/ sq m for the upper 200 meter column.

The general pattern of zooplankton biomass (estimated as displacement volume and computed for 100 cubic meter) of the Andaman Sea (92-98°E; 5-17°N) reveals that the area on the eastern side of the island was generally poor in zooplankton standing crop (average 3.6 ml/100 cubic m). An isolated patch of fairly high standing crop (average 11.8 ml/100 cubic m) was observed at few places around the Andaman Island. The biomass was moderate in the other areas (average 6.3 ml/100 cubic m) except for a few isolated patches of poor biomass. The biomass values reported from the east coast of India ranged from 7.8 to 8.2 ml/100 cubic m during south west monsoon period, 2.5 to 15.4 ml/100 cubic m during late south west monsoon period, 8.9 to 32.2 ml/100 cubic m in June and 8.9 to 10 ml/100 cubic m from the west coast of India during post-monsoon period. In comparison, the average zooplankton biomass from the Andaman Sea (5.6 ml/100 cubic m) is generally poor. The biomass in Andaman Sea sampled in February-March, 1963 also showed poor biomass ranging from 0.5 to 5.5 ml/standard IOSN haul. The total zooplankton counts ranged between 760/100 cubic m in deeper places and 13900/100 cubic m at shallower regions of 12°32 minute N and 92°30 minutes E.

Components

Copepoda formed the dominant component of zooplankton forming 53.9 per cent of the total counts. They were fairly uniformly distributed at almost all places of sampling. Highest numbers of them (7150/100 cubic meter) were recorded in 12°32 minute N; 92°30 minute E, but their counts were far less compared to reported figures in earlier studies from east and west coasts of India.

Cheatognaths formed the next abundant group (12.9 per cent). They were poorly represented in the around the islands, except for a patch of higher density in the eastern side of Middle, South and Little Andaman and at 12°32 minute N; 92°30 minute E, where their highest density (1380/100 cubic m) was observed. The density of Chaetognaths were relatively low compared to IIOE data available in 1973 for the seas around Andaman Island.

Pelagic tunicates were found at all places around Andaman Island. Appendicularians formed 8 per cent of the total counts with peak density (1290/100 cubic m) at 13°30 minute N; 91°55 minute E. Thaliacea, represented by both salps and doliolids formed a less abundant component (1.8 per cent) and the maximum density of them was observed at shallower region (530/100 cubic m).

Ostracods were fairly abundant around the Andaman and Nicabar Islands. They formed about 4.7 per cent of the counts and their maximum abundance (2100/100 cubic m) was observed at 12°32 minute N; 92°30 minute E. Maximum population range was scattered between 92 to 94°E longitude. Their total distribution in the surveyed area was very similar to earlier observations in 1972.

Euphasiids occurred at all places in small numbers forming 4.1 per cent of the total counts. Their maximum density (1700/100 cubic m) was observed at 13°N; 91°58 minute E.

Decapod larvae and adult decapods were in general poorly represented around the islands. They contributed to 2.5 per cent of the total zooplankton counts with maximum density (540/100 cubic m) at 12°32 minute N; 92°30 minute E.

Fish eggs and larvae together formed only 1.4 per cent of the zooplankton counts. Fish eggs were represented only at 39 per cent of the places (maximum density 250/100 cubic m) at shallower regions. Fish larvae were better represented around Andaman Islands, their highest density being 250/100 cubic m at shallower regions of 12°32 minute N; 92°30 minute E.

Apart from these, other groups, such as, Hydromedusae, Scyphomedusae, Siphonophora, Pteropoda (both Thecosomata and Gymnosomata), Heteropoda, Cladocera, Amphipoda, Mysidacea and larvae of Polychaeta, Echinodermata, Cirripedea, Bryozoa, Stomatopoda and Cephalopoda also occurred at several places, but contributed only a small percentage of the total counts.

Secondary Production

Large amount of data on the zooplankton standing crop in terms of biomass, estimated as displacement volume is available from the Indian Ocean. Several attempts have been made to relate this to the other trophic levels to project the energy transfer.

It was felt desirable to conduct some observations so as to facilitate more direct methods for these calculations.

Large variations are possible in the percentage of water content and organic carbon content depending on the composition of zooplankton, especially the dominance of gelatinous organisms. Earlier estimates from estuarine and off-shore waters of Goa had yielded values ranging from 61.9 to 81.7 mg dry weight per ml of zooplankton biomass. In the Andaman Sea, it varied between 68.7 to 82.2 mg dry weight per ml and an average of 75.4 mg dry weight/ml zooplankton biomass was used for further estimates. Estimates of organic carbon for mixed zooplankton from estuarine and off-shore region of Goa ranged from 30.4 to 41 per cent of dry weight. Studies from Cochin backwaters had yielded the organic carbon content of non-gelatinous zooplankton as 33.1 per cent and gelatinous zooplankton as 5.1 per cent dry weight. In the present estimates of Andaman Sea, it ranged from 27 per cent to 38.4 per cent of dry weight with a mean value of 34.2 per cent.

Using these mean values, the average zooplankton standing crop in the Andaman Sea was calculated, which varied between 185.6 mgC/sq m in areas having low biomass, 324.9 mgC/sq m areas having medium biomass and 608 mgC/sq m in regions of high biomass for the upper 200 meter column depth. The average for the entire area obtained was 288.8 mgC/sq m for the 200 m column depth.

Distribution of Zooplankton in Relation to Thermal Barrier

Vertical distribution of temperature had been found to have definite influence on the vertical distribution of zooplankton. Earlier studies from Indian waters have shown that zooplankton have a tendency to accumulate above the discontinuity layer. Thermocline had been found to act as a barrier in vertical movement of some species of ostracods and siphonophores.

However, the effect of discontinuity layer on the distribution of species of many groups of zooplankton in Andaman Sea from 200 meter to surface and from thermocline depth to surface reveal that similar to earlier findings, biomass distribution in general, showed higher values above the thermocline. The general increase was by 60 per cent compared to values for the 200 m column depth. Highest biomass value recorded was 23 ml/100 cubic m at 10°55 minute N; 93°30 minute E for the water column above thermocline.

The distribution of other major groups showed that in general, higher abundance occurred above the thermocline for Chaetognatha (average increase 126 per cent), Appendicularia (34 per cent), Thaliacea (33 per cent), Decapoda (158 per cent) and Euphasiacea (42 per cent), although at few places, the abundance was higher in the 200 m to surface column. For Copepoda greater aggregation above thermocline was found at 9°32 minute N, 93°30 minute E; 11°25 minute N, 93°E to 13°30 minute N, 91°55 minute E. But on the average their total abundance was more or less even in both above thermocline and 200 m to surface column depth. Ostracoda, on the other hand, showed a decrease (29 per cent) in the layer above thermocline compared to 200 m column depth. At some places, they were not represented in collections made from the thermocline layer.

Among the common copepod species, *Undinula vulgaris, Euchaeta marina, Eucalanus monachus, Eucalanus mucronatus, Corycaeus* sp. and *Oncea* sp., on the average showed higher aggregation above the thermocline, although at some places the distribution was more or less similar for thermocline and 200 m depth. Species like, *Eucalanus attenuatus, Pleuromamma indica* and *Aetedus* sp. were not present at some places above the thermocline, but when present, were observed in higher numbers above the thermocline layer. This could be attributed to the vertical migration of these species and that discontinuity layer did not affect their vertical migration. *Candacia* sp. and *Copilia* sp. showed more or less even distribution above and below thermocline, whereas, species like, *Scolicithrix dana, Rhincalanus cornutus* and *Saphrina* sp. had greater abundance in the layer below thermocline. Some of the less common species, like, *Euchaeta corcinna, E. wolfendeni, Scolicithrella nicobarica, Scaphocalanus* sp., *Scotocalanus* sp., *Pleuromamma xiphias, Gaetanus armiger* and *Euchirella amoena* were not collected above the discontinuity layer.

Among chaetognaths, *Kronhitta pacifica, Sagitta cuflata, S. ferox, S. neglecta, S. pacifica* and *S. robusta* were present in higher numbers above the thermocline at most of the places. *S. regularis* was always represented in higher numbers in this layer. *S. bipunctata* and *Pterosagitta draco* showed higher aggregation below thermocline. The distribution of *Sagitta bedoti* was more or less even above and below thermocline. *S. hexaptera* and *Kronhitta subtilis* were sporadically found in the samples and confined to the stratum below thermocline.

Many of the ostracod species like, *Euconchoecia aculeate, Halocypris brevirostris, Conchoecetta gigesbrechti, Metaconchoecia rotundata, Ornithoconchoecia striola* and *Spinoecia porrecta* were better represented in the 200-0 m hauls as observed earlier (George *et al*, 1975). *Cypridina dentate* showed accumulation above thermocline. Distribution of *Ornithoconchoecia atlantica* did not seem to be influenced by thermocline. *O. bispinosa* and *Conchoecetta accumunata* were present only in samples from 200-0 m, suggesting that thermocline acted as a barrier in their vertical movements.

Among the decapods, *Lucifer typus, L. hanseni, Leptochela aculeocaudata, Thalassocaris* spp. and *Alpheus* spp., in general showed higher aggregation above the thermocline. On the other hand, *Sergestes orientalis, S. crassus, S. cornatus, S. atlanticus* and *Hippolyte* spp. were encountered more below the discontinuity layer suggesting that thermocline acts as a barrier for the distribution of these species. *Brachyuran zoea*, although were collected from 200 m column, their density was higher above the thermocline.

The numerical abundance of other groups and species was scant to draw any definite conclusions on their distribution in relation to thermocline. The vertical migration of zooplankton is to be considered, while comparing their distribution above and below discontinuity layer. Variations both in biomass and number of organisms occurred in both the layers. But spectacular increases above thermocline was observed both in day and night. Hence, it would seem that vertical migration would not be a serious factor affecting distribution of species in relation to thermocline, except in species where discontinuity layer acts as barrier. Also many species seem to be able to migrate to optimum light intensities within the thermocline layer during day time and variations occurring in various places would be mostly a characteristic of zooplankton abundance of the area itself.

Crustaceans and Total Zooplankton Diversity in Andaman Sea

Crustacea formed the bulk of zooplankton collected from Andaman Sea. The major crustacean groups are Copepoda, Ostracoda, Euphausiacea and Decapoda.

Copepoda dominated the zooplankton and contributed to the bulk of biomass. 47 species belonging to 33 genera were collected in February, 1979. Of them, 42 species belonged to the Suborder Calanoida and the rest to Cyclopoida. No Harpacticoides were encountered in the sample. Fifteen species, namely, *Undinula vulgaris* (21 per cent), *Euchaeta marina* (11.5 per cent), *Pleuromamma indica* (9.3 per cent), *Eucalanus monachus* (7.4 per cent), *Aetedus giesbrechti* (5.8 per cent), *Oncea* sp. (5 per cent), *Corycaeus* sp. (3.9 per cent), *Oithona* sp. (3.7 per cent), *Candacia* sp. (3.7 per cent), *Eucalanus attenuatus* (3.3 per cent), *E. mucronatus* (3.1 per cent), *Saphrina* sp. (2.9 per cent), *Rhincalanus cornutus* (2 per cent), *Copilia* sp. (2 per cent) and *Temora stylifera* (1.9 per cent) contributed to about 86.5 per cent of the total copepod counts. These species are generally present in most of the places, although there are variations in their count.

Some species, like, *Canthocalanus pauper, Calocalanus pavo, Euchaeta corcinna, Rhincalanus nasutus, Eucalanus crassus, Centropages tenuiremis, C. gracilis, Temora turbinate, Calanopia* sp., *Tortanus forcipatus, Acartia erythraea* and *Labidocera acuta*, which are common in the coastal waters of Indian peninsula were represented only sparsely at a few places. Some of the oceanic forms, like, *Euchaeta longicornis, Scolicithrella nicobarica, Scahocalanus* sp., *Scotocalanus* sp., *Pleuromamma xiphias, Gaetanus armiger* and *Euchirella amoena* also come under this category. Other species, which exhibited a fair but discontinuous distribution are *Eucalanus elongates, Euchaeta wolfendeni, Centropages furcatus, C. calaninus, Pontellina plumata* and *Acartia amboyenensis*. In general, the composition of copepods were constituted by a mixture of neritic and oceanic species with euryhaline forms dominating the counts. In the coastal waters although species like, *Undinula vulgaris, Euchaeta marina* occur fairly commonly, the copepod counts are usually dominated by smaller forms like, *Acrocalanus* spp., *Paracalanus* spp., *Acartia* spp., *Pleuromamma indica* and *Aetedus* spp., which are common in the Andaman Sea, rarely occur in neritic waters.

Ostracods were fairly abundant in Andaman Sea. In total, 17 species were identified from the collections. Of these 16 belong to the family Halocyprididae and one to Cypridinidae. *Halocypris brevirostris* (39.5 per cent), *Spinocea porrecta* (15.9 per cent), *Metaconchoecia rotundata* (14.4 per cent), *Conchoecetta giesbrechti* (9.7 per cent), *Ornithoconchocia striola* (5 per cent), *O. atlantica* (4.7 per cent) and *Euconchoecia aculeate* (4.7 per cent) were the common species.

Cypridina dentate, which is common form in the east and west coasts of India was only occasionally represented and was confined mainly to places located to the eastern side of the Little Andaman Island. Of the 16 species of ostracoda, recorded earlier from the Andaman Sea, *Microconchoecia curta* and *Platyconchoecia prosadena* were not represented in February, 1979 collections. *Paraconchoecia discophora, Conchoecetta accumunata* and *Orinthoconchoecia bispinosa* are new records for the Andaman Sea. *Para conchoecia elegans, P. procera, P. decipiens, Spinoecia parthenoda, Conchoecia magna* and *Conchoecilla daphnoides* were other species represented.

Euphausiids occurred in all places of Andaman Sea, although generally in low numbers. Juveniles of euphausids formed about 25 per cent of their total numbers. *Euphausia* sp. (22.2 per cent), *E. diomedia* (8.2 per cent), *Thysonopoda* sp. (7.7 per cent), *Namatocylis gracilis* (7.6 per cent) dominated the counts. *N. microps, Stylocheiron affine, S. carinatus, S. elongatum, S. suhmii, S. longicorne* and *Tysanopoda monacantha* were the other species occurred in the collections.

Members of the family Sergestidae contributed to the majority of decapods (41.5 per cent) followed by carideans (31.3 per cent), brachyurans (15.6 per cent), penaeids (8.7 per cent) and other decapod larvae. A similar trend had been observed in the general composition of decapoda in IIOE samples also. Among sergestids, the oceanic form, *Lucifer typus* dominated (18.3 per cent) followed by *L. hanseni* (8.5 per cent) and larvae of *Sergestes* spp. (7.9 per cent). Both the species of *Lucifer* were distributed around Andaman and Nicobar Islands. *L. penicillifer* was collected from a few places in small numbers. The larvae of other species of *Sergestes* recorded were *S. orientalis, S. crassus, S. cornatus* and *S. atlanticus*. *Sergestes orientalis* has been reported from the plankton of southeast and southwest coasts of India and *S. crassus* from the southwest coast of India. The occurrence of the other two species of *Sergestes* larvae appear to be new records from Andaman Sea as well as Indian waters. *S. cornatus* was collected at 10°22 minute N, 92°12 minute E, whereas, *S. crassus* and *S. atlanticus* were recorded from a single station each (10°32 minute N, 92°12 minute E and 9°50 minute N, 92°40 minute E respectively) in small numbers.

Metapenaeus spp., *Parapenaeopsis* spp., *Penaeopsis rectacuta, Solenocera indica* and *Gennadas* spp. were collected around Andaman Island in small numbers. Of these, *Solenocera indica* (4.5 per cent) and *Penaeopsis rectacuta* (2.8 per cent) were the main forms, the later had been reported to be a major constituent of penaeids of Andaman Sea.

Thalassoearis spp. (10.8 per cent), *Alpheus* spp. (7.9 per cent) and *Hippolyte* spp. (3.3 per cent) were the main carideans that contributed to the decapod composition. Other carideans, *Leptochela aculeocaudata, Proeessa* spp., and *Periclimenes* spp. occurred in small numbers at one or two places.

Non-caridean forms, such as, *Axius* spp., *Calocaris* spp., *Callianassa laticauda* and *Callianassa* sp. were collected from mostly a single place.

Brachyuran zoea were fairly abundant andwere present in all places. Their maximum density 127/100 cubic m was observed in shallow place (12°32 minute N; 92°30 minute E). Larvae of Anomura were rather rare, confined to a few places on the western side of Andaman Island.

Species Diversity

Species diversity index (D) varied between 3.2 at 8°12 minute N, 93°36 minute E and 8.1 at 10°22 minute N, 92°12 minute E. The diversity was fairly uniform around the islands except for another low patch on the eastern side of the Andaman Island (11°50 minute N, 93°08 minute E and 13°N, 93°05 minute E). Evenness in distribution of species (E) was usually high as typical of the oceanic waters. It was generally high on the eastern side of the islands and showed an inverse relationship with biomass. Evenness was quite high compared to values generally obtained in coastal waters and quite comparable to data available from other oceanic stations along the east coast of India.

Chapter 14
Corals of Andaman Sea

Andaman Sea have the richest coral diversity. Most of the coral reefs of Andaman Sea are of fringing type, colonizing nearer to the coast line on east and west coast of Andaman Island. A total 0f 177 species of hard corals, falling under 57 genera have been reported from the Andaman Sea.

Coral reefs are providing habitat for numerous species, serving important ecological functions (Precht, 2006). Although are located in nutrient poor crystal clear waters, yet they support a very high biodiversity. The presence of corals increase the associated faunal communities in and around coral reefs. Pillai (1983) listed 135 coral species from Andaman and found Andaman Islands were less diverse (31 genera with 82 species) than the Nicabar Islands (43 genera with 103 species). The diving studies of Government of India and UNDP GEF field mission in 2001 reported a total of 198 species of scleractinian coral from Andaman groups of islands of which 111 are supposed to be new records to India. But, on verification with other studies only 94 species are found to be new records and this also include some non-scleractinian corals. Venkatraman *et al* in 2003 described 228 species of corals belonging to 58 genera and 15 families from Andaman Sea. A total of 223 species of scleractinian have been reported from the Andaman Islands by Turner *et al* in 2009. Reporting of seven newly recorded species were also made by Tamal *et al* during 2010 from the islands of Andaman and Nicobar. In 2011, seven species of scleractinian corals were added by Mondal *et al* from Ritchie's Archipelago. However, the species composition of corals in Andaman Sea increased up to 418 as reported by Ramakrishna *et al* (2010). Further search may discover more new coral species in years to come.

Coral reefs are distributed in restricted places of the sea. Andaman Sea has both fringing and barrier reefs. Both stony and soft corals are found in the subtidal regions of the reefs. The genera of stony corals found in Andaman reefs are *Favia, Favites,*

Goniopora, Montipora, Sinularia etc. The genera of soft corals include, *Nephthya, Dendronephthya* etc.

The biodiversity of Indian coral reefs is unique. They contain 23 species of ammonifying bacteria, 15 species of nitrifying bactera, 3 species of nitrogen fixing bacteria, 23 species of phosphate producing bacteria. About 50 species of diatoms, 55 species of phytoplankton and 20 groups of polychaete worms have been recorded in Indian reefs.

The composition of coral reefs include about 185 species of benthic algae, 15 species of sea grass, 15 species of sea weeds, 110 species of porifera, 6 species of crustaceans, 105 species of echinoderms, 599 species of bony fish, besides different species of crabs, bivalves, gastropods and cephalopods.

Coral reefs are hard limestone structures built up by the cementing process and depositional activities of the species of class Anthozoa, Scyphozoa, Hydrozoa and also the calcifying algae. In some areas of coral reefs, rich algal diversity is seen, such as Sargassum, Ulva, Cladophora etc.

Seventy species of sponges, 27 species of prawns, 200 species of mollusks, 3 species of marine mammals has also been recorded.

Ritchie's Archipelago is one of the most diverse areas in Andaman Sea, and consists of islands like, Sir William Peel Island, Havelock Island, Outram Island, Nicolson Island, Wilson Island, Henry Lawrence Island, Inglis Island, South Button Island, North Button Island, Middle Button Island, Sir Hugh Rose Island, John Lawrence Island and Neil Island. Seven scleractinian coral species have been recorded from this archipelago.

Colonies of *Lobophyllia diminuta*, belonging to family mussidae were collected at Elephant Beach (Lat. 12°582 minute N, Long. 92°5838 minute E) of Havelock Island at a depth of 4 meter. Colonies are small, flat or dome-shaped. Corallites are phaceloid and usually monocentric, but may have three centers. Corallites average 16 mm diameter. Septal teeth are few in numbers, but are long and conspicuous. Columella are circular or oval and are well developed. It is mottled orange, green or white, usually with pale mouths. The species is included as vulnerable in the IUCN red list category and its occurrence is uncommon.

At a depth of 7 to 11 meters in Henry Lawrence Island of Ritchie's Archipelago (Lat. 12°12.598 minute N, Long. 93°o.03883 minute E), *Acanthastrea regularis* an uncommon coral species belonging to Mussidae family have been collected, which are vulnerable listed under IUCN red list category.

Colonies are found to be massive, and subplocoid. Septa are uniformly spaced with 8 to 10 evenly spaced rounded teeth. Teeth on adjacent septa are often aligned, forming concentric circles. Some septa are more prominent than others. Columellae are weakly developed. Colonies do not have thick tissue over the skeleton. Colour is variable, brown and yellow brown, usually with contrasting corallite walls and centers.

Commonly occurred species of *Scolymia australis* of Mussidae family have been collected at Henry Lawrence Island from a depth of 5 to 24 meters of Andaman Sea (Lat. 12degree 12.598 minute N and Long. 93°0.03883 minute E).

Usually solitary, but sometimes two or four centers occur in one corallite, or occasionally in separate corallites. Corallites are saucer shaped and less than 60 mm in diameter. Septa are sturdy with blunt saw like teeth. It is usually a mixture of cream, red, blue and green in colour. Enlisted as least concern in IUCN red list category.

At Henry Lawrence Island (Lat. 12°12.598 minute N; Long. 93°03.883 minute E), from a depth of 5 to 20 meters, colonies of *Montatrea cavernosa* of family Faividae have been collected, which occur commonly in Andaman Sea.

Colonies are massive, forming boulders or domes or are flat plates. Corallites are variable, but are usually conical and exert. Long and short septa strongly alternate, with alternate septa joined to the columella. It is commonly seen in green, brown, grey or orange in colour.

A good number of colonies of *Plerogyra simplex*, belonging to family Euphyllidae, have been located in Andaman Sea, off Henry Lawrence Island at a depth of 11 to 23 meters. This, near threatened uncommon species are IUCN red list category.

Colonies are branched and the branches are of uniform length and uniformly spaced. Branches are monocentric, forming phaceloid colonies. The coenosteum is smooth, with costae only found below corallite rims. During the day, vesicles sometimes form a continuous mass, although the underlying phaceloid shape can be distinguished. At night, long tentacles are extended from the base of the vesicles. It is cream, grey or pale brown in colour.

Colonies of *Siderastrea sidereal* of family Siderastreidae have been located at a depth of 4 to 16 meters of Andaman Sea, adjacent to Henry Lawrence Island (Lat. 12°12598 minute N, Long. 93°03883 minute E).

Colonies are encrusted to massive structure and are usually over one meter across. Corallites are 3 to 4.5 mm diameter, with numerous septa, tightly compacted giving a small surface. Septa are uniformly separated and reduced in number from the wall to the columella without forming many fan like fusions. Colour is usually light reddish brown. A commonly occurring species and listed as least concern in IUCN red list category.

Colonies of usually uncommon, *Dichocoenia stokesi* of Meandrinidae family have been collected from 5 to 14 meters depth of Andaman Sea, near Henry Lawrence Island. Colonies of this species are massive, often spherical or may form thick sub-massive plates. Corallites are evenly spaced, plocoid or ploco-meandroid. Septo-costae are usually in two neatly alternating orders. It is orange brown with white septo-costae, rarely green in colour.

Commonly known as Colt coral, blushing coral or blanching soft coral, *Cladiella australis*, is distributed at the reef area of Andaman Sea at 3 to 5 m depth from the island ecosystem of Little Andaman. The colony in live condition is beautifully chocolate brown colour and is encrusted with broad and low stalk. When disturbed, the live coral will reduce its biomass and instantaneously change into lavender-violet with the rapid retraction of polyps. The colony looks pinkish white with polyps as dark dots just after alcohol preservation. Published informations show that extracts from *Cladiella australis* are capable of inhibiting human oral carcinoma cells.

The Gorgonian

The Gorgonians or horny corals form one of the major tropical communities of appreciable economic importance. Their graceful and soft colouration constitute one of the chief attractions of the " sub-marine garden " of Andaman Sea. They inhabit through a range of depth from the tide mark to over 300 meters. Most of them are found attached to hard objects, like the corals, while a few are encountered in soft bottom to which they fasten by tufts of stolons. The gorgonians include the sea fans, sea whips and sea feathers. Although more than 15 species have been identified, only a few species, such as, sea fans – *Gorgonia* spp., *Plexaura* spp., *Paragoria* spp; sea whips – *Barireum* sp. and sea feathers-*Muricae* spp. are exploited for commercial purposes.

The gorgonians range from low forms to immense branching colonies of 2 to 3 m height. Yellow, orange, red, black and purple are among the most common colours. They are used for the interior decoration and pharmaceutical preparations. As regards its nature, it is protein yielding typical amino-acid on decomposition, but is lower in sulphur content than true horn. Its composition differs in different species, but frequently contain bromine and iodine, united to amino-acid tyrosin. The black variety of *Gorgonia* sp. can be distinguished from the tar coated ones by the gorgonian of the axis and also by the basal plate, strengthened by a lamella or network of gorgonin.

The gorgonians are known to contribute to considerable biomass of the littoral zone. Because of their sessile habits and firmness, they provide shelter for a variety of sessile animals and passage-ways for polychaetes and thus serving a good feeding ground for many of the commercially important fishes.

Biological and Physical Features of Coral Reefs

It is emphasized in the literature,that coral reefs are complicated systems. They are complex as geological structures (Pichon, 1982) as self building and eroding constructions (Scoffin *et al*, 1980) and as trophic organizations (Grigg *et al*, 1984). Their diversity is not only in the variety of structures and processes, but is also highlighted by the great densities and high numbers of species. The hermatypic stony corals (Scleractinia) take up a keystone position as the main builders of the coral reef habitat. Their diversity on reefs varies with the geographic distribution of coral reefs from some 50 species in the Caribbean Sea (Wells and Lang, 1973) to about 330 recorded from eastern Australia.

In surveying reefs, identification of coral species and coral communities is used to classify the reefs and reef habitats. Factors, such as, coral colony shape, colony size, interspecific interactions, development of algal vegetations, presence of sea urchins, have all been used to explain reefs ecologically. The ecology of coral species and their interaction with biotic and abiotic environment seems to indicate an enormous array of possible expressions of reefs conditions apart from the traditional species list. These "expressions of reef conditions" are ecological variables, more widely applicable than the presence or absence of particular coral species.

There are several terms in use for habitat-related shapes of organisms, namely, ecad, ecophene, ecophenotype and habitat form. In reef coral studies, especially, the additional terms ecotype and ecomorph have been applied, with a preference to the last one.

List of Ecological Variables

Living Coral - Tabular – Table-shaped

☆ Branching thick – Branches thicker than 2 cm (index finger)

☆ Branching thin

☆ Massive

☆ Foliose - Erect blades

☆ Encrusting – Colony shape follows bottom contours, under side completely attached

☆ Fungiid (mushroom) – Loose mushroom coral

☆ Fragment loose massive – Living, broken-off pairs of colonies

☆ Fragment thin branching

☆ Fragment plating – flattened blades parallel to bottom or horizontal plane, underside or periphery only partially attached

☆ Fragment foliose

☆ Fragment attached – Living, broken-off pairs of colonies, but cemented to bottom

Dead Coral

☆ Dead piece – Dead loose piece of coral, maximum diameter less than 25 cm

☆ Bottom - Dead, bare, attached coral rock

☆ Out living old – Old (discoloured) scar (bare skeleton) outside, periphery of living tissue of coral colony old wound

☆ Out living new – New (white) scar (bare skeleton) outside periphery of living tissue of coral colony new wound

☆ In living old – Same as out living old, but in the living tissue surface of the colony

☆ In living new – Same as in living old, but in the living tissue surface of the colony

Sediment - Sediment + Rubble

☆ Fine sediment – piece, larger than 1 cm in maximum diameter

☆ Rubble – Dead loose piece of coral, smaller than Dead

Other - Crevices – All sediment smaller than rubble

☆ Crustose corallines – All inaccessible holes and all overhangs all surfaces with coralline encrusting algae with a diameter less than 1.7 cm (length chain link).

☆ Algae – With a diameter less than 1.7 cm

☆ Excavating sponge – All filamentous and fleshy algae surfaces dominated by inhalant/exhalent tubes of excavating sponges, *e.g.* Clionidae

☆ Soft corals – All signs of interaction between coral and alcyonarian neighbour including overgrowth, toppling and aggression

☆ Organ-pipe (*Tubipora*)

☆ Other fauna

Interaction Coral

☆ Overgrowth – (Includes area between tube openings) coral grows, physically in contact, over neighbouring coral. No signs of antagonistic reactions

☆ Toppling – Coral colony has fallen on other colony, all signs of antagonistic reactions, *e.g.* dead

☆ Aggression – Margins between neighbouring colonies such as caused through extra-mesenterial digestion, sweeper tentacles etc.

Interaction Other

☆ Soft coral – All signs of interaction between coral and alcyonarian neighbour including overgrowth, toppling and aggression

☆ Fire coral (*Millepora*) – Similarly defined

☆ Other – Similarly defined

Predation

☆ Fish, Crown of thorns (*Acanthaster*) – All marks and scrapes in living coral tissue

Maximum Size

☆ Maximum diameter and height of 10 largest individually recognizable colonies

Preliminary Indicators

☆ *Acropora* tables, *Acropora* branched, *Acropora palifera*, *Porites* massive, *Diploastrea, Seriatopora, Pocillopora eydouxi*, foliose colonies. Included were easily recognized colony shapes and species listed as or found to be typical for certain environment.

Miscellaneous

☆ Depth, Angle slope, Relief, Urchins, Mushroom corals, Crown of thorns

Scleractinian Coral Species Recorded from Andaman Sea

Family – Acroporidae

1. *Acropora nasuta* (Dana), *A. grandis* (Brook), *A. humilis* (Dana), *A. palifera* (Brook), *A. rectina* (Nomenzo), *A. pacifica* (Brook), *A. digitifera* (Dana), *A. canalis* (Quelch), *A. clavigera* (Brook), *A. calamaria* (Brook), *A. hyacynthus*

(Dana), *A. polymorpha* (Brook), *A. corymbosa* (Lamark), *A. echinata* (Dana), *A. robusta* (Dana), *A. clathrata* (Brook), *A. squarrosa* (Ehrenberg), *A. cancellata* (Brook), *A. botryoides* (Brook), *A Formosa* (Dana), *A. Variabilis* (Klunzinger), *A. palmerae* (Wells), *A. monticulosa* (Bruggemann), *A. diversa* (Brook), *A. intermedia* (Brook), *A. armata* (Brook), *A. pulchra* (Brook), *A. brueggemanni* (Brook), *A. surculosa* (Dana), *A. conigera* (Dana), *A. irregularis* (Brook), *Montipora fruiticosa* (Bernard), *M. hispida* (Dana), *M. composite* (Crossland), *M. foliosa* (Pallas), *M. florida* (Nemenzo).

Family – Pocilloporidae

2. *Pocillopora elegans* (Dana), *P. damicornis* (L), *Stylophora mordax* (Dana)

Family – Agariciidae

3. *Pachyseris gemmae* (Nomenzo), *Pavona obtusta* (Quelsh), *Leptoseris papyracea* (Dana)

Family – Mussidae

4. *Lobophyllia hemprichii* (Ehrenberg), *Mussa angulosa* (Pallas), *Euphyllia glubrescens* (Chamisso and Eysehardt), *Euphyllia* sp., *Symphyllia recta* (Dana)

Family – Merulinidae

5. *Merulina ampliata* (Ellis and Sol), *M. laxa* (Dana)

Family – Faviidae

6. *Favia abdita* (Ellis and Solander), *F. halicora* (Ehrenberg), *F. speciosa* (Dana), *F. pallida* (Dana), *Diploastrea heliopora* (Lamark), *Platygyra daedalea* (Ellis and Solander), *P. sinensis* (Milne Edwards and Haime), *Goniastrea pectinata* (Ehrenberg), *G. planulata* (Milne Edwards and Haime), *G. benhami* (Vaughan), *G. retiformis* (Lamark), *Goniastrea* sp., *Coelaseris magiri* (Vaughan).

Family – Oculinidae

7. *Galaxea fascicularis* (L)

Family – Thamnasteriidae

8. *Psammocora contigua, Psammocora* sp.

Family – Poritidae

9. *Alveopora* sp., *Porites porites* (Pallas), *P. tenuis* (Verill), *P. nigrescens* (Dana), *Goniopora columma* (Dana), *G. tenuidens* (Quelch), *G. stokesi* (Milne Edwards and Haime), *G. peteolata* (Bernard).

Family – Pectinidae

10. *Echinophyllia aspersa* (Ellis and Solander)

Family – Fungiidae

11. *Fungia echinate* (Pallas), *F. fungites* (L), *F. horrida* (Dana), *Herpolitha limax* (Esper).

Fungiids

It was found that free living fungiid corals, which did not live in well-illuminated reef parts are usually thin and flat. At reef zones, which receive light of high intensity, the fungiid corals are usually dome-shaped. For both types of corals, attached or free living, the shape is most clearly influenced by light.

In areas where water can be very turbulent, several fungiids are represented by relatively many thick and heavy specimens, which will more likely remain unbroken than thin and light corals. They are also less likely to be washed ashore or to be transported to deeper reef zones. If such a coral is dome-shaped instead of flat, it is even more likely to remain in an upward position as shown for *Fungia scutaria*.

Fungiid corals lying on soft substrata at sheltered locations, usually had a thin and flat shape. Such corals gives less pressure to the soft bottom and run a smaller risk of sinking down into the mud than thick and arched corals. Fungiid corals optimally adapted to silty environments have a flat under-side. They are thin at the margin and convex around the mouth, so that sediment can easily be rejected.

The variation in shape of a fungiid coral species is primarily related to light intensity, which depends on depth, turbidity and bottom gradient. The occurrence of a coral shape depends further on the ability of corals with such a shape to endure the turbulence, sedimentation and substratum type in their habitat.

The reef flat coral coverage is limited by waves. Just above the drop-off, at depths between 6 to 10 meters, coral coverage reach its maximum. The highest concentration of fungiid corals is also found here.

Below the drop-off coral coverage decreases strongly with increasing depth, due to poor light conditions. Attached fungiid corals belonging to genera, *Lithophyllon* and *Podabacia* are rarely found. Free living fungiids, like, *Fungia echinate* (Pallas), *F. fungites* (L), *F. horrida* (Dana) are only observed on ridges protruding from the vertical wall. The steep slope below the drop-off makes the site unsuitable for free living corals.

The highest number of fungiid species is recorded at places, where coral coverage is very high.

The coastal areas in general have higher species numbers than the open sea areas. This can be explained by the great variety in coastal reef habitats. Coastal reefs can differ from each other in several variable physical conditions (*e.g.*, currents, salinity, sedimentation, wave action), whereas reefs more remote from the shore vary only in oceanic influences, like currents and wave energy.

The scleractinian species are colonial and living attached to a hard substratum.

The best known environmental factor influencing growth and form of zooxanthellate corals appears to be light intensity. Additional important factors in this respect are turbulence and sedimentation.

Reefs in open sea undergo least terrigenous influence and are exposed to waves during long periods of the year.

Many free living fungiids found in shallow reef zones are dome-shaped. Their skeletons are usually thick and heavy. The septa and costae are also thick in some species, namely, *Fungia scruposa* and *F. horrida*, the septal and costal ornamentations are coarse.

A thick and archedshape protects corals in turbulent conditions against fragmentations and helps them to remain in an upright position. The large septal dentation may give protection against predation and the long costal spines may operate as anchors to stabilize the position of the corals. Sand is relatively easily rejected from their sloped surfaces. The specimens living relatively deep are in general flatter and thinner than those from shallow water habitats. In shallow reef parts, reflected light comes abundantly from many directions. In deeper and, consequently darker zones, light mainly comes from above, so that the flatter the animal, the more light it can receive.

In relatively quite environments, many thin and fragile corals are encountered, some of them belonging to species usually having thick forms.

The effects of environmental conditions on eco-morph variations in fungiids are not always distinct, because these conditions vary in time and because the position of individual corals may change as a result of wave action.

Rainforest of the Sea

Coral reefs, the rain forest of the sea, are the living and dynamic ecosystems, harbouring about 25 per cent of all marine fish species, though the coral reefs cover only about 1 per cent of the earth's surface. Reef ecosystem, a self-sustainable one, bear rich species diversity and high productivity. It provides innumerable services, as nursery grounds for marine invertebrates and fishes in addition to providing natural barriers for coast line protection from storms and cyclones. Nearly 500 million people depend on coral reefs for food and livelihood. It is a potential source for development of new medicines. Corals play a significant role in purifying carbon dioxide in sea waters, thus playing an important role in global warming

Coral reefs in India is met with high species diversity, especially in Andaman Sea, besides Gulf of Mannar, Gulf of Kutch and Lakshadeep.

Most of the coral reefs of Andaman Sea are fringing type, cololizing nearer to the coast line of east and west coasts of Andaman Island. Venkatraman *et al* in 2003 reported the presence of 177 species of hard corals falling under 57 genera.

Corals are tiny marine animals, belonging to the Phylum Cnidaria and class Anthozoa. They mostly live in colonies and each colony consists of thousands of small individuals and genetically identical polyps. Polyps are multicellular organisms, usually few millimeters in diameter. They are radially symmetrical with tentacles surrounding a central mouth. They have only one opening, through which food is ingested and waste is expelled out. Corals are largely divided into two groups. Hard corals, found in the shallow waters, made up of stony calcium carbonate, which are the principal reef builders. Soft corals are found in deeper waters and are made up of protein and calcium carbonate.

Symbiotic relationship exist between corals and zooxanthellae, a single-celled dinoflagellates. These zooxanthellae live in the endoderms of corals. Several different species of zooxanthellae exist in corals, and many corals contain as many as two or three species in each host. Corals host a wide range of symbiont associations, which may have an adaptive advantage during environmental disturbance. These algae produce food by photosynthesis in the presence of sunlight, but uitlize only a part (more or less 5 per cent) of food produced by them, providing the rest to the host corals. This excess organic energy is translocated to corals, which are rich in carbohydrate, but low in nitrogen and most coral supplement the nitrogen by actively feeding on zooplankton or assimilate dissolved organic nitrogen. With the help of nematocysts on the tentacles of most corals, they catch zooplankton from the water. In return, the zooxanthellae get the nitrogenous nutrients from the corals, which would otherwise be scarcely available to them in the oligotrophic sea with coral reefs. Zooxanthellae are also protected from the solar radiation. Although the algae-coral relationship is characterized by their specificity, this does not mean that there is exclusively one-to-one association of one host with a specific alga. Rather it is an association resulting from several mechanisms that prevent the promiscuous random formation of combinations, while at the same time, while allowing for the probability of the occurrence of others.

Corals reproduce both sexually and asexually. Sexual reproduction occurs by either internal or external fertilization. Though most corals are hermaphrodite, sexes are separate in some coral species.

Inside each polyp of hermaphrodite coral eggs and sperms develop and mature. Many of the factors stimulate maturation and spawning of corals. In Kaneohi Bay, Hawaii, many corals spawn during summer months from May onwards. Gradual rise in sea temperature triggers maturation of the eggs and sperm in the adult corals. After maturation of sperms and eggs, most of the corals follow the lunar cycle to coordinate exact time of release of eggs and sperms or larvae to the surrounding waters. Apart from this, diurnal cycle, tide and weather condition is also important in spawning process.

Most corals around the world ocean broadcast spawners. Polyps extrude their eggs and sperms into the water concurrently for external fertilization and development. *Fungia scutaria* and *Montipora capitata* participate in mass spawning. In normal circumstances, *Fungia scutaria* spawns in one to four late evenings following the full moon between 5 to 7 pm, from the months of June through October, whereas *Montipora capitata* spawn 0 to 4 days of new moon at around 9 pm. In *Fungia scutaria*, sexes are separate. It produces gamets that lack zooxanthellae initially and take their symbionts from their environment, whereas, *Montipora capitata* releases both eggs and sperms together in bundles. Many corals release the planula larvae directly into the water, as is the case with *Pocillopora damicornis*. Spawning time of corals generally depends upon time of the year, lunar cycle, temperature, tide etc. It was noticed that lower water movement facilitates a better rate of fertilization, specially in the case of broadcast spawning. Planula larvae have to settle on a hard substratum after fertilization to grow and multiply. Generally larvae would have to settle within a period of few days. During this period, they are very much vulnerable to predation and environmental stress. Settled larvae also have to face severe competition for food and space.

Coral species throughout the world ocean are now facing severe environmental and human interferences. Coral bleaching in 1997-98 caused extensive reef damage in many regions of the world. In Indian Ocean, coral mortality reached over 90 per cent in some areas, are now at risk of loosing this valuable ecosystem and associated economic benefits from fisheries and tourism. Reasons for the destruction and degradation of corals throughout the world are;

Mass bleaching of corals due to global climate change associated with increased atmospheric and sea surface temperature (SST) caused large scale mortality of corals. During April 2010, 36.5 to 69.5 per cent coral bleaching in Andaman waters was reported due to the rise of sea surface temperature.

Rise in acidity of the sea water due to high rate of dissolution of carbon dioxide from atmosphere in the water causes not only reduced rate of shell formation, but also facilitates dissolution of the calcium carbonate shell of the corals.

Destructive fishing methods (dynamite and cyanide fishing), overfishing and selective fishing cause immense damage to the coral reef ecosystem. In an undisturbed ecosystem, all the component and organisms play an important role in maintaining a healthy and sustainable environment. Removing large quantities of fishes or carnivorous fishes will facilitate higher population density of coral eating fishes and other organisms that prey upon coral colonies, as well as coral larvae. The species diversity of reef fishes will also alter. Blasting in a reef area destroys not only the corals, but also kills fish and invertebrates in a large surrounding area.

Application of cyanide to harvest ornamental and food fishes (groupers) affects photosynthesis of the symbiotic zooxanthellae and cause considerable coral destruction. Over a time, these types of fishing damage the whole reef ecosystem that affect the livelihood of many subsistence fishermen.

Poor land use, deforestation, coral and sand mining, dredging etc cause significant damage to the coral reef areas by the way of sedimentation. So also, sewage and industrial effluent and influx of nutrient rich water from agriculture run-off impart significant damages to the fragile coral reef ecosystem. Oil spills are also highly detrimental to the coral reef organisms.

Development of coastal areas, construction of roads, industry and tourism produce much stress to the coral ecosystem directly or indirectly. Besides natural calamities, like, tsunami, earthquake, hurricanes and tropical storms are detrimental to the corals.

Diseases, plagues and invasion of aquatic invasive species like, *Kappaphycus alvarezii* and crown-of-thorns starfish are also causing deleterious effect to the self-sustained ecosystem.

Discolouration of symbiotic zooxanthellae or their pigments is called coral bleaching, which can occur due to several reasons, like exposure to higher sea surface temperature and excessive solar radiation for longer period, reduced sea water salinity, high rate of sedimentation, pollution by pesticides, bacterial diseases and destructive method of cyanide fishing.

Corals become colourful due to the presence of symbiotic micro-algae, zooxanthellae. When they bleach, the corals become white exposing internal white skeleton through the transparent tissue. A reduction or loss in the density or pigment content of symbiotic algae can be elicited by a variety of environmental factors. When unfavorable condition persist for a longer period, mass bleaching might occur. After the whole area gets bleached, the affected corals become white in colour owing to loss of zooxanthellae. Bleaching can be partial, if stress responsible for bleaching is removed or the intensity is lost, and most corals can recover by re-colonizing with new micro-algae. If not, they will start to degrade. After death, other algae will slowly invade upon the corals and envelop the whole corals. The most frequent environmental stress responsible for bleaching corals is a rise in water temperature of 1 to 2°Celsius above the seasonal maximum.

Andaman Sea also witnessed bleaching events during 1998, 2002, 2005 and 2009. In 1998, the bleaching in corals was observed in 2 to 39 per cent and death in corals, especially branching types varied between to 3 to 55.4 per cent. In 2005, the bleaching was more severe, and maximum bleaching was observed in Havelock Island (69.49 per cent) followed by South Button Island (67.28 per cent), Nicolson Island (56.45 per cent), Red Skin Island (43.39 per cent), North Bay (41.65 per cent) and Chidiyatapu (36.54 per cent) and branching corals were most affected, while the massive corals were found to have withstood the elevated SST relatively. Coral communities of Andaman Sea were significantly damaged by mining of coral boulders and coastal sand siltation and deforestation, discharge of effluents from timber and match factories and infestation of crown-of-thorns star fish.

Conservation of reef ecosystem is a global issue. It has been estimated that 10 per cent of the world's reef has already been lost, and 60 per cent are threatened by bleaching, disease and by several human activities, like shore-line development, polluted run-off, ship groundings, over harvesting, destructive fishing and global climate change. However, given the right environmental conditions and a sufficient period of time, reefs can regenerate themselves and even during bleaching events. If some corals or patches remain unaffected, these can provide a basis for asexual reef generation. Thus to conserve coral reef ecosystem, impact of causative agents responsible for bleaching or degradation of corals has to be minimized or removed.

Continuous monitoring of the reef ecosystem is necessary to generate detailed database on coral reef biodiversity and on associated community living in those areas for conservation and management of reef ecosystem. Use of remote sensing, identification through spectoral signature and validation through ground scouting are some of the recent technologies widely being used to monitor the health of the reef ecosystem and to generate data base.

Marine protected areas play an increasingly important role in sheltering biodiversity in coral reef ecosystems from climate related impacts and in the recovery of corals from massive bleaching events.

Transplantation is an important method which can be used for restoring degraded ecosystem. Transplanted corals grow faster and mature, and adult corals can be used as a source for coral larvae. Larvae of *Pocillopora damicornis, Montipora capitata* and *Fungia scutaria* are produced in the laboratory after rearing the host coral for several months. Settled larvae can be transplanted at the required places, and if general condition is optimum, larvae will establish and grow in that environment. Coral seeding and transplantation are some of the direct methods, which can be used to rehabilitate and restore degraded reefs. Cryopreservation of coral gamets will also be very important for conservation of corals. The first frozen bank for Hawaiian corals is an attempt to protect them from extinction and to preserve their diversity.

Chapter 15
Sponges

Sponges are asymmetrical benthic animals, and are strikingly coloured. They represent a major component of reef communities. So far,4562 species of sponge have been described in the world. A total of 275 species has been recorded from Gulf of Mannar and Palk Bay. Other areas with numerical abundance are Gulf of Kutch (25 species), Andaman Sea (112 species) and Lakshadweep (109 species). Recent studies have revealed that sponges contain several peculiar chemical compounds that are not found in any other animals. Arabinose nucleosides isolated from *Tethya crypta* have proven cancer inhibiting properties and this discovery has triggered a worldwide interest in the biochemistry of this group. In India too, several institutions have taken up the extraction and characterization of several pharmacologically active compounds from sponges. So far, no species of sponges are exploited commercially in India.

A list of 83 species, belonging to 8 orders, 32 families, 51 genera, recorded from Andaman Sea are given below;

Phylum – Porifera, Class - Demospongiae

Family – Tetillidae

Cinachyra arabica, C. australiensis, Paratetilla bacca, Tetilla cranium, T. dactylloidea

Family - Ancorinida

Ecionemia accrous, Myriastra clavasa, M. purpurea, Rhabdastrella globostellata, Stellata cavernosa, S. haeckeli, S. orientalis, S. validissima

Family – Geodiidae

Erylus lendenfeldi

Family – Pachastrellidae

Dercitus simplex, Poecillastra eccentrica, P. tenuillaminaris

Family - Theneidae

Thenea andamanensis

Family - Chondrillidae

Chondrilla australiensis

Family - Clionidae

Cliona carpentari, C. ensifer, C. kempi, C. lobata, C. mucronata, C. quadrata, C. vastifica,Thoosa (Cliothosa) lancocki

Family - Spirastrellidae

Spirastrella andamanensis, S. inconstans

Family - Tethyidae

Tethya andamanensis, T. diploderma, T. repens, T. robusta

Family - Theonellidae

Discodermia gorgonoides, D. papillata, Theonella swinhoei

Family - Azorickidae

Leiodermatium pfeffarae

Family - Microcionidae

Clathria atrasanguinea, C. vulpine, Echinochelina barba

Family - Raspailiidae

Echinodictyum asperum, Raspailia viminalis

Family - Rhabderemiidae

Rhabderemia prolifera

Family - Anchinoidae

Ectyobatzella enigmatica

Family - Crambidae

Psammochela elegans

Family - Myxillidae

Damiriopsis brondstedi, Iotrochota baculifera

Family - Phoriospongiidae

Tedania anhelans

Family - Guitarridae

Biemna lipsasigma, B. tubulata

Family - Halichondriidae

Amorphinopsis foetide, Petromica massalis, Spongosorites halichondrioides

Family - Chalinidae

Gellius flagellifer, G. megastoma

Family - Niphatidae

Gelliodes fibulatus

Family - Phloeodictyidae

Calyx clavata

Family - Petrosiidae

Xestospongia testudinaria

Family - Spongiidae

Phyllospongia foliascens

Family - Clathrinidae

Clathrina coriacea

Family - Leucettidae

Pericharax heteroraphis

Family - Hyalonematidae

Hyalonema aculeatum, H. affine, H. indicum, H. lamella, H. martabanense, H. masoni, H. nicobaricum, H. rapa, Lophophysema inflatum, Pheronema raphanus, Semperella cucumis

Family - Aphrocallistidae

Aphrocallistes beatrix, A. bocagei, A. ramasus

Family - Farreidae

Farrea occa

Family - Tretodictyidae

Hexactinella minor

Family - Euplectellidae

Dictyaulus elegans, Regadrella decora, Euplectella aspergillum, E. regalis, E. simplex

Family - Rossellidae

Lophocalyx spinosa

Sponges along Indian coastline are bathymetrically distributed between 3 to 60 meter. Sponges (phylum Porifera) are sessile marine filter feeders that have efficient defence mechanisms against viruses, bacteria, or eukaryotic organisms. Being the richest source of pharmacologically-active chemicals from marine organisms, atleast some of the bioactive secondary metabolites isolated from sponges are produced by functional enzyme clusters, which originated from the sponges and their associated micro-organisms. More than 5300 different products are known from sponges and their associated micro-organisms, and more than 200 new metabolites from sponges are reported each year (La Port *et al*, 2009). Marine sponges are a treasured store of pharmacy on a centennial outlook with respect to their complex diversity of secondary metabolitic compounds. Though the molecular activity of a host of metabolites is still far from a clear answer, the reported mechanism of interferences onto the pathogenesis of varied diseases, forms the prime factor of significance to transform these bioactive compounds into useful medicines.

Culture of Marine Sponge

Among all metazoan phyla, marine sponges are considered as the richest source of biologically and pharmacologically active chemicals. More than 5300 different products are recorded to have obtained from sponges and their associated micro-organisms. Every year, about 200 new metabolites are reported from sponges. The detection of novel metabolites from sponges gains importance and also provides scope for developing new drugs against disease causing bacteria, virus, fungi and parasites. In nature, the chemical interactions in the marine habitat of sponges suggest that products from them function as defense tools to protect them against predators including fish. Sponge product ara-A (vidarabine), the anti-viral drug used against the *Herpes simplex encephalitis* virus has advanced to the late stages of clinical trials. Others, such as, manzamine A (active against malaria, tuberculosis, HIV and others), lasonolides (antifungal activity) and psammaplin A (antibacterial activity) are considered as promising leads. However, most sponges contain only trace quantity of the bioactive molecules. The increasing demand for initial experimental trials, possible success and subsequent industrial use for scaling up will lead to severe pressure on the wild population and hence the possible overexploitation and extinction of the target species as such. In view of the limited availability of larger quantities of defined source material, and to cater the requirements without loss of bioactive potential, culture of marine sponges could be considered as one of the options.

As such, mariculture of two species, namely, *Callyspongia subarmigera* (Ridley) and *Echinodictyum gorgonoides* was attempted at Vizhinjam Research Center of CMFRI.

All-glass aquarium tanks of 60x45x30 cm fitted with filtration unit and perforated Perspex panels served as bioreactors for sponge culture in the laboratory. Compressed air with air-water lift was provided for re-circulating sea water and for development of ammonia oxidizing bacteria in the bioreactor.

Freshly collected sponge *Echinodictyum gorgonoides* from Kanyakumari were washed and explants of about 4 to 6 g size were prepared by cutting with sharp knife.

Individual explant was weighed and kept in sea water without allowing them to dry. The explants were fixed to the Perspex panel.

The sponge explants fixed in Perspex sheet in the aquarium were fed with micro-alga, *Nanochloropsis* sp. at the rate of 3.5 x 1000000 cells initially. The cell density was increased gradually and the algal feed was provided twice per day. The wastes remaining in the culture system were removed every day and water was exchanged on alternate days. At different intervals of time, sponge tissues from the bioreactor were aseptically removed for evaluating the bioactivity using standard microbiological and other bioassay methods.

In open sea sponge culture, the sponge masses (*Callyspongia subarmigera*, *E. gorgonoides* and *C. diffusa*, collected off Kanyakumari and Vizhinjam) were cleaned with fresh sea water, placed in plastic circular fruit baskets (closed) and held at varying depths, one above the other at 1 m depth intervals and suspended in the vertical plane, at Vizhinjam Bay, Trivandrum coast. Fouling organisms attached to the baskets were periodically removed. The average pH of the water varied between 7.60 and 8.10. The dissolved oxygen content ranged from 3.6 to 4.17 mg/l, while the salinity was between 29.5 and 35 ppt.

The sponge species that produce bioactive compounds can be cultivated in the laboratory with sea water re-circulatory system as well as in open sea. In both the systems, the sponges retained their health state to a large extent and also their potency to produce bioactive compounds.

It could be inferred that the bioactivity potential of the sponge in culture conditions is determined by the survival and growth, which are influenced by the farming environment. The marine environment influences the biosynthesis and yield of target metabolite. In order to achieve the maximum production of specific metabolites or molecules from marine sponges or any other organism with bioactivity potentials, it is essential to devise and develop novel culture methods with considerable flexibility. For these, sponges having different bioactivity patterns are to be collected from different locations and cultured in different depths of a selected marine habitat.

Chapter 16

Coelenterata

Ocean's Gelatinous Inhabitants

Jelly fish and other gelatinous creatures from Cnidarians (Jelly fish) to Ctenophore (Comb jelly fish) with waiving cilia that propel them along to a mass of other creatures including gelatinous snails, worms and larvaceans, tiny tadpole like filter feeders that spin mucous webs to catch their food.

Jellies seems to have bigger role in the ocean system and could even provide the missing piece in a long standing puzzle about how carbon cycles from the ocean's upper layers to its floor.

Mass blooms of jelly fish, which appear when the creatures reproduce in spring or summer are a perennial nuisance to fishermen and coastal dwellers. They decimate fisheries by munching on fish larvae and eating all of the shrimps and plankton, the same food that fish rely on.

The fragile rarely survive the temperature changes and physical buffeting they experience during the trip to the surface. Many jellies eat prey that glow and to avoid becoming a beacon to their own predators, they camouflage their guts with red or black pigment. These pigments can become lethally toxic on exposure to light, including the head lamps of submersibles.Lindsey, the marine biologist caught one red jelly fish after four year's trial and brought to the surface in a protective case and when he was about to start filming, the light on the video camera inadvertently switched on at full strength, the jelly spat out its gut and disintegrated within minutes. About 2000 species of jelly (a new species almost in every dive) are known, many more lurk undiscovered in the deep. There were so many jelly fish in the water, during blooming, they fill the nets before targeted fish could be caught. Jellies are major players in the ocean's carbon biomass.

Jellies have been perceived as carnivores, that drift passing around, snaring prey when their chance upon it. Some jellies come up from the deep water under the cover of night to gorge on plankton and krill in surface waters before sinking back down to digest their meal at depth. Other jellies have evolved specific features to pery on their gelatinous cousins. Some jellies of the genus *Beroe*, for example, are made of nothing, more than a mouth like sac, with tooth-like macrocilia that chomp exclusively on other jellies. A substantial percentage of oceanic biomass is tied up in the bodies of jellies that are feeding on each other.

Jellies diverse role in the food web coupled with their immense numbers could answer some outstanding questions about how carbon-cycles through marine ecosystems.

Jellies can provide shelter and food for a huge range of creatures, who hide in jellies fold and even nibble on the jellies themselves.

In the Gulf of California, it was discovered that a crustacean, called, *Oxycephalus* nurses her young atop a comb jelly fish. And many deep sea explorers have observed that *Deepstaria enigmatica*, one of the largest midwater jelly fish, which looks like a thin sac of white jelly almost always has one or two crustaceans, called, *Anuropus* living with it.

Jelly Fishes

Jelly fishes, members of the class Cnidaria of phylum Coelenterata, are fascinating creatures that dwell im marine aquatic ecosystems. Jelly fishes are found in all oceans round the world, at all levels, starting from the surface to the deep sea. They are believed to be among the oldest inhabitants of the earth. Jelly fishes are free-swimming members and possess different morphological characteristics that represent several Cnidarian classes including Scyphozoa (about 200 species), Staurozoa (about 50 species, Cubozoa (about 20 species) and Hydrozoa (about 1000 to 1500 species). The jelly fishes are found distributed in oceans from the surface to the deep sea. The life span of these jelly fishes is about a few hours to several months.

Jelly fishes are generally translucent aquatic creatures. Jelly fishes are kept captive by humans for their display in aquaria and zoos. Jelly fishes, however, is not adaptable to closed spaces. The photo protein causes the bioluminescence effect in jelly fishes. As such, the jelly fish has become an interesting aquatic animal for its green fluorescent protein as marker of genes, and also for evolving methods allowing these proteins to be inserted into the cells and organisms. Many different fluorescent colours of jelly fish can be used as gene markers

Jelly fish have been used for extracting bioluminescent proteins. Many species of jelly fish, namely, *Aequoria Victoria* found in North Pacafic Ocean emit bioluminescent glow produced by the green fluorescent protein (GFP) present in them, which startles their predators.

Molecular biologists have taken advantage of this protein by splicing it to produce luminous genes that can be easily observed in living cells.

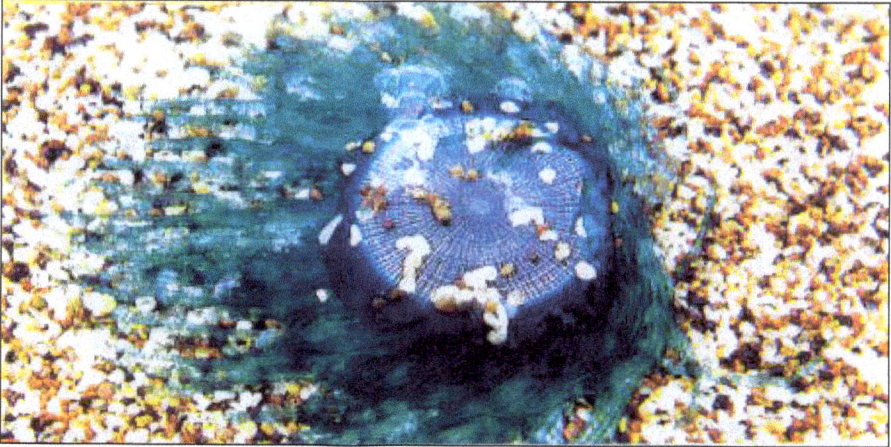

Figure 16.1: *Porpita porpita*

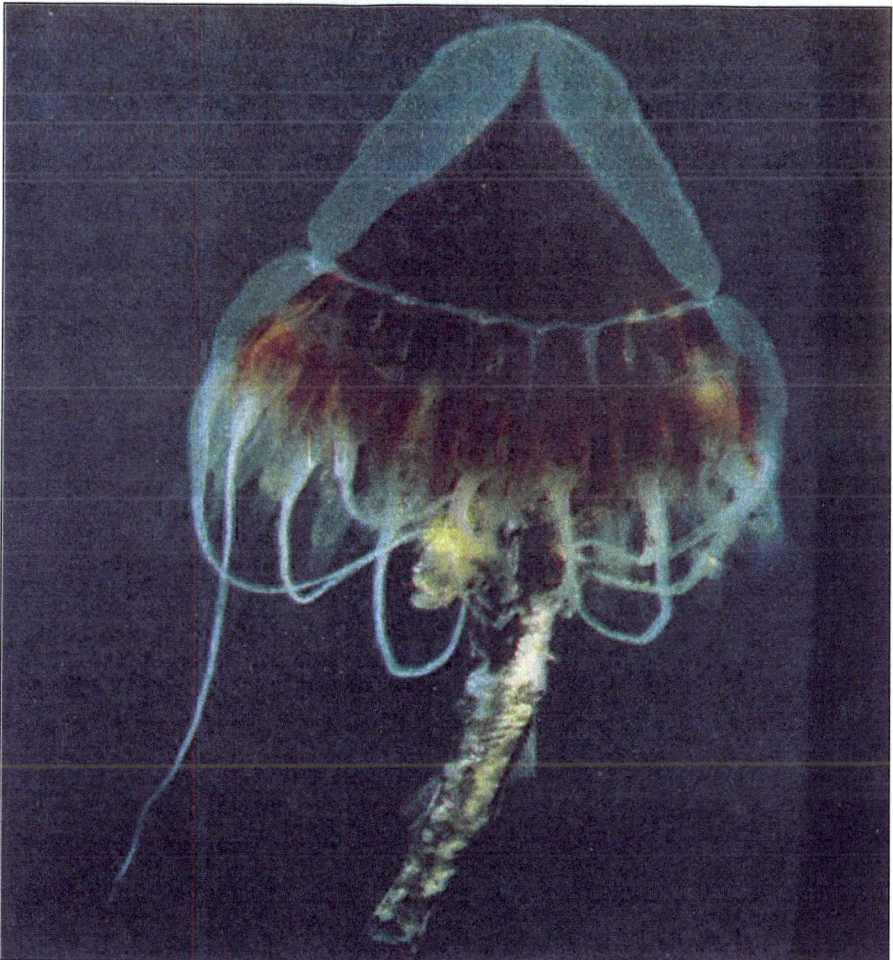

Figure 16.2: Deep Sea Jellyfish, *Periphylla*

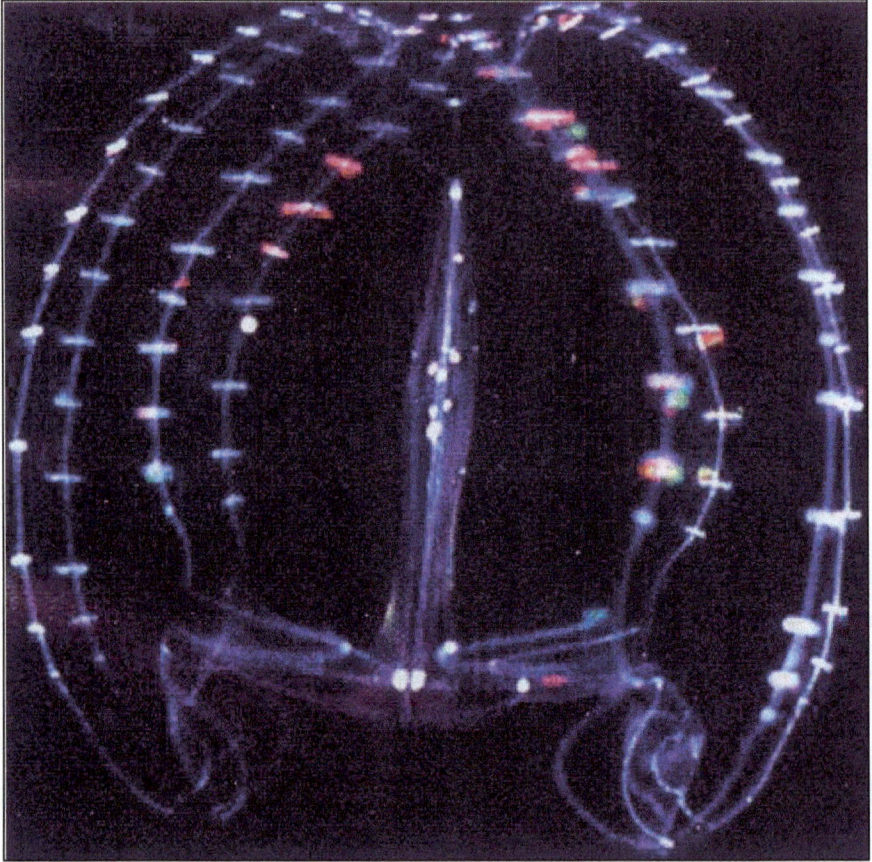

Figure 16.3: Polyp (Free swimming)

Jelly fishes have another part of their life-cycle, polyp phase. They do not have specialized digestive, osmoregulatory, central nervous, respiratory or circulatory systems. They do not need a respiratory system, since their skin is thin enough that the body is oxygenated by diffusion. The body has more than 90 per cent water. They do not have either a brain or central nervous system. The main predators of the jelly fish are tuna, shark, swordfish, sea-turtles and others.

The abundance of many jelly fish species may increase in warm conditions. An extraordinary mass landing of jelly fish is observed in the Cuddalore district of Tamil Nadu during the last week of March, 2009, the majority of landing being *Crambionella* sp.

The jelly fishes were floating on the sea surface about 10 to 15 km away from the sea shore. They were caught by scoop nets. Around 3000 tonnes of jelly fishes have been landed during the last week of March, 2009. Later, strangely, there were no jelly fish catches.

Global warming have caused proliferation of jelly fishes. After tsunami, the fishermen had been getting large quantity of jelly fishes.

Seasonal jelly fish bloom occurred in Jakhau coast, Gujarat. Of late, backed by the heavy demand of jelly fish from China and South East Asian markets, 75 fibre glass canoes with inboard engines are involved in jelly fish collection using multifilament gill nets. Each boat brings harvest of around one tonne of jelly fish per day. The fishery has two seasons, namely, November-December and April-May. Initial processing is carried out in the landing center itself. They are then transported to Veraval for further processing and export. The landings of 800 tonnes of jelly fish was observed in 2010, which is more than three times of the previous year's landings of 250 tonnes.

Chapter 17

Sea Anemones

Symbiotic relationships are of great ecological importance. Symbiotic relationship exists on land as well as under water. In the densely packed environment of a coral reef, there is a concentration of symbiotic relationships far greater than any other habitat. According to Roughgarden (1975), the most common and well known associations that have prompted much of investigations are those between sea anemones and the coral associates.

The relationships between sea anemones and other symbionts are the type of symbiosis, which basically mean "living together". For both the species benefit, symbiosis takes place between sea anemones and clown fishes. The degree of specificity in symbiotic relationships depends on the number of host species that a symbiotic will live and interact within the nature (Miyagawa, 1989).

The first evidence of direct transfer of photosynthetic carbon from zooxanthellae to the host was obtained for the temperate anemone, *Anthopleura elegantissima*. The sea anemone has a symbiotic alga, called zooxanthellae, which are found in the anemone's tissues. These algae are necessary for the long time survival of the anemone. Anemones are exposed to pronounced fluctuations in sea water temperature and they host multiple symbionts with different temperature tolerances. The distribution pattern of zooxanthellae and a green algal symbiont within the host are strongly correlated with physiological tolerances of the algae to the temperature, suggesting environmental control of algal complements.

The associations occurs between different species of crabs and anemones are not well understood. Many harmit crabs have anemones attached to their shells, some of which are anemone shrimps. Anemone shrimps have been known even to share anemones with or without clown fishes. If an anemone shrimp and a clown fish must share a host, then large anemones make better hosts than small ones. In the absence of an anemone, most anemone shrimps will accept other hosts. The anemone shrimp will have particular need of a hiding area directly after moulting, while its new exoskeleton remains soft and sensitive to toxins in the host anemone. After the exoskeleton gets hardened, the anemone shrimp slowly readjust itself to the toxin of its host.

Out of 28 species of clown fish found in the Indo-Pacific, 10 species of them are found living with anemones. Sea anemones provide shelter and a variety of other benefits to the fishes. In turn, the anemone fishes provide the anemone with certain advantages. It is believed that the clown fish has a special mucous coating that prevents animal's stinging capsule from firing. Since clown fishes are poor swimmers, they would become an easy prey without the protection of the host anemones. The anemones are equipped with microscopic stinging structured called, nematocysts located at the end of their tentacles.

The main advantage that the fish gain is that of protection. The stinging tentacles of the anemone provide a safe place for the fishes from predators. During day time anemone fishes may venture short distances away from the anemone to find food. At night, the fishes settle deep into the tentacles and remain there until the morning. This protection is not only beneficial to adult anemone fishes, but to eggs and juveniles as well. The clown fishes select spawning sites for the deposition of eggs near the anemone and in contact with tentacles, for extra protection from predators, which avoid coming into contact with the tentacles. Fish may also gain some nutrients from feeding on wastes around the anemones or even on the anemone's tentacles. The stinging tentacles of the anemone remove external parasites from the fish. It has been hypothesized that the tactile stimulation that the fish receive from the tentacles of the anemone may be beneficial to their health and well being.

The anemone fishes rarely occur in nature without being associated with a host anemone. Removing either the anemone fish from the anemone and vice versa, the anemone fishes almost always become prey for larger fish. Anemones, on the other hand, are often found in nature without fishes, hypothesing that the fishes are not essential to the anemones for their survival, fish provide to them some benefit. Most members of clown fish live with only a few host anemone species, but in one complex, *Amphiprion clarkia* has been found living with up to ten different host anemone species.

Chapter 18
Meiofauna

Interstitial Meiofauna

The archipelago of Andaman and Nicobar Islands offer a rich variety of habitats for the colonization of fauna. Thus the biogeographic importance of these islands lies mainly in providing isolated biotopes, favoring evolution of a variety of species. In many of the islands, the coast is mostly rocky and sandy or muddy. In some areas, the sandy beaches are extensive, while in other places, they are restricted to patches between cliffs. Many of the beaches are exposed to surf action, while some are sheltered. The sands are mostly fine and medium, silicious or coralline, and grain shape varied from spherical to angular. Temperature and salinity of interstitial water in the habitat varied between 26 to 30°Celsius and 29 to 34 ppt respectively. The intertidal sands are sufficiently rich in organic detritus.

The fauna collected from coarse sandy substrates to soft sediments, showed widely distributed species in different kinds of substratum. Sheltered beaches with adequate coarse particle size generally harbored rich population of fauna, probably due to less disturbed habitat that permits an uninhibited proliferation of the fauna. Beaches with very fine and muddy sand yielded poor faunal concentration. Bulk of the fauna occurred near half-tide level 10 to 40 cm below surface, depending on the distance from low water level.

The density and diversity of fauna varied from island to island and from locality to locality around Andaman Sea. Thus a considerable percentage of the species are stenotopic with restricted distribution, while the others appeared to have a wider distribution. Quantitatively, the density of the fauna is comparatively low and a maximum of only 600 to 800 specimens were obtained in 100 cubic cm. Almost all the typical groups of interstitial meiofauna, excepting, Mystacocarida and Bryozoa were collected in the sandy beaches of Andaman Sea. Quantitatively, the copepods formed a dominant element of the total meiofauna. Annelids and Nematodes were generally second in overall abundance. The Ciliata, Turbellaria, Gnathostmulida, Gastrotricha, Isopoda and Amphipoda are moderately represented. All other groups of animals occurred in very small numbers.

Out of the total 324 species collected from 24 shore stations of Andaman Sea (collection made 3 to 5 places depending on the exposure between the low and high tide levels of the intertidal zone) 15.4 per cent appear to be cosmopolites, 14.1 per cent widely distributed species, 10.2 per cent Indian Ocean forms and 59.6 per cent uncertain endimics.

Table 18.1: Summary of the Zoogeographical Relationships of the Interstitial Meiofauna of Andaman Sea

Group	Percentage of Total Collection			
	Cosmopolitan	Widely Distributed	Indian Ocean Form	Uncertain Endemics
Ciliata	38	19	–	45
Cnidaria	–	–	–	100
Turbellaria	3	10	3	84
Gnathostomulida	–	–	–	100
Nematoda	28	20	7	45
Gastrotricha	10	5	10	75
Kinorhyncha	–	–	50	50
Nemertina	–	–	–	100
Rotifera	–	–	–	100
Archiannelida	8	16	16	60
Polychaeta	33	8	14	45
Oligochaeta	–	50	–	50
Ostracoda	–	37	–	63
Copepoda	11	21	17	51
Isopoda	–	–	–	100
Amphipoda	–	–	–	100
Tardigrada	50	–	25	25
Collembola	–	–	–	100
Acarina	–	17	–	83
Mollusca	–	17	17	66
Echinodermata	–	–	–	100
Tunicata	–	–	–	100

The composition and abundance of genera and species of Andaman Sea are more or less in agreement with the other regions of the world. In all the major groups of the fauna encountered, quantitative abundance of few dominant species is observed. Thus about 20 per cent of the total species encountered comprised about 75 per cent of the total number of meiofauna individuals, while bulk of the remaining species occurred in small numbers. Many of the interstitial meiofauna groups with considerable number of genera and species are known to have a wide geographical distribution. Thus qualitatively the groups, Ciliata, Turbellaria, Gnathostomulida, Nematoda, Gastrotricha, Arciannelida, Polychaeta and Copepoda are best represented in the Andaman Sea with many a variety of texa, while the remaining group have only moderate to poor representatives.

Meiobenthic Fauna

Meiobenthos play an important role in benthic food chain. Meiobenthic fauna in the depth range of 30 to 2750 m, of Andaman Sea ranged between 68 and 430 per 10 cm square. Faunal components were maximum in clayey deposits and minimum in the coralline sand. The population was generally dominated by the nematodes. Faunal density showed unusually low value at intermediate depths (200-500 m). At all the depths, the fauna was mostly present in the upper 2 cm of the sediments and only nematodes and foraminiferans were found below 4 cm depth. Changes in the population density appear to be related to the changes in the hydrography, biotic feature and bottom characteristics.

The sampling area lies between Lat. 8 and 15°N and Long. 91 and 94°E. Bottom deposits in the shallow regions of Andaman Sea (above 200 m) is characterized by sand followed by coralline sand up to 500 m and mostly clayey below 500 m. Bottom temperature ranged from 27.9°at 30 m to 5.02°Celsius at 2000 m. A steady decrease in temperature from shallow to a depth of 2150 m was observed. Changes in temperature were more pronounced between 200 and 300 m. Salinity did not show much change and the values ranged between 31.88 ppt at 30 m and 35 ppt at 500 m. Dissolved oxygen was found to be very low and the value ranges from 4.82 ml/l at 50 m to 0.45 ml/l at 300m.

A total of 7 taxonomic groups and nauplii were identified in 10 cm square surface area. Total meiofauna numbers ranged from 68 to 438/cm square. Nematodes, foraminiferans, harpactacoid copepod, polychaetes and turbellarians were both the most abundant taxon at all places except at 13°N and 92°20 minute E, where benthic foraminifera dominated the fauna. Average biomass of meiofauna in Andaman Sea was 14.26 mg/10 cm square. Maximum biomass was observed at 1500 m and minimum at 850 m.

Changes in the population density of meiobenthos in relation to depth revealed that the fauna were highest in the shallow region (0-100 m) and decreased with increasing depth up to 500 m. Below 500 m and up to 1500 m, the faunal density again increased and subsequently decreased to minimal number at high depths (2000 m).

As regards the vertical distribution of meiofauna, about 68 per cent of the total fauna were found in the upper 0-2 cm of the sediment, regardless of sediment type. The fauna progressively decreased with increasing depth and no fauna was observed below 6 cm in the sediment.

As regards to the vertical distribution of the major texa (nematodes, foraminifera and others) found in the sediment at 85 m, 650 m and 800 m depth, all texa decreased both qualitatively and quantitatively with the increasing depth. Nematodes and foraminiferons were the only texon encountered in the entire 6 cm deep sediment. Among other groups, only polychaetes and turbellarians were observed below 2 cm in the sediment.

The influence of sediment type on meiofauna is well documented. In the 1979 February investigation, the faunal changes coincided with nature of the substratum and hydrographic changes. The shallow region (0-100 m) dominated by sand with little admixture of silt harbours maximum fauna. Between 200 and 500 m, the sediment is coralline sand and the fauna was poor, with increasing depth the sediment changes to clayey deposits, which follows marked changes in faunal density and at deeper depths (more than 2000 m) as expected, the fauna reduced to only one taxonomic group, the nematode.

It has also been observed, that total disappearance of some of the meiobenthic group towards increasing depths was due to the changes in sediment type. Surprisingly, the total meiofauna number increased considerably between 500 and 1500 m. This is due primarily to changes in sediment from coralline sand to clayey deposits and also hydrographic conditions. Similar pattern and high meiobenthic population in clayey deposit have been observed in the Bay of Bengal. This suggest that changes in sediment type is important for meiofauna.

Marked changes in bottom temperature and dissolved oxygen were observed in the Andaman Sea. Both temperature and oxygen have shown sharp decrease below 100 m and a steady trend between 500 and 1200 m. Thus in the region of maximum faunal change between 200 and 500 m, at least three ecological factors, temperature, dissolved oxygen and sediment composition are responsible for this change. Though it has been correlated that changes in meiofauna with changes in bottom temperature, sediment composition and accompanying changes in the availability of food, yet interestingly, the place (14°N, 93°E), where the dissolved oxygen value at bottom were lowest (0.45 ml/l), no fauna in the sediment was observed, which indicate that oxygen is a limiting factor in the distribution of meiofauna.

The vertical distribution of the number of organisms in the sediment decreased with increasing depth. Concentration of fauna at the sediment surface has been attributed to the availability of oxygen, food supply and living space.

For deep sea benthos, two main sources of food, detritus and bacteria exists. As far as the first source, the available information suggest high amount of detritus in Andaman Sea. High number of heterotrophic bacteria are also present in the sediment. But the productivity of the water column, which form the main source of food for benthos is lower than other areas of Bay of Bengal. Similarly, organic carbon in the sediment is also low (0.45 to 1.50 per cent). Limited mixing, sharp stratification and low productivity of the water column may reflect on low benthic production of the Andaman Sea.

Table 18.2: Meiofauna Density per 10 cm Square, Biomas Values in Parenthesis in mg/10 cm sq. (Wet weight)

Locations	Nema	Fora	Cope	Poly	Turb	Lame	Kino	Nauplii
8.12' N, 93.36' E	102 (10.2)	68 (2.02)	–	–	34 (0.17)	–	–	–
11.50' N, 93.08' E	152 (15.2)	102 (3.06)	–	34 (1.12)	34 (0.17)	34	34 (0.14)	–
12.30' N, 93.30' E	152 (15.2)	136 (4.08)	34 (0.24)	–	–	–	–	–
14. N, 93. E	–	–	–	–	–	–	–	–
13. N, 93.25' E	68 (6.8)	–	–	–	–	–	–	–
12.05' N, 93.30' E	102 (10.2)	102 (3.06)	–	–	–	34	–	–
11.25' N, 93. E	232 (23.2)	34 (1.02)	68 (0.47)	102 (3.36)	–	–	–	–
10.22' N, 92.12' E	68 (6.8)	102 (3.06)	–	34 (1.12)	34 (0.17)	–	–	–
10.16' N, 92.25' E	34 (3.4)	–	–	–	34 (0.17)	–	–	–
10.18' n, 92.28' E	34 (3.4)	–	34 (0.24)	–	–	–	–	–
10.14' N, 92.32' E	152 (15.2)	–	34 (0.24)	–	34 (0.17)	–	34 (0.14)	–
10.28' N, 93. E	302 (30.2)	34 (1.02)	68 (0.47)	34 (1.12)	–	–	–	–
10.28' N, 92.40' E	102 (10.2)	–	–	–	–	–	34 (0.14)	34 (0.06)
10.38' N, 92.43' E	152 (15.2)	–	34 (0.24)	–	68 (0.34)	–	–	–
10.38' N, 92.40' E	34 (3.4)	–	–	34 (1.12)	34	34 (0.17)	–	–
10.30' N, 92.18' E	68 (6.8)	34 (1.02)	34 (0.24)	34 (1.12)	–	–	–	34 (0.06)
10.40' N, 92.05' E	204 (20.4)	34 (1.02)	34 (0.24)	68 (2.24)	–	–	–	–
10.40' N, 92.20' E	102 (10.2)	68 (2.04)	34 (0.24)	34 (1.12)	34 (0.17)	34 (0.06)	–	34
11.20' N, 92.08' E	68 (6.8)	–	–	34 (1.12)	34 (0.17)	–	34 (0.14)	–
12.32' N, 92.30' E	68 (6.8)	68 (2.04)	34 (0.24)	34 (1.12)	–	34	–	34 (0.06)
13. N, 92.20' E	136 (13.6)	186 (5.58)	34 (0.24)	–	–	–	–	34 (0.06)
13.31' N, 92.25' E	102 (10.2)	34 (1.02)	34 (0.24)	34 (1.12)	–	–	–	–

Nema: Nematoda; Fora: Foraminifera; Cope: Copepoda; Poly: Polychaeta; Turb: Turbellaria; Lame: Lamellibranchiata; Kino: Kinorhyncha.

Table 18.3: Number of Total Meiofauna and Dominant Group per 10 sq.cm of Sediment Recorded at different Places

Places	Depth (m)	Total Fauna	Dominant Group and Density/10 cm sq
Off Massachusettes, (1964)	40-58	127-988	Nematoda, 50-924
Off Massachusettes, (1964)	69-179	237-537	Nematoda, 202-507
Off Massachusettes, (1964)	366-567	117-127	Nematoda, 110-117
Off East Africa, (1966)	1045-5030	15-170	Nematoda, 14-128
Off North Carolina (1971)	250-500	138-1174	Nematoda, 60-1026
Off North Carolina (1971)	800-2500	40-149	Nematoda, 32-134
Bay of Bengal (1977)	100-500	260-1630	Foraminifera –
Bay of Bengal (1977)	500-1000	122-1050	Nematoda –
Loch Nevis, Scotland (1964)	101	541-2224	Nematoda, 405-2072
Buzzards Bay (1960)	18	276-1861	Nematoda, 250-1860
Bermuda coast (1970)	2-13	123-1333	Nematoda, 88-958
Kerala coast (1972)	4-12	78-1455	Nematoda, 42-1146
Andaman Sea (1979)	100-500	68-340	Nematoda, 34-102
Andaman Sea (1979)	501-2000	68-438	Nematoda, 34-302

Table 18.4: Distribution of Meiofauna in Relation of Depth of Andaman Sea

Depth Range	No. of Observations	Average Density per 10 sq cm
0-100	6	288
101-200	3	136
201-500	4	176
501-1000	3	236
1001-1500	3	220
1501-2000	2	134
> 2000	1	68

Chapter 19

Chaetognaths

Chaetognaths inhabiting in the sea around Andaman Islands up to 48 km from the coast, during February, 1979 was studied from zooplankton samples from 34 places of which 6 stations (1967-1975) were located enroute to the Andaman Islands.

Chaetognaths occurred at all places of Andaman Sea. Maximum density of 1959 per 100 cubic meter was recorded at 11°50 minute N, 93°08 minute E. Stations located between southeast of Andaman Island and north of Nicobar Island had relatively greater population density. Low values were continuously recorded at places enroute to Nicobar Islands (6°N, 92°52 minute E and 8°N, 93°38 minute E). The mean density of chaetognaths in the area was 659/100 cubic meter.

In all, 13 species belonging to three genera, *Krohnitta*, *Pterosagitta* and *Sagitta* were found to inhabit in the island waters. *Sagitta enflata* was the most abundant chaetognath in the sea around Andamans. Other common species are *S. bipunctata*, *S. pacifica* and *K. pacifica*.

Distribution

S. enflata was the dominant chaetognath species found in most places and it was absent only at 9°50 minute N, 91°58 minute E. Percentage incidence of the species in the collection was 44.7. Maximum density of the species was found in the region southeast of Andaman Island. At places, 6°32 minute N, 94°05 minute E and 6°N, 92°50 minute E also they were relatively abundant.

S. bipunctata is second in abundance and constituted 15.3 per cent of the total chaetognath population. Their maximum density was noted towards the north of the Andaman Sea, and at all places, numerically they were more than *S. enflata*. Maximum density of 577/100 cubic meter was obtained from 10°55 minute N; 93°30 minute E.

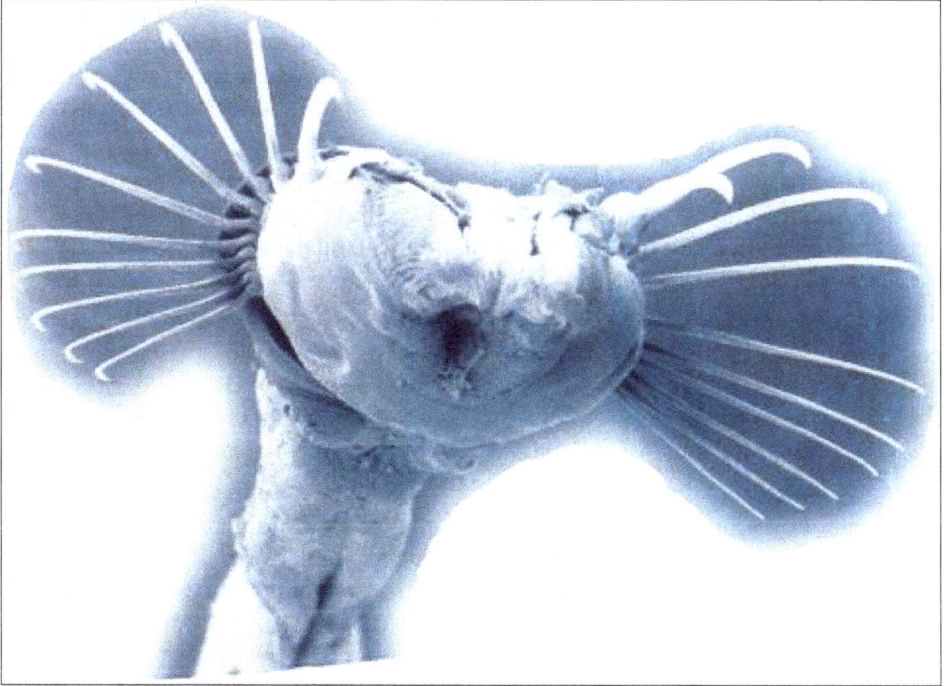

Figure 19.1: Head of *Sagitta* sp.

Eventhough percentage of incidence (12.3) of *S. pacifica* was only next to *S. bipunctata*, they were very common in the collection and was the dominant species at 12°30 minute N, 93°30 minute E. Higher density was observed southeast of Andaman Island. The highest recorded density of 337/100 cubic meter was at 11°50 minute N, 93°08 minute E.

K. pacifica was commonly encountered in the collections. Their incidence in the Andaman Sea was 8 per cent. They are found all around the Andaman Island with scattered areas of high density. Maximum numer of specimens (209/100 cubic meter) was obtained from the 9°50 minute N, 92°40 minute E.

In mature specimens of *K. pacifica*, the arrangement of ova in the ovaries appeared to be different from the normal pattern. Usually the ova are small, round and closely arranged within the ovary. In all the mature specimens (stage iv maturity) of Andaman Sea, ova were elliptical with distinct spacing in between. This type of arrangement of ova is similar to that described for *S. oceania*. Such variations in the disposition of ova had not been for any of the species of *Krohnitta*.

P. draco was commonly encountered in the collections. All the negative stations were located east of Andaman Islands and north of Nicobar Islands. *P. draco* formed 5.7 per cent of the total chaetognath population. The highest density of 172/100 cubic meter was recorded at 6°32 minute N, 94°05 minute E.

S. neglecta was not very common in the Andaman Sea. Where ever present, they were represented in fairly good numbers and hence the percentage of incidence amounted to 5.3 per cent. Maximum number of 126/100 cubic meter was obtained from 13°N, 93°05 minute E. Relatively higher density of the specimens was noted close to the island.

S. bedoti was occasionally found in the collections and their incidence was 4 per cent. The highest density was recorded at 9°32 minute N, 93°30 minute E (114/100 cubic meter). Stations located south of Andaman Island sustained higher density of the species.

Among rare species, the two closely related species, *S. ferox* and *S. robusta* were confined mainly to the sea surrounding the Andaman Island and percentage of incidence of each of them was 1.3. Placewise representation was more for *S. ferox*, than *S. robusta*, *S. pulchra* and *S. regularis* were found only at a few places. *Krohnitts subtilis* was obtained mostly from places located away from the land. *Sagitta hexaptera* was limited to the places south of Andaman Island.

Eventhough chaetognaths are present at all places of Andaman Sea, numerically they are not very abundant as can normally be expected from a tropical sea. The International Indian Ocean Expedition data on chaetognaths from the Andaman Sea gave a mean value of 2000/100 cubic meter for the northeast monsoon period during 1973. The investigation of 1979 gave the number of 659/100 cubic meter only, one third of the IIOE record. The different types of net used for collecting zooplankton samples may probably account for this variation.

During the IIOE seven samples were collected from the Andaman Sea and 12 species were recorded. In the IIOE collections, *K. subtilis* was absent and *K. pacifica* was very rare. *S. regularis* and *S. bedoti*, which were very rare in 1979 samples were reported to be common in IIOE samples. Of the 13 epiplanktonic species recorded from the sea around Andaman Island, *K. subtilis*, *P. draco*, *S. bipunctata*, *S. enflata* and *S. hexaptera* are cosmopolitan forms of warm and temperate oceanic waters. *K. pacifica*, though widely distributed is restricted to the equatorial belts of the Atlantic, Pacific and Indian Oceans, the remaining species are typical Indo-Pacific forms.

The Lakshadweep atolls and Andaman Islands are situated approximately along the same latitudes. However, the sea around the Lakshadweep appears to sustain more number of chaetognaths. The average number of chaetognaths was 3184/100 cubic meter for the northeast monsoon period, which is about four times higher than the mean value obtained for the Andaman Sea. The different species encountered at both the seas are almost the same with certain amount of variation in their percentage of incidence. *S. regularis*, *S. neglecta* and *S. bedoti* were common in Lakshadweep, while their abundance is relatively low in the Andaman Sea. *S. bipunctata*, which was second in the order of dominance in the Andaman Sea is rarely found in the Lakshadweep Sea.

Depending on the pattern of abundance of various species, different groups could be recognized within the area of sampling.

1. *S. enflata, S.pacifica* and *S. bedoti,* which were common to the region sustained maximum density south of Andaman Island.

2. *K. pacifica* and *S. bipunctata,* which were also common in the collections exhibited higher abundance towards the northern part of the surveyed region.

3. *P. draco* and *S. neglecta* showed a patchy distribution.

4. Incidence of the rare species was observed mainly north of 10°N.

Chapter 20
Holothurians

Echinoderms, spiny skinned marine animals live in rocky, sandy, muddy and coral environments. One thousand one hundred fifty species of holothuroids available, world wide. About 160 species under 62 genera are available in India. Chinese are the traditional consumers of the processed sea cucumber products, called beche-de-mer and they consume it for the treatment of weakness, impotency, debility of the aged, constipation, due to intestinal dryness and frequent urination. Sea cucumbers are important in biomedical research and from the nutritional point of view, the derivatives are ideal tonic in high protein, and in low fat. In the western countries, it is used as a rich source of polysaccharide, "Chondroitin sulphate" and "glucosamine". The bioactive substances with anti-inflammatory and anti-tumour activity are well known for the therapy of arthritis and HIV, several commercial products, Arthi Sea, Sea Cu Max, Sea Jerky (arthritis medicines, nutritional supplements) are also derived from sea cucumber.

Sea cucumbers is an important component of the food chain as deposit and suspension feeders, and are also called earthworms of the sea. Important species of sea cumbers are *Holothuria nobilis*, which grow to a length ranging from 50 cm to over one meter, The other important species are, *Holothuria scabra, H. spinefera, H. atra, Actinopyga militaris, A. echinites, S. hermanni* and *B. marmorate*.

Sea cucumbers are caught by skin diving, trawl nets, push nets and are exported to Singapore and Hongkong after processing. In Gulf of Mannar, sea cucumber habitats in 15 fishing areas support 20000 fishermen and around 500 kg of beche-de-mer is processed every month.

Sea cucumbers (Holothurians), a delicacy for the Chinese, the Japanese and the Koreans, are consumed in fresh, chilled, frozen, processed in brine, occurs in all the oceans, including Andaman Sea. In Andaman Sea, 34 species, belonging to 11 genera

Figure 20.1: Starfish and Sea Urchin (Echinoderms)

Figure 20.2: Starfish, *Protoreaster lincki*

have been reported to occur from shallow to deeper waters. Any sea cucumber that is large with a thick body wall can be used for processing. In the world, about 30 species of sea cucumber are used for edible purpose. From the Indian seas, 15 species can be used for processing. *Holothuria scabra*, the most valuable species are available in Andaman Sea along with the following other species;

* ☆ Genus – *Cucumaria* (1) *C. alcocki* Koehler and Vaney, (2) *C. bacilliformis* Koehler and Vaney

* ☆ Genus – *Actinopyga* (1) *A. booleyi*, (2) *A. lecanora* Jager, (3) *A. mauritiana* Quoy and Gaim, (4) *A. miliaris* Selenka

* ☆ Genus – *Labidodemas* (1) *L. semperianum* Selenka

* ☆ Genus – *Holothuria* (1) *H. albida* Bell, (2) *H. cadelli* Bell, (3) *H. exilis* Koehler and Vaney, (4) *H. papillata*, (5) *H. prompta*, Koehler and Vaney, (6) *H. pyxis* Selenka (7) *H. rugosa* Ludwig, (8) *H. atra* Jager, (9) *H. edulis* Lesson, (10) *H. maculate* Brandt, (11) *H. marmorate* Jager, (12) *H. monacaria* Lesson, (13) *H. pardalis* Selenka, (14) *H. scabra* Jager, (15) *H. vagabunda* Selenka

* ☆ Genus – *Pelopatides* (1) *P. gelatinosus* Walsh, (2) *P. mollis* Koehler and Vaney, (3) *P. ovalis* Walsh

* ☆ Genus – *Synallactes* (1) *S. woodmasoni* Walsh

* ☆ Genus – *Protankyra* (1) *P. errata* Koehler and Vaney, (2) *P. innominata* Ludwig, (3) *P. timida* Koehler and Vaney

* ☆ Genus – *Amphideima* (1) *A. investigatoris* Koehler and Vaney

* ☆ Genus – *Stichopus* (1) *S. chloronatus* Brandt (2) *S. variegates* Semper

* ☆ Genus – *Synapta* (1) *S. beselli* Jager

* ☆ Genus – *Chiridota* (1) *C. rufescence* Brandt

Holothuria scabra has two breeding peaks, the major peak in March-May and the minor one during October-December. The sea cucumber are subjected to thermal stimulation during the breeding season by raising the temperature 5°Celsius. They can be induced to spawn by feeding them on rice bran; *Sargassum* spp. and soya bean powder. The larvae are fed by a combination of *Isochrysis galbana* and *Chaetoceros calcitrans*. The doliolaria can be induced to settle by the addition of powdered algae at a concentration of 0.5 g/500 l. This will induce the doliolaria to settle and also serve as food for the newly settled pentactula. One month old pentactula are given *Sargassum* spp. extract for one month in a tank without sand. This consisted of fine mud with good organic matter with addition of Algamac. This improved the growth rate. After two months, when they reach a length of 20 mm, the juveniles can be transferred to shrimp farms. This was tried in Vietnam and it was found that the weight of the sea cucumbers increased from 30 g to 300 g in 3 months time. In shrimp ponds, much of the feeds goes as waste and settles down at the bottom of the tank polluting the environment. *H. scabra* is a good candidate species to rear them in shrimp ponds utilizing the waste feed of shrimp pond.

Sea cucumbers are soft worm-like marine animals, commonly found in shallow water areas of the sea to deep ocean floors. Most of the sea cucumbers are deposit feeders, living on organic matter and associated microorganisms. Although there are more than 1250 species of sea cucumber exist throughout the world ocean, only a couple of dozen species are considered commercially important.

The dried body wall of the sea cucumber is one of the most important and highly priced sea food product in the international market and is marketed as *beche-de-mer* (trepang or hai-som). Besides the reputation as an aphrodisiac, *beche-de-mer* is also said to cure low blood pressure, kidney disorders and impotence and prevents aging. *Beche-de-mer* is one of the four "sea treasures" of Asian cuisine, the other being shark fin, abalone and fish bladder.

Commercialization of sea cucumber aquaculture in Australia has taken a pioneering step with Bluefin Seafoods Sea cucumber culture project. The sea cucumber culture system consists of a microalgal culture uint, brood stock collection and management, induced spawning, larval culture, juvenile growout and sea ranching.

The tropical sea cucumber species, *Holothuria scabra* (sand fish) and *H. versicolour* (golden sand fish) are mass produced in the hatchery for sea ranching.

Among the tropical *beche-de-mer* varieties, these two species consistently fetch the highest price in the international markets (Hong Kong and Singapore).

Chapter 21
Echinoderms

Starfish – Crown-of-Thorns

Acanthaster planci, an obscure member of the coral reef fauna is a spectacular starfish. It can grow to more than 60 cm in diameter and has over a dozen of arms. The whole upper surface is covered with 2.5 cm long spines, coated with a toxic mucus. Any one injuried by these spines is likely to suffer severe pain and nausea. As early as 1931, T. Mortensen recorded that the crown-of-thorns fed on madreporian corals, the major reef builders.

It feeds by everting its stomach, spreading the filmy gastric membranes over the coral and applying them closely. The flesh of the coral polyps is then digested by the gastric juices and absorbed. When the starfish withdraw its gastric membranes from the coral, the flesh is gone and all that is left is a dead, clean, white skeleton. At its normal low densities, the crown-of-thorns is usually a nocturnal feeder and during the day it retreats under ledges, where it is not easily seen.

On the coral reefs, the crown-of-thorns breeds in the summer months, males and females releasing sperm and ova into the water. It has been estimated that a single female releases between 12 and 24 million eggs in one season. The fertilized eggs become young larvae, which float as plankton for a period and then settle. During this early planktonic stage they must have many predators. Small fishes, in fact, been seen feeding on the eggs as they are being released. It is likely that any of the normal plankton feeders of the reef will consume crown-of-thorns larvae. The corals themselves are major feeders on plankton and may be important predators on the larvae.

After it has settled, the crown-of-thorns larva undergoes metamorphosis. During the very small early stages, immediately after settlement, it probably falls prey to a variety of reef animals. These early stages have not so far been found in quantity on the reefs, but E.C. Pope has discovered newly settled larvae of sub-tropical starfish in short algal turfs and it may well be that recently settled crown-of-thorns larvae will be found in similar situations, as this kind of environment is common on the reef.

Once its metamorphosis has been completed, the starfish appears to be immune from attack by most of the other reef species. The metamorphosed crown-of-thorns contains a toxic material related to holothurin, a toxin found in some sea-cucumbers, and even at the small juvenile stages it seems to deter the majority of predators. It has been reported that when small crown-of-thorns, up to 12.5 cm in diameter, were offered to carnivorous snails of various families, none of them would feed on the starfish. Nor have predators on small juveniles been found during field observations.

The rate of growth of the starfish has been estimated at about 1.25 cm per month, which means, that it takes two years for a starfish to reach diameter of 30 cm. The adult is a formidable animal. Its large size and array of poisonous spines make it understandably unpalatable and at present, only one predator is definitely known to feed on it. This is the large carnivorous mollusk, *Charonia tritonis*. Popularly known as triton shell, it reaches 40 to 45 cm in length and with its powerful radula, it is able to tear open and ingest even the crown-of-thorns. An adult triton takes between 12 and 24 hours to consume an adult starfish. Starfish as small as 17.5 cm in diameter was eaten.

There has been a report that another animal preys on crown-of-thorns. This is the shrimp, *Hymenocera picta*, 4 cm in length. It is claimed that the shrimp and its mate dance about on the starfish, until it retracts its tube feet, thus releasing the grip on its arms on the substrate. The shrimp then lifts up the unattached arms and feed on the flesh of the starfish through the grooves on the under surface.

The crown-of-thorns starfish is present on most coral reefs in Indian Ocean, including Andaman Sea. It is not known in large numbers until infested by the

Figure 21.1: *Nassaria nivea*

Figure 21.2: Gastropod, *Tibia fusus*

plague. Under normal circumstances, the small numbers and cryptic day time behaviour of the crown-of-thorns mean that it is seldom seen by the divers. During normal conditions, at most, a dozen of crown-of-thorns are seen on the coral reef. But in infested area, with a short period many hundreds can be seen. Pearson has defined the plague area as one in which more than forty starfish are counted in a twenty minutes search. Endean counted 5750 starfish in a hundred minute swim on Green Island (Plague infested area) in 1967.

This is somewhat surprising in a coral reef system, the most complex of all the marine ecosystems. Current ecological theory maintains, that in such a situation, where there are large numbers of interdependent species, energy flow has many alternative pathways, implying that an increase in the numbers of one species would be damped by others, which impinge on it. The result should be a tendency towards stability, but perhaps sufficient account of the possibility have not been taken that any complex system might include some quite simple sub-units. Lacking the checks and balances conferred by complexity, these simpler ecological systems could be prone to fluctuations. The crown-of-thorns starfish appears to belong to a simple sub-system of this type, at least after it has undergone metamorphosis. The adult is an extremely specialized animal, with only one type of food and one definitely known predator.

Crown-of-thorns have been reported from Andaman Sea, in Labyrinth Island near Wandoor. Besides, *Acanthaster planci* (Linn), the following forteen species of Sea Urchin have also bee found in Andaman Sea. (1) *Lagamum depressum* (Lesson), (2) *Prionocidaris verticillata* (Lamarck), (3) *Tripneustes gratialia* (Linn), (4) *Stomopneustes variolaris* (Lamarck), (5) *Diadema setosum*, (6) *Prionocidaris baculosa* (Lamarck), (7) *Echinothris calamaris*, (8) *Echinostrephus molaris* (Blainville), (9) *Temnopleurus toreumaticus* (Leske), (10) *Stylocidaris tiara* (Anderson), (11) *Echinothrix diadema*,

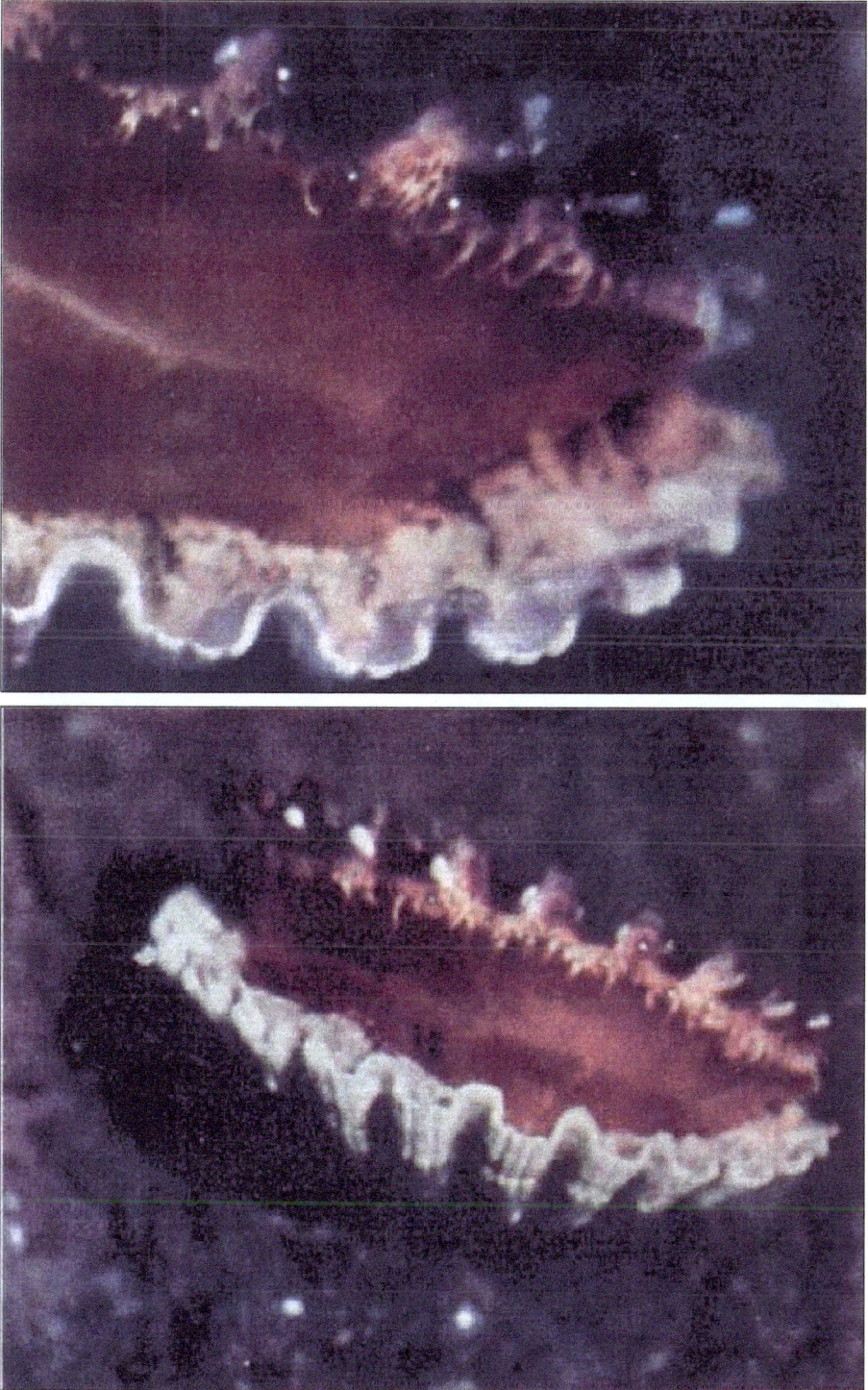

Figure 21.3: Scallop, *Pecten jacobaeus*

(12) *Echinomitra mathaei* (Blainville), (13) *Eucidaris metularia*, (14) *Toxopneustes pileolus* (Lamarck).

Two sea urchins, *Diadema savigni* and *Diadema setosum* are distributed in Andaman Sea. The sharp spines of sea urchins can inject poison into the flesh if caught by bare hands. As the spine has backwardly directed barbs, it is not easy to remove those spines except with the help of needle. The human contact with pedicellariae of the sea urchin *Tripneustes gratilla* was responsible in Japan for swellings of lips or mouth and the ovaries of this sea urchins also produce the same reaction, when they are not sufficiently washed before consuming. The roe or gonad of sea urchins are consumed as delicacy in European and Indo-Pacific regions.

During the reproductive season, generally in spring and summer months, the ovaries of certain sea urchins are reported to develop toxic properties injurious to humans.

Another coral dwelling sea urchin, *Echinothrix calamaris* possess with banded slightly thicker and shorter spines. In between there are slender reddish brown spines ensheathed by a thin membrane with a sac like venom gland at the tip, the pain is immediate, but does not last as long as that caused by *Diadema*.

The list of starfishes found in seas around Andaman and Nicobar Islands are given below.

(1) *Patiniella pseudoexigua* (Darlnall), (2) *Fromia milieporelia* (Lamarck), (3) *Ogmaster capelia* (Muller and Treschel), (4) *Culcita novaequineae* (Muller and Treschel), (5) *Astropecten polycanthus* (Muller and Treschel), (6) *Ophiomastix annulosa* (Lamarck), (7) *Ophiocoma scolopendrina* (Lamarck), (8) *Ophiarthrum pictum* (Muller and Treschel), (9) *Ophioloma pica* (Muller and Treschel), (10) *Ophiocoma erinaceus* (Muller and Treschel), (11) *Ophioplocus imbricatus* (Muller and Treschel), (12) *Craspidaster hasperus* (Muller and Treschel), (13) *Ophidepis cinctata* (Muller and Treschel), (14) *Metrodira subulata* (Gray), (15) *Astropecten monocanthus* (Slanden), (16) *Echinothrix calamaris* (Pallas), (17) *Archaster typicus* (Muller and Treschel), (18) *Echinaster luronicus* (Gray), (19) *Ophiomyxa bengalensis* (Kochler), (20) *Steliaster equestris* (Retzius).

Eleven species of echinoderms for the first time were recorded from the west coast of India (Kerala coast).

The records of *Ophiocoma erenaceus* Muller and Troschel and *O. pica* Muller and Troschel from Vizhinjam are interesting. Other species like, *Ophiocoma brevpes, O. scolopendrine* (Lamarck) are likely to be available along Kerala coast. The species of *Ophiocoma* occur gregariously in the Andamans and also in the islands of Lakshadweep. The occurrence of *Actinopyga mauritiana* is also interesting. *A. eclinites* (Jaeger) and *A. miliaris* are likely to be recorded on the Kerala coast, when intensive collections are made. Vizhinjam being a rocky coast, members of the sub-genus *Semperothuria* found attached to the rocks near the shore are bound to yield more species under this subgenus.

Family – Astropectinidae sp. *Astropectin hemprichi* (Muller and Troschel) collected from Neendakari

Family – Asterinidae – spp. – *Asterina sarasini* (de Loriol), *A. coronata* (V.Martens) from Vizhinjam

Family – Ophiodermatidae – sp- *Ophiaracnella gorgonis* (Muller and Troschel) from Cochin

Family – Ophiocomidae – spp- *Ophiocoma pica* (Muller and Troschel), *O. crinaccus* (Muller and Troschel)

Family – Holothuroidea – spp. – *Holothuria edulis* (Lesson), *H. flavomaculata* (Semper), *H. imitans* (Ludwig), *Actinopyga mauritiana* (Qouy and Goimard) from Vizhinjam

Family – Cucumaridae – sp. – *Actinocucumis typicus* (Ludwig) from Neendakari

Protoreaster lincki, commonly called red general star or red spine star is a beautiful, decorative species, which is relatively large in size upto 30 cm, with five short triangular arms having bright red or orange reticulate pattern on the dorsal side.

Distributed throughout the Indo-Pacific region as a sand or sea grass bed dweller, this species is never encountered in intertidal areas. It is a voracious scavenger, mainly feed on sponges, sea anemones, soft corals etc. It is an enemy of pearl oysters along the Gulf of Mannar area. The species attracted the attention of aquarium keepers for its bold shape and bright colour.

At Tuticorin, the species is being collected by the skin divers engaged in chank fishery during off season. After collection, the specimens are washed thoroughly in sea water to remove adhering sand and other extraneous particles and sun dried. Some traders are giving formic acid treatment before drying to avoid off odor, while storage and also colour and beauty for long time. During peak season, on an average, 200-300 pieces are being collected. Like other echinoderms, star fish constitute a major benthic community structure.

Culcita novaeguinea (pin cushion star fish) reoccurred after five decades in the Palk Bay area of Mandapam, which generally feeds on young polyps of coral reefs.

Chapter 22
Molluscs

A variety of molluscan resources, such as, cephalopods, edible oysters, pearl oysters, mussels and gastropods are distributed along the coastal areas of Indian Ocean, including the Andaman Sea.

In Andaman Sea, mollusks constitutes one of the major fauna. Topography of the sea and tropical climate provide optimum conditions for their rapid growth and multiplication. Submerged coral reefs and rocky beds provide them proper shelter to flourish and reproduce.

Turbo marmoratus, Turbo intercostalis, Turbo petholatus, Trocus niloticus, Trocus radiatus, Xancus pyrum, Lambis lambis, Lambis truncate, Cassis cornuta, Hemifusus cochlidium, Hemifusus pugilinus, Melo spp., Murexramosus, Purpura persica, Bulla ompulla, Tonna dolium, Cypria tigris, Cypria Arabica, Cypria argus, Conus litteratus, Conus miles, Conus textiles, Babylonia canaliculata, Nautilus pompilus, Pinctada fucata, Pinctada maragatifera, Pinna bicolour, Haliotius sp. and some species of Oliva and Cockles are commercially important shells available in Andaman and Nicobar Sea.

Cephalopods

The group known as Cephalopods (Class- Cephalopoda) consists of bilaterally symmetrical mollusca, with a well developed head that contains circumoral (surrounding the mouth) crown of mobile appendages, that bear suckers and/or hooks (except in *Nautilus*). The mouth has chitinous beak like jaws and a chitinous tongue-like radula (band of teeth). The shell is variously modified, reduced or absent and is enveloped by the mantle; an external shell occurs only in the primitive form *Nautilus*. Cephalopods are soft bodied animals with their primary skeletal features, a cranium, and in most forms, a mantle/fin support (cuttle bone or gladius). One pair of ctenidia (gills) is present (two pairs in *Nautilus* only). The central nervous system is highly developed, especially, well-organized eyes. A funnel or siphon (tube) expels

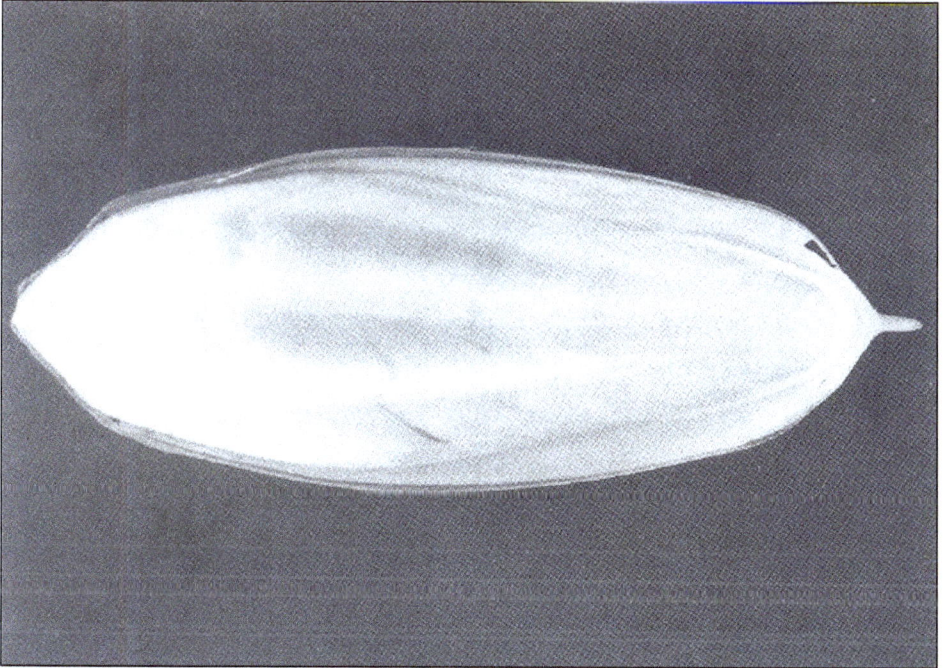

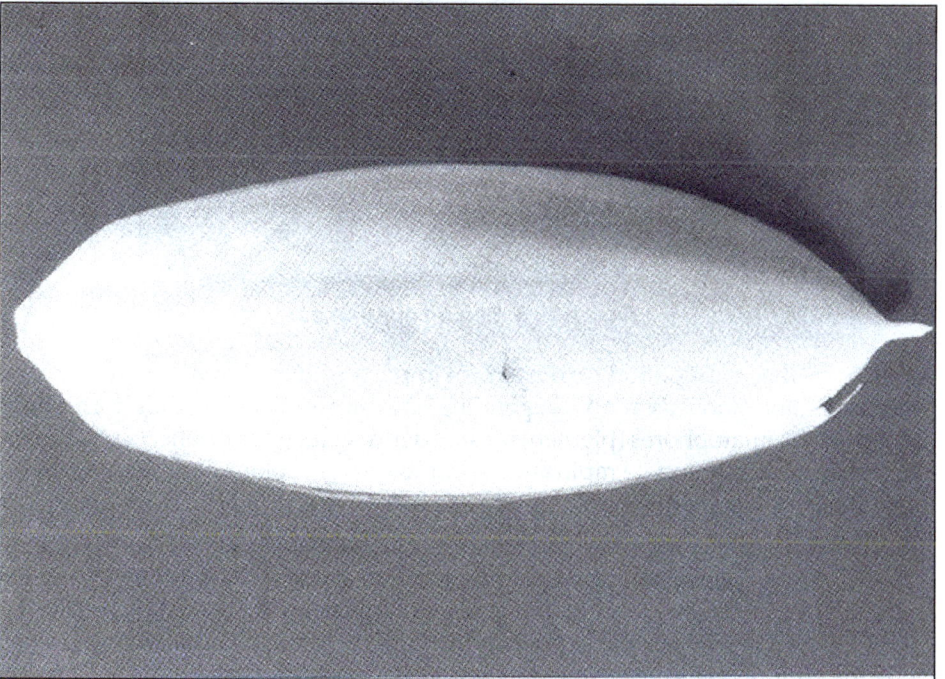

Figure 22.1: Cuttle Bone

Figure 22.2: Egg Cluster of *Sepia pharaonis*

Figure 22.3: Sepia Young Ones

water from the mantle (body cavity) providing propulsion and expelling waste products. Colouration is variable depending on group and habitat; most forms have numerous chromatophores (pigment sacs) and iridocytes (shiny, reflective platelets) in the skin, to accommodate rapid changes in colour and colour patterns, that are an integral part of their behaviour.

The size of adults ranges from 2 cm to 20 m in total length; the largest specimen may weigh over one ton (average size of commercial species is 20 to 30 cm total length and 0.1 to 1 kg in weight). Locomotion is achieved by drawing water into the mantle cavity followed by its jet-like expulsion through the funnel, and also by crawling along the bottom on the arms (mostly Octopods, occasionally sepioids). Fins on the mantle provide stability, steering and secondary locomotion. The sexes are separate, eggs are heavily yolked and development is direct without metamorphic stages.

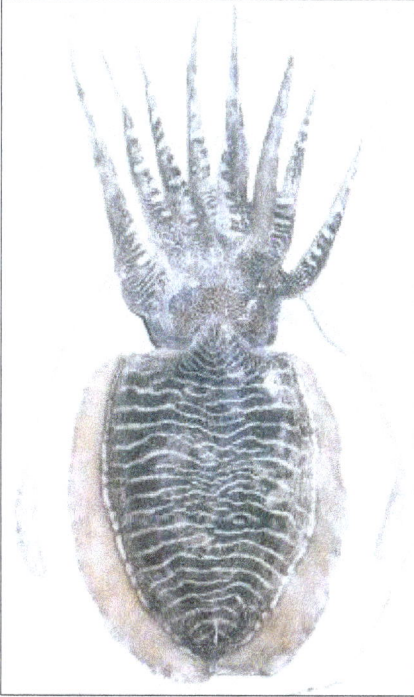

Figure 22.4: Cuttle Fish, *Sepia* spp.

Figure 22.5: Cuttle Fish, *Sepia elliptica*

Figure 22.6: The Copulating Cuttle Fish, *Sepia officinalis*

The total number of living species of cephalopods is fewer than 1000 distributed in 43 families. Cephalopods occur in all marine habitats of the world; benthic, cryptic or burrowing on coral reefs, grass flats, sands, mud and rocks; epibenthic, pelagic and epipelagic in bays, seas and open ocean. The range of depth extends from the surface to over 5000 m. Abundance of cephalopods varies (depending on group, habitat and season) from isolated territorial individuals (primarily benthic octopoda) through small schools with a few dozen individuals to huge schools of neritic and oceanic species with millions of specimens.

The four groups of cephalopods, squids, cuttle fishes, octopuses and chambered nautiluses are easily distinguished by external characteristics. The squids have an elongated torpedo-like body with postero-lateral fins and eight circumoral arms, not connected at bases with a web, with two (occasionally more) rows of stalked suckers bearing chitinous rings (and/or hooks) running the entire length, plus two longer tentacles with an organized cluster (tentacular club) of two or more rows of suckers (and/or hooks) at the distal end. The cuttle fishes have broad sac-like bodies with lateral fins that either are narrow and extend the length of the mantle (Sepiidae) or are short, round and flap like (Sepiolidae). In either case the posterior lobes of the fins are free (subterminal) and separated by the posterior end of the mantle. Ten circumoral appendages, the longest (4th) pair (tentacles) retractile into pockets at the ventro-lateral sides of head. The remaining arms frequently with 4 rows of stalked suckers with chitinous rings, never hooks (otherwise 2 rows). Eyes covered with a transparent membrane, and eye lids present. Shell is thick, chalky, calcareous (cuttle bone of *Sepia*) or thin chitinous (Sepiolidae). The octopuses have a short sac-like body, either with no lateral fins, or with separate paddle-like fins in some deep sea forms and 8 circumoral arms only (no tentacles) with bases connected by membraneous web and unstalked suckers, without chitinous rings along the length of the arms. The chambered nautiluses are characterized by an external shell, coiled, simple suture lines, smooth external surface. They have two pairs of gills, funnel bilobed, two circumoral appendages. Nautilus ows its survival largely to its camouflaged, pressure-resistant shell, tentacles and air-tight chambers that act as ballast tanks. As the nautilus grows, it moves forward in the expanding shell, and natural secretions form a partition or septum behind its fleshy body. Thus the nautilus creates a series of ever-larger chambers, at an estimate rate of one every few weeks. With a master builder's skill, the mollusk fashions as many as 38 chamber, most of them increasing in size with a mathematical consistency. The final chamber, inexplicably is smaller.

A thin tube, the siphuncle, links all the chambers to the body like a life line. It is thought to control buoyancy by regulating the ratio of gas to liquid in the chambers. The shell displays a spiral symmetry found often in nature. But even the most perfect spider web can't match the flawless nautilus.

Squids, cuttle fishes and octopuses, an important marine resource has been neglected until the stagnation in marine fish production despite various development measures. Beyond the continental shelf in tropical seas, major conventional commercial resources, at present, boil down to two important groups, namely tunas, squids and cuttle fishes. Squids are also present in the continental shelf waters, where cuttle fishes occur as demersal resource.

The cephalopods are prey of a great variety of fishes, cetaceans and cephalopods themselves. Octopods form a major part of the food of yellowfin tuna. Cephalopods, especially squids are a very favourite food of sperm whales. Thus it is seen that cephalopods form an important forage for several species and a very significant constituent of the food of tunas and billfishes.

Cephalopods have been classified into the following sub-classes.

1. Ammonoidea – It had external shell, coiled, complex suture lines, sculptured external surface. All of them extinct now. Have been found Upper Silurian to Upper Cretaceus age.

2. Nautiloidea – The animals have external shell, coiled, simple suture lines, chambered, smooth external surface (Some extinct forms with straight shells). They have two pairs of gills, funnel bilobed, two circumoral appendages. Extinct except Nautilus.

3. Coleoidea – They have internal shell, enveloped in tissue, calcareous, chitinous, or cartilaginous. One pair of gills and funnel is tube like. Eight to ten circumoral appendages.

The sub-class Coleoidea is divided into following orders.

(a) Belemnoidea – Shell is located internally, unchambered, conical, strongly developed. Found in Upper Mississipian to Cretaceous, now all extinct.

(b) Sepioidea – Shell are calcareous (cuttle bone) or reduced, chitinous. Ten circumoral appendages. Fourth pair (tentacles) retractile into pockets, posterior fin lobes are free.

(c) Teuthoidea – Shell (gladius) chitinous, rod or feather shaped. Ten circumoral appendages. Fourth pair (tentacles), contractile, not retractile, no pockets, suckers (hooks) stalked (tentacles secondarily lost in few groups), posterior fin lobes fused.

(d) Vampyromorpha – Shell chitinous, thin broad, ten circumoral appendages. Second pair sensory filaments only. Arms with papilla-like cirri; tentacles absent. Two pairs separate, paddle-like fins; all black pigmentation.

(e) Octopoda – Shell reduced, cartilaginous. Eight circumoral appendages, arms only, suckers unstalked, fins absent or single pair, widely separated, paddle-shaped.

The order Teuthoidea is divided into two sub-orders;

(i) Myopsida – Eye covered with corneal membrane; gonoducts single, suckers on buccal lappets and hooks absent.

(ii) Oegopsida – Eyes open, no membrane, gonoducts paired; No suckers on buccal lappets; hook present in many groups.

The order Octopoda is divided into two sub-orders;

(i) Cirrata – Arms with papilla-like cirri; one pair widely separated paddle-like fins; shell a single saddle-shaped structure (or absent).

(ii) Incirrata – Arms without cirri, fins absent, shell vestage, a pair of curved rods (or absent).

Cephalopods in the Andaman Sea consists of;

Sepia aculeate

The mantle is broadly ovate, width three-fifth in length; fins moderately broad and extend along the entire length of the mantle. The head is large,about as long as wide and narrower than mantle opening.

The oral arms are short, sub-equal in length in the order of 3,4,2,1. The dorsal arems are rounded on the outer side, the lateral arms are keeled and the ventral ones broad at the base with strong swimming membrane. The buccal lappets bear a few minute suckers with smooth, chitinous rings. The arm suckers are uniformly arranged in four rows and bordered by protective membrane on either side. The left ventral arm in male is hectocotylized with modification at the proximal half of the arm. At the base of the arm, there are about 12 rows of normal suckers followed by about 6 transverse rows of very minute suckers, with a pit like excavation in the middle of the modified portion and this is distinct in adult males.

The tentacles are comparatively very long, slender and keeled on the outer side; the stem are triangular in cross section. The tentacular clubs are long, about one third mantle length and slender but not much expanded; beyond the base of the club the protective membrane run as two ridges on the oral side of the tentacular stem. The club suckers are minute, sub-equal in size and arranged in about 10-12 longitudinal series in males, whreas females possess 13 or 14 series; the numerous minute suckers on the tentacular clubs give them a spongy appearance.

The cuttle bone is elongate oval in shape with granular rough dorsal surface and has three low longitudinal ribs; the chitinous margin is very narrow. The ventral surface has a slightly convex striated zone anteriorly and is concave at the posterior end, where the posterior innercone has a thick rounded ridge in contrast to the distinct plate-like form in *S. pharaonis*. On the ventral surface, in the striated zone, there is a longitudinal medial ridge with a faint groove running medially; the striae are notched in the middle. The last loculus is short and its medial portion is slightly concave. Spine is small, strong and not keeled.

The maximum size recorded is 230 mm. The colour of the dorsal mantle is very variable and consists of different shades of grey or brown with white spots, streaks or patches in varying patterns with numerous chromatophores.

Sepiella inermis

The species is readily recognized by the smaller size, absence of spine in the cuttlebone, the oval outline of the shell and the presence of a distinct glandular pore at the extreme posterior end of the mantle.

The mantle is broadly oval, the dorsal margin is angularly rounded mid-dorsal projection is not much pronounced; the ventral margin of the mantle is emarginated; the mantle has a pigmented gland and an orifice at its posterior end ventrally. The

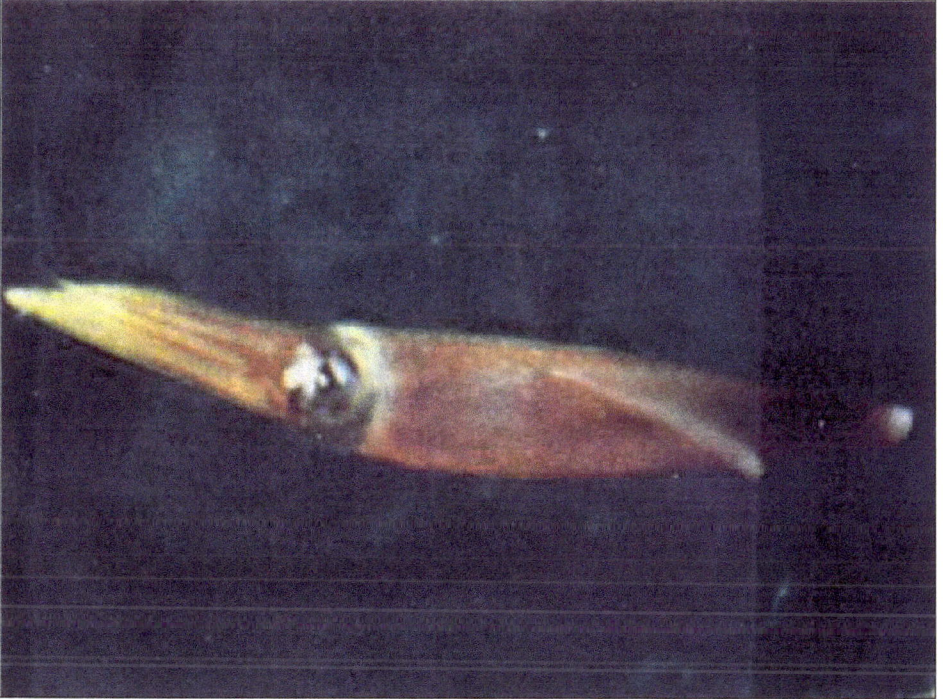

Figure 22.7: The Squid, *Loligo vulgaris*

Figure 22.8: Cranchid Squid

Figure 22.9: Squid, *Loligo* spp.

funnel is short and thick. The fins begin slightly behind the anterior mantle margin, are narrow anteriorly and are broader posteriorly. The head is rather short and broad. The arms are stout, sub-equal in length, laterally compressed and tapering to slender tips. The arm suckers are uniformly minute and arranged quadri-serially. The basal suckers are larger and are progressively reduced in size towards the distal end. The protective membranes of all arms are well developed. In females the arm suckers are provided with smooth rings, whereas those of males have strong dentate ring.

The tentacles are very long and slender. The swimming membrane of the tentacular clubs is slightly shorter than the club. The clubs are provided with minute, sub-equal, numerous suckers arranged in 16 to 24 transverse rows. The clubs are long, but not very much expanded.

In males, the left ventral arm is hectocotylized and in its proximal half, there are ten rows of minute suckers set widely apart in 4 rows and transverse ridges are also

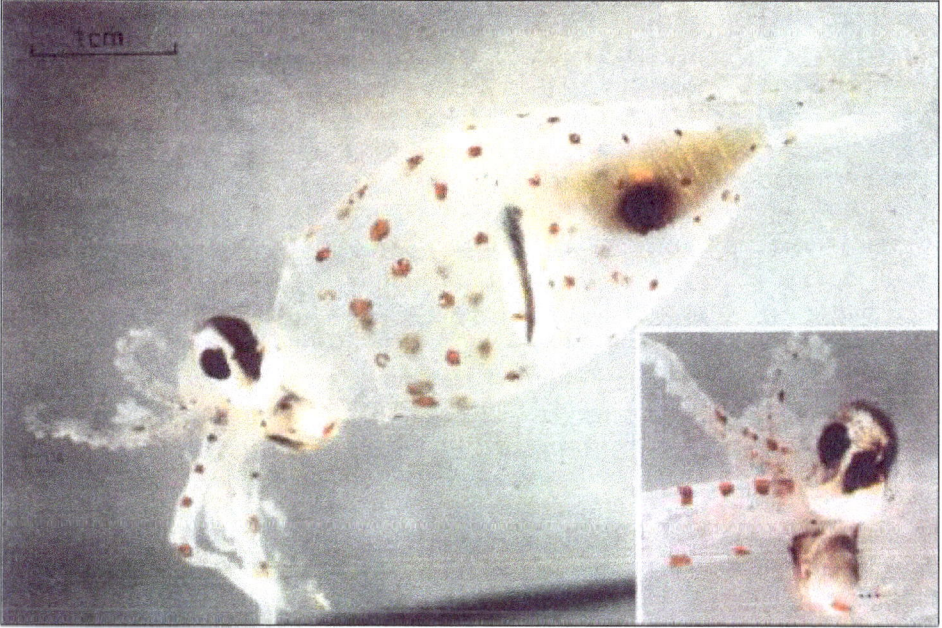

Figure 22.10: A Cranchid Squid, *Cranchia scabra*

present. The distal half of the arm is narrow and quadriserially arranged normal suckers.

The cuttle bone is oval in shape and without spine. The dorsal surface is granulose and has a low mid-rib and the ventral surface has wavy striae with a distinct median narrow groove and many jointed radiating furrows on the striated area. The last loculus is short and concave. The inner cone has "V" shaped limbs and a small thick rounded knob at the end. The outer cone is broad and extends beyond the inner cone

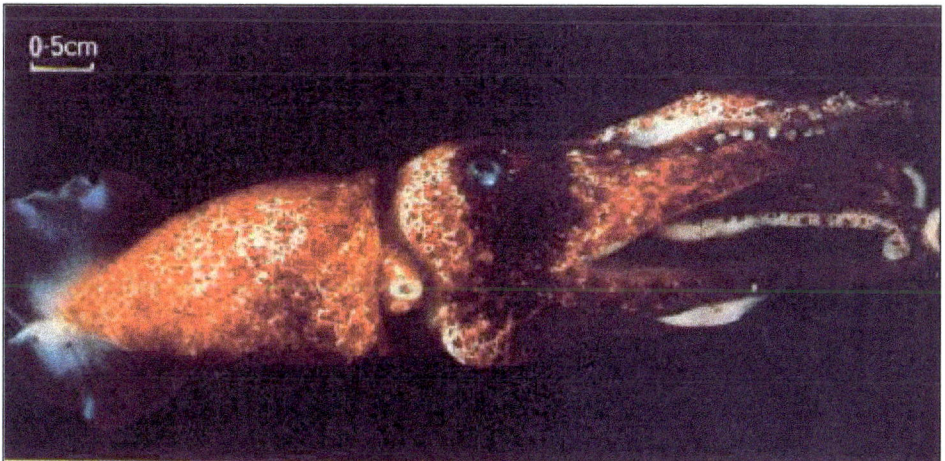

Figure 22.11: Benthic Crinoid, *Pentacrinus*

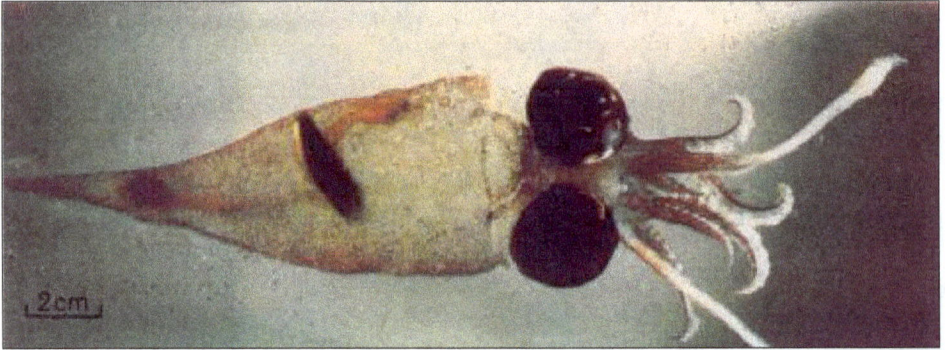

Figure 22.12: Crinoid

and is round at the posterior end with a slight marginal notch on either side. The outer cone is brittle and thin.

It is a moderate size cuttlefish with a maximum mantle length of 124 mm. The dorsal surface of head, mantle and arms is grayish with numerous melanophores, in the fresh condition. Faint longitudinal stripes extending from the base of the arms to their tips on the aboral side are seen. There is a row of dark ornamental ocelli on either side of the fins on the margins on the dorsal side in males.

Sepia elliptica

It is a tropical Indo-Pacific species, extending from northern to western Australia, Exmouth Gulf, Queensland, Capricorn Island group, Gulf of Carpentaria and Vietnam and occur mainly in coastal waters at a depth range of 16 to 142 m (Jereb *et.al.*2005). the occurrence of this species in Indian waters were reported by Silas *et.al.* (1985). Sivasubramanian (1991) has reported this species from Bay of Bengal to a depth of 100 m. *S. elliptica*, the neritic species exhibits bathymetric distribution and are caught along Cochin and Verabal coasts by trawling beyond 40 m depth.

The mantle of *S. elliptica* is oval with the triangular dorsal anterior margin. The arm length is sub-equal and the arm suckers are tetra serial. Club sucker-bearing surface flattened with 10 to 12 minute suckers in transverse rows. Swimming keel of the club extends well proximal to carpis. The cuttle bone is oval and angular V-shaped anteriorly, bluntly rounded posteriorly and the dorsal surface is greyish white. The maximum dorsal mantle, recorded for this species is 175 mm. The largest sizes recorded for males and females in Cochin area are 129 mm and 119 mm respectively. The species seems to be extensively feeding on prawns. Males and females in Cochin area attain sexual maturity at a minimum size of 75 mm and all individuals of both sexes mature when they reach 115 mm.

Doryteuthis singhalensis

Mantle long, thick, muscular and narrow tapering posteriorly to a sharp end. The anterior mid-dorsal projection is rounded at the tip. The ventral margin is emarginated at the middle. Fins large and rhombic in shape extending along more than 60 per cent of the length of the mantle. They are widest at about the middle; the anterior margin of fin lobes is slightly convex and the posterior margins are concave. Head relatively small, eyes large. The buccal membrane has given projections and each possess about eight minute suckers. The funnel is small, stout and placed in a deep furrow beneath the head.

Arms are rather short, 20 to 30 per cent of the mantle and in the order of 3,4,2,1 or 3,2,4,1. The third pair of arms are longer and stout, well compressed and have poorly developed swimming membrane, the protective membrane on either side is well developed. The horny rings of the larger arm suckers bear on the distal side 6 to 11 slender teeth; the proximal side is smooth.

The distal half of the left ventral arm of males is hectocotylized and the proximal portion of the arm bears about 15 pairs of normal suckers, which are more or less equal in size, but smaller than those on other arms. The pedicels of the suckers in the modified distal portion are fleshy and enlarged and those on the ventral row are greatly enlarged. The pedicels in the modified portion bear minute suckers without horny rings.

The tentacles are moderately long, slender and compressed. The clubs are short and slightly expanded with a trabeculate protective membrane on either side. The club suckers are arranged in four rows; those on the two median rows are slightly larger than the lateral suckers. The rings of the suckers bear about 20-22 sharp curved teeth. The teeth are not of uniform size; a few of them are smaller and arranged alternately with long curved teeth. An oval shaped light organ is present on either side of the rectum. The gladius is narrow with nearly straight margins gradually tapering to a point. The maximum mantle length recorded for this species is 50 cm in males and 31 cm in females.

Symplectoteuthis oualaniensis

The mantle is long, muscular and cylindrical up to the point of origin of fins and tapers abruptly to a narrow point at the posterior end. The dorsal margin is slightly produced in the middle. The fins are short, muscular and broad with convex anterior margin and form an angle of about 65°laterally. The head is large and as wide as mantle and bears comparatively short arms. The funnel is short, compact and set in a deep pit present on the ventral side of the head; faveola with 7 to 9 longitudinal folds in the central pocket and 3 to 5 lateral pockets on either side. Funnel locking apparatus is shaped and fused in its middle portion with the mantle groove, so that separating the funnel free from mantle-lock is difficult.

The arms are large, strong in the order 3,2,4,1 and compressed with the third pair, strongly keeled. Arm suckers are biserial; the protecting membrane have prominent trabeculae; the larger arm suckers are provided with about 12 sharp teeth around the entire rim of the horny rings.

The left ventral arm in males is thicker and longer than the right ventral arm and hectocotylized. About one half of its distal part is devoid of suckers and papillae. In the basal portion of the hectocotylized arm 14 or 15 suckers are present in two rows protected by flap-like membranes. There is a series of pits in a single row along the base of the protective membranes.

The tentacles are short and muscular and laterally compressed. The clubs are small and slightly expanded; the suckers are quadriserial with the inner rows on the manus layer. The larger suckers of the club bear about 20 sharp teeth on the rims of which four are larger and located one in each quadrant. An oval photophoric patch is present on the antero-dorsal surface of mantle; there are also two photophores on the intestine.

This is a large species of oceanic squid, attaining a size of 35 cm. The head, dorsal mantle, fins and arms are uniformly chestnut brown colour.

Diamond back squid, *Thysanoteuthis rhombus*, is an epipelagic oceanic rhomboid squid of bright red colour having a thick cylindrical muscular mantle, wide anteriorly and tapering gradually posteriorly to a blunt end. The fins are long occupying the entire length of the mantle on lateral side. The fin is diamond shaped, being broader in the middle and tapering at both anterior and posterior ends. The head is shorter and eyes are prominent. The outer lateral arms are the longest and the inner arms are the shortest, provided with a crest like muscular projection at the base of each arm. The other arms are intermediary in length and devoid of any structures. All the arms possess two rows of suckers and sucker rings with sharp teath. The morphometric measurements are; mantle length-630 mm (dorsal) and 610 mm (ventral), mantle width- 550 mm, fin width – 200 mm, head length – 110 mm, width of head – 150 mm, funnel length – 75 mm, eye diameter – 40 mm and tentacle length – 600 mm.

The squid inhabit warm tropical and partially sub-tropical waters. It occupies near surface waters during night and migrates to mid-waters during day time and often occurs alone or in pairs. The species is distributed in Atlantic, Pacific and Japanese waters.

Heavy landings of *Loligo singhalensis*, Ortmann, 1891 was observed at Pamban along the Gulf of Mannar coast, Tamil Nadu during the first fortnight of November, 2010, as by catch from trawl net fisheries, mainly targeting medium to large pelagics off Pamban. The landing of this squid was ranging from half to one tonne per boat. Their length and weight range were 102 to 216 mm and 27 to 94.4 gram respectively. Their sex ratio (Male: Female) was found to be 1:0.66. The gonads of both sexes was observed to be fully ripe conditions. The size of male was relatively larger than female.

Gonatus anyx, one of the most abundant Cephalopods in Pacific and Atlantic Oceans and is an important prey for a variety of vertebrate species.

A pelagic egg brooding habit have had been questioned because some aspects of the squid's biology seem to preclude brooding, principally, the degeneration of musculature following sexual maturation was presumed to limit locomotion and render squids unfit for egg protection.

It has been observed five squids, each holding an egg mass in its arms at depths between 1539 and 2522 m in Monterey Canyon, off California. The squids used hooks on their arms to hold the egg mass, which consists of two thin membranes in a continuous flat sheet, open at both the distal and proximal ends. The egg mass forms a hollow tube, that extends from the mouth to well beyond the end of the arms and contain about 2000 to 3000 eggs.

Repeated extension of arms at about 30 to 40 seconds interval flushed water through egg mass. This behaviour served to aerate the eggs in the hypoxic mid-waters found off California. Aggressive arm movements and escape swimming caused partial disintegration of the more mature egg masses and hatching of the released eggs.

A squid bearing undeveloped eggs made a vigorous escape by using fin and mantle contractions, whereas those with advanced embryos showed only respiratory mantle contractions and did not move away, which suggest that with the apparent stage of egg development, the squid may have undergone a gradual degeneration of locomotory capacity.

Low temperature and large eggs will prolong development in *G. anyx*. The abundance of juveniles in near surface waters peaks seasonally from April through July, which may indicate a yearly cycle with an egg development period of 6 to 9 months. Gonatid squids have sufficient lipid stores to fuel metabolism for a long brooding period and the gradual decline in mantle muscle and digestive gland conditions indicates a progressive use of energy stores of long egg brooding period.

Despite retaining some capacity for swimming, the relatively immobile brooding squids are found within the usual diving range of whales and elephant seals and so may provide as easy target for such mesopelagic mammals

Gonatid squids and other ontogenetic migrators, therefore, represent a direct energetic, as well as trophic link between deep and shallow biomass.

Mantle length of an adult *Gonotus anyx* (Cephalopoda) is about 145 mm. The squid was found holding a tubular egg mass in a horizontal resting position at 2522 m depth. Hatched embryos were about 3 mm in length. The *in situ* temperature was 1.7 to 3.0°Celsius and oxygen concentration 45-90 micro mol/l.

Octopods

Octopuses, popularly known as devil fishes, are marine benthic animals found to live from the sea coastal water down to 1000 m of its depth. The major species of octopuses, which contribute to the global fishery come under the genera *Octopus, Cristopus* and *Eledone*. As many as 200 species of Octopodidae are known to occur in the world oceans (Worms, 1983) and of these, about 60 species are known to occur in the Indian Ocean.

The octopuses, considered as delicacy, are also caught during low tide from oceanic regions by adopting simple fishing methods, like trap setting, harpooning or poisoning the coral rock pools, which they inhabit. In shallow areas they are caught by setting traps and also by using long lines, hand lines and spears. Once caught, octopuses are killed by fishermen and their bodies are turned inside out, thereby forcing gills, heart, viscera out through the wide opening of the branchial chamber. This is known as turning its cap.

Thirty eight commercial species of octopus have been reported from the Indian seas. *Octopus membranaceous, Octopus dollfusi, Octopus lobensis, Cristopus indicus, Octopus globosus, Octopus cyaneus, Octopus aegina* are the main species contributing to the fishery along Indian coasts.

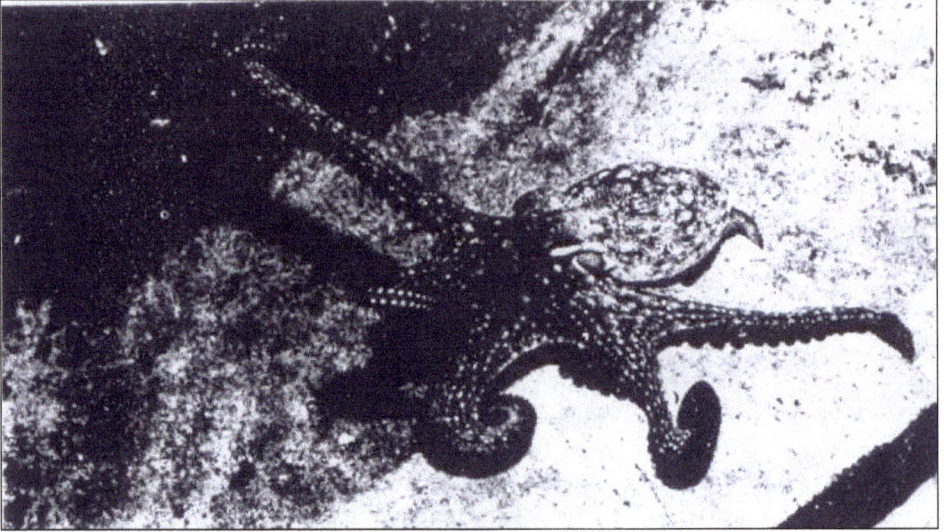

Figure 22.13: Cephalopods Regular Components of Patterning

Figure 22.14: Octopus, *Octopus* spp.

The occurrence of *Octopus vulgaris* and *Cristopus indicus* have been reported from Andaman Sea.

Unlike squids and cuttle fishes, octopuses lead a solitary life and do not form schools. Some octopuses are known to make seasonal migration, which are influenced by breeding activity. Octopuses are exclusively carnivorous and they feed on crustaceans, fishes and mollusks.

Octopus has two types of chromatophore, that produce different colours depending on their degree of expansion; black to red-brown from the dark ones, and red to pale orange-yellow from the ligher ones. Below them is a layer of tiny reflecting bodies, iridocytes, that give green and blue by refraction and white by reflection.

At hatching there is a larval set of about seventy red-brown colour cells, which is replaced after the octopus settle out of the plankton at 4 to 6 weeks of age by an adult set of smaller brown-black and orange-red chromatophores together with the iridocyte layer. By the end of a year, the adult possesses one or two million of the colour cells, most of them on the upper surface of the body, there being 100 to 200 of them per sq mm. Despite enormous growth of the animal during the life span, the size of each of these organs remains the same.

The skin surface carrying the colour elements is broken up into patches joined by infoldings that give a matt texture. The roughness is greatly increased by papillae, particularly large above the eyes and over the mantle. The papillae often appear

Figure 22.15: *Octopus macropus*

white or green, because the iridocytes are concentrated within them. Other concentrations of iridocytes, on the front and back of the head, on the mantle and along the tops of the arms, produce the white patches and spots. They remain even after death. Black or brown bars across the front of the first, second and third pairs of arms and across the eye, extending the line of pupil are produced by preferential expansion of the dark chromatophores.

The possession of a body musculature with fibres running in all three main directions in space allows great postural flexibility and much of the behaviour repertoire is expressed through this means. The arms may be extended as loose as tentacles, on the bottom, their tips gently explore the surrounding surfaces, or they may be curved under the body or curled back stiffly over the head as armour. A set of subcutaneous muscles at times raises folds of loose skin that run together to form an interbranchial web, when the arms are spread out. Each sucker is able to move independently. They may be tucked below the arms and invisible or extended, supplying the arm with a wavy outline, or else raised so that their conspicuous white under surface, lacking chromatophores, can be seen. The mantle after being rounded and flattened like a partly distended ballon, may be recast into an ornamented to give. At rest on a flat surface, the octopus head is raised a little above the level of the mantle. It may, however, be flattened into the body, or raised so high that the animal can best be described as standing.

Octopus dollfusi

The mantle is elongately oval in shape, head small, eyes well developd but not conspicuous, there are no cirri over the eyes.

The arms are moderately long, sub-equal and stout at their base; the first pair of arms are shortest. The intrabranchial membrane is well developed between all the arms except in between the first pair, where it is narrow. The arm suckers are cup-like and fleshy without chitinous rings and are directly set on the arms uniformly, in two alternating rows. The normal suckers are large at the base, become smaller progressively along the length of the arms and minute at the tips. There are few slightly enlarged suckers at the base of the second and third arms in males.

In males, the third right arm is hectocotylized. This arm is slightly shorter in length than the left third arm, the ventral margin of the web is rolled into the form of a deep groove, which serves as the spermatophore groove and extends up to lingual. The lingual and calamus are well developed and the former is about 8 to 10 per cent of the arm. There are eight or nine lamellae in each demibranch of the gills. One of the characteristic anatomical features of this species is that the male reproductive system consists of an elongate and slender penis and a large coiled diverticulum together forming a reverse 6-shaped structure. The spermatophores are moderately long and distinctly arms with a series of teeth-like cirri.

This is a moderate size octopod with a maximum mantle length of 9 cm. The body and arms are light greyish brown colour with a reticulate pattern all over the body.

Octopus aegina

Mantle rounded to elongately oval in shape and closely covered with small tubercles or fine papillae and fine reticulate pattern on the dorsal surface. Head is small with a narrow neck region. Eyes are prominent and have a single large cirrus near the base of each eye; fresh specimens with a narrow distinct white band across the base of head between eyes. Arms are long and stout, the dorsal pair distinctly short than others. The web between dorsal arms is very shallow, depth between other arms moderate; on the dorsal side of each arm a cluster of dark chromatophores present at the base of each sucker. The right third arm in males is long, slender and hectocotylized; lingula is short and about 5 to 8 per cent of the hectocotylized arm; groove very shallow and without ridges, calamus very small and distinct. In males, abruptly enlarged suckers may or may not be present. Male genital organ consist of a characteristic long, slender penis and a long diverticulum together forming an "U" shaped structure terminating in an additional secondary loop. Spermatophores are long and unarmed.

The maximum size recorded is about 10 cm. The species is odd dull brown colour with a fine reticulate pattern on the dorsal surface of mantle, head, arms and web. The oral surface of arms is whitish. At the base of each sucker on all arms there is dark pigmentation.

Octopus, which were earlier discarded as "devil fish" are now being exploited in commercial quantities. The major species of octopus, which contribute to the world fishery come under the genera Octopus, Cistopus and Eledone. Octopus in continental shelf and oceanic region are caught mainly as by-catch in the bottom trawl.

Cephalopods form about 10.5 per cent in trawl landings of Maharashtra (north west region) with Octopus dominated by the species, *Cistopus indicus* contributing 7.1 per cent toward the cephalopod catch.

Thirty eight species of octopods belonging to the family Octopodidae, Thermooctopodidae and Agronautidae exists around the Indian Seas including Andamans and Lakshadweep.

A new entrant to the octopus fishery is *Octopus membranaceus*, commonly known as "web foot octopus" in Maharashtra coast during December to February. The mantle length of the species, ranged from 50 to 80 mm. The depth of operation was 30-40 m at 70-80 km off north west coast.

Some of the important distinguishing characters of *O.membranaceus* a saccular to elongate mantle with small close-set tubercles over head, mantle and arms. Two cirri or warts observed over each eye, arms are moderately long, robust and the web low. The right arm III is hectacotylised in males. Lingul is slender and long with 4 to 6 per cent of arm length. The presence of a conspicuous dark ringed ocellus on the web base of arm II, antero-ventral to the eyes.

It is a benthic shallow water species occurring down to about 60 m depth. It shows strong cryptic behaviour and usually hides in holes on flat bottom. The spawning season extends from December to February. It is an Indo-Pacific species extending from Indian Ocean to Japan, China, Philippines and southward to

Australia.In Kerala, the species contribute 82 per cent towards the octopus catch. The length range of species landed at Cochin is 20-90 mm in mantle length. The maximum mantle length of the species is 80 mm and maximum total length 300 mm in Mumbai waters. The size range of the species is smaller ranging between 30 and 80 mm and the landings are comparatively of lesser magnitude.

Two octopods, namely, *Argonauta argo*, Linnaeus and *Berrya annde*, new to Arabian Sea were obtained from the collections of vessels of Integrated Fisheries Project.

Mantle of *A. argo* is ovoid with a broad blunt apical point. Head small, almost merged into the mantle. Eyes prominent. Mantle opening very wide with thick ventral edge. Arm formula 1234. First pair of arms thicker at their bases and bear the characteristic membrane, which extends nearly their whole length. Suckers closely set and alternately arranged. Distal suckers very minute. Remaining three pairs of arms show no specialization. They taper gradually to slender extremities. Arm bases joined by a very low web with a formula of ABDEC.

Funnel is long with a broad base and projects beyond the eyes. It opens in advance to the edge of the web sector E. Funnel organ shaped with slender limbs.

Gills with 10 to 12 filaments in each demibranch. Adhesive apparatus present. Pallial organ with a round knob and a narrow transverse depression at the base. Cephalic element with an appropriate shaped organ, which can be locked with the pallial element to secure tight closing.

Colour of the preserved specimen brownish dorsally and paler ventrally. Chromatophores found densely distributed on the dorsal surface of the mantle, head and first two pairs of arms. Ventral surface more whitish with widely scattered chromatophores.

Shell compressed. A narrow keel with double row of tubercles present. Shell surface smooth, but traversed by a number of ribs like projections. They form secondary ribs. Shell opening rather narrow with convex rims. Colour whitish with blackish keel and tubercles. Distributed in warm and temperate seas.

Berrya annae has rounded body, pouch like covered with soft gelatinous tissue. External surface of mantle very smooth except for some small soft scattered warts on the dorsal side of the eye openings and body surface

Head rather broad, a little narrower than mantle, weakly constricted behind forming a moderate neck. Eyes somewhat prominent with small eye openings.

Mantle aperture wide. Funnel organ consists of two V-shaped pads, entirely similar to one another. Limps of funnel organs narrow with curved base and acute extremities.

Arms sub-equal in length, slender, short and tapering to the points. Base of arms connected with a soft semi-transparent muscular web and extended between arms to nearly 36 to 48 per cent of their length. Suckers relatively small, little raised, widely set, regularly alternating in two rows. No enlargement of suckers in male. Third right arm hectocotylized, shorter and stouter than its mate of left side. The species is distributed in Arabian Sea and Indian Ocean.

Cephalopod landings by trawlers in Mumbai consists of *Loligo duvauceli, Sepia aculeate, Sepia inermis, Sepia pharaonis* and Octopus species. They are landed almost throughout the year. The fishing season is from September to May with peak during September to January. Among the landings of cuttle fish at Mumbai, *Sepia aculeate* formed the dominant species, but from 2000 onwards, a new entrant *Sepia prashadi* has shown its occurrence in cephalopod fishery. *S. prashadi* occurs from September and landing lasts upto December.

Though there was no change in fishing grounds, gear or effort, a definite pattern in the landings was evident. The landings of *S. aculeate* was unusually low in September, which gradually picked up by December and then again it became dominant species from January onwards.

The estimated average catch of *S. aculeate* for the years 2001-2004 ranged from 2.95 tonnes in September to 91.82 tonnes in December and the average catch of *S. prashadi* decreased from 54.40 tonnes in September to 10.20 tonnes in December.

The mantle length of *S. aculeate* landed in September and October ranged between 60 to 130 mm, while *S. prashadi* ranged between 55 to 110 mm. From November, the size of *S. aculeate* ranged between 70 to 180 mm and *S. prashadi* ranged between 70 to 170 mm.

The two species could easily be identified, since the cuttle bone of *S. prashadi* is very distinct with its dorsal surface being pink in colour. Adult *S. prashadi* develop prominent lateral stripes on its dorsal surface and the size ranges from 130 to170 mm.

Cephalopods, comprising of mainly by *Loligo durauceli, Sepia pharaonis* and *S. aculeate* represent one of the most important marine fishery resources off Verabal between Lat. 21°59 minute to 21°30 minute N and Long.57°49 minute to 69°03 minute E. The high incidence of berried cephalopod females in the catch give an indication that the fishing ground would be a probable breeding ground for cephalopods. The operation of multi-day trawlers (5 to 8 days) in the deeper unexploited waters (80-100 m) for cephalopods resulted in this huge increase in catch (5692 tonnes) for the first quarter of 2009. The species dominating were *Octopus* sp., *Sepia pharaonis, Sepiella inermis* and *Loligo* sp.

Cephalopods comprise about 10.5 per cent of trawl landings of Maharashtra with octopus contributing 7.1 per cent. *Cistopus indicus* dominates the octopus fishery in Mumbai waters. Apart from *C. indicus* and *Octopus membranaceus, Octopus dollfusi,* commonly known as "marbled octopus" is observed regularly.

O. dollfusi have elongated oval mantle and inconspicuous eyes. The arms are moderately long and stout with dorsal arms being the shortest. The mantle, head and arms are covered dorsally with numerous large reticulate warts, with each unit bordered by a darkly pigmented line. Some of the larger suckers at the base of the arms are bluish in colour.

The species occurred among other species of octopus during December-May. The depth of fishing operation was about 30-40 m, 70-80 m off the north-west coast. The maximum mantle length of the species is 90 mm. Among 18 specimens, 2 were females.

Cephalopods, which once treated as discardable by-catch, have now been upgraded for their export potential. In India, 80 per cent of the landings of cephalopods takes place along its west coast, and the latest estimate of cephalopods landing in India for the year 2007 were 94804 tonnes against the global catch of 3.512 million tones in 2003.

The flesh of cephalopods are firm and it blends to provide, a variety of processed and preserved products, owing to absence of bones, ease of cleaning and leaving behind a marginal quantity of wastage. They have also high food value. The meat contents of cephalopods is much higher than fish, containing 40-70 per cent consumable fraction. Due to high protein content (16-21 per cent), low carbohydrate and lipid content (1.0-1.5 per cent), cephalopods are widely accepted as a choice of food in various parts of the world. The cephalopod meat is prepared for food by cutting them into slices and treated them with spices and frying or cooking into curries, cutlets and soups.

There are about 80 cephalopod products exported from India and the quantity exported from India during 2006-2007 was 102953 tonnes, contributing about 17 per cent to the total marine products export from India. Some of the exported products are whole squids, stuffed tubes, squid rings, frozen tentacles, fillets, roe, squid rings battered and frozen.

Non-edible parts of the cephalopods are converted to meal, similar to fish and prawn meal.

Cuttle bones are commercially used in preparing fine abrasives and dentitrices.

Artists have used the ink of cuttle fish as a natural "sepia" pigment. Medicinal value of cuttle fish is attributed to the ink. Imitation caviar, named "Cavianne" is made up of a mix of squid ink, pectin from apples, extract of sea urchin, oyster and scallop, as well as a type of gum derived from kelp.

Ambergris (obtained from sperm whales, which is used as a fixative in perfumery) is formed around beaks of squid consumed as food. Squid liver extract are used for human consumption and in dehydrated form they are used as food for livestock.

Nautilus pompilius

Nautilus, popularly known as pearly nautilus or the chambered nautilus in contrast to the other interesting remarkable cephalopod, the paper nautilus, *Argonauta* is the only living representative of the sub-class Nautiloidea. The remarkable feature of the pearly nautilus is its beautiful coloured external shell which distinguish this from other cephalopods *Nautilus pompilius* is extensively distributed in Indian Ocean, including Andaman Sea.

The body is enclosed in an external calcareous, spirally coiled and multichambered shell. The animal occupies the last chamber, which is the largest. The inner surface of the chamber is pearly and the outer surface orocellanous and pigmented with wavy reddish brown bands on whitish background. The colour markings on the shell are very conspicuous.

The successive chambers of the shell are separated by a system of concave septa, which are perforated in the middle and a shelly tube is formed running through the chambers. Through this the vascular prolongations of the tip of the mantle, the siphuncle runs. Although the tubular siphuncle runs through the chambers to the apex of the shell, it does not open into the chambers. Except the outermost recently formed chamber in which the body is present, all the other chambers are filled with gas, which gives buoyancy to the animal.

The body of the animal consists of head, which bears eyes, a system of tentacles and a sac-like visceral mass. Arms are absent, funnel is bilobed. Mouth is situated at the end of the head and surrounded by foot with two sets of lappets forming an inner and outer circle. The edges of the circles are beset with numerous small, thin and annulated retractile tentacles. The tentacles are devoid of suckers and are adhesive in nature. A special fleshy hood used for closing the last chamber as an operculum is present above the head region. Arms and suckers are absent. Unlike other cephalopods, *Nautilus* possesses two pairs of gills, two pairs of kidneys and two pairs of auricles. Sexes are separate. A portion of the tentacle crown, which involves four tentacles of the lappets on the right side is modified to form the spadix.

Gastropods

Commercially important marine gastropods are fished by skin diving to depths ranging from 12 to 26 meters. A part of the collection is also made from intertidal zones during neap tide. Live shells are found hidden under rock, coral reefs and in-rock crevices. *Lambis, Strombus, Murex, Conus, Oliva* are the major genera, which thrive well in the intertidal zones. Some shells are also collected by mechanized vessels during bottom trawling. Flat bottom dinghies, keeled bottom dinghies (of Andaman group of islands) and dug-out canoes (of Nicobar group of islands) are the crafts employed in shell fishing. Various types of knives and iron rods are used during fishing to remove the animals from rock crevices and to detach them from substratum

Gastropod shells are utilized for several purposes since historical times. The sacred chank (*Xancus pyrum*) was in use even thousands of years before for worship and producing war calls. Gastropod shells are the raw material for shell craft industry, like the manufacture of items like, rings, bangles, chains etc out of the shells of *Trochus niloticus, Turbo marmoratus, Turbo intercostalis* and different species of cowries and olive shells. Most of the gastropod shells, available at present, are utilized for the manufacture of various types of ornaments and decorative pieces. But trochus and turbo shells are more commercially important.

Shell fishing in the waters around Andaman and Nicobar Islands is regulated by Fisheries Regulation, 1938 and Shell Fishing Rules, 1978.

The shell fishing can be undertaken by the licencee in the sea around Andaman and Nicobar Islands to distance of three miles from the coast of the island, measured from low water mark. May 1st to 30th September every year are closed season. Shell fishing season starts from 1st October and continues up to 30th April every year. License for shell fishing is issued for any of the nine zones to the highest bidder. During a fishing season, collection should be restricted to 15 tonnes of shell from a zone.

Shells after collection from the sea, the animals are killed either in boiling water in case of small shell, or kept inside pits and covered by soil for atleast a week. By this time due to bacterial activities, the muscular part of the animal disintegrates leaving empty shells.

Strombus sp., the queen conch of Andaman Sea is impressive and prized as holiday souvenirs. At the microscopic scale, they are one of the nature's greatest engineering masterpiece, a stunningly intricate hierarchical architecture of inorganic crystals interwoven with organic molecules.

In natural shell growth, an organic outer layer (the periostracum) is deposited first. It remains unmineralized but provides a base on which the mineralized shell is deposited. First a micrometer-thick layer of perpendicularly oriented elongated crystals froms, abutting the periostracum. Then comes the body of the shell, which grows to a thickness of a few mm and has a three-layer crossed-lamellar structure.

In a wounded conch, 24 hours after damage, a transparent organic membrane is deposited over the wounded area. Once the membrane is in place, fine crystallites of aragonite-a particular crystal form of calcium carbonate nucleated rapidly on the organic membrane. Each conch generate a 100 micron depth of such abnormal tissue in the damaged area. After 6 to 8 days, elongated crystals are deposited, orient perpendicular to the direction of shell growth. As in normal shell development, the setting down of these elongated crystals preceded the formation of the crossed-lamellar microstructure.

Fasciolaria trapezium, golden brown shell, commonly known as trapezium horse conch, is species of marine gastropods in the family Fasciolariidae, occur in Andaman Sea. They are traditionally exploited for shell craft industry. Being banned under Schedule I of the Wild Life Protection Act, the fishermen throw these shells into the sea, when caught in the net. Sometimes shells are thrown removing the operculum, which is used to produce an adhesive matrix for coating incense sticks with powdered sandle wood and other sweet smelling materials. The average weight of an operculum is 1.8 gram. This operculum is sold in the local market at the rate of Rs. 1800 per kg.

Trochus niloticus of Andaman Sea is in great demand in shell craft industry. This animal attain full growth within three months. The developing young feed on soft algal filaments.

Gastropods, like, *Tibia maculata* and *Tibia curta* occurred in large quantities as by-catch along with the lobsters in gill nets along the rocky coastal belt of Mangrol, Porbunder and Muldwarka of Gujarat. Other gastropods, such as, *Archipecten* sp. and *Murex* sp. also make their occurrence in small quantities. The gastropods living in the muddy or nearer to rocky area also get entangled in the gill net. These gastropods are available in large numbers during high tide after post-monsoon season. The catch rate was 15-25 kg per unit during September-December and considerably lower (5-8 kg per unit) during January-March. The gastropods are mainly constituted by *T. maculate* and *T. curta* forming about 94 per cent, while the sun-shell, *Archipectin* sp. and *Murex* sp. form 3 per cent and 2 per cent respectively. Majority (71 per cent) of the gastropods (*Tibia* sp.) were alive and 29 per cent were dead and were mostly occupied by the hermit crabs. The size range of the gastropods varied between 71 and 136 mm with a size group 80-84 mm forming 17 per cent and the size range 80-104 mm contributing 68 per cent of the population.

Two species of gastropod, namely, *Nassaria nivea* and *Tibia fusus* were identified from the trawler landings (20-40 kg) from deep sea, 40 miles from the shore, north of Tuticorin at a depth of 200-400 meter.

The maximum shell diameter of *N. nivea* ranged from 0.81 to 1.40 cm and weighing 1.1 to 5.2 gram.

After the tsunami, there is an increasing trend of gastropod population, especially, *Babylonia spirata*. It is found in the sea at a depth of 10 meters. The snail, *Babylonia spirata*, is known to be distributed in Indo-Pacific region. Among 14 species, *B. spirata* and *B. formosae* are considered to be polytypic, that is, in Indian waters only two species have been recorded along the east and west coasts of India. *B. spirata* is common along the east coast, whereas, *B. zeylanica* is dominant in west coast. *B. spirata*, a common gastropod species occurring along the south east and south west coast of India, has recently gained importance for its commercial and food value.

The marine gastropod, *B. spirata* is distributed in the sandy mud bottoms upto 5 to 10 fathoms depth on the south eastern coast of India. They are carnivorous feeders, feeding on the dead and decaying animal matter and living flesh. The osphradium is always well developed and is used in detecting food.

Generally the mollusks breed in summer and winter continuously. Breeding of *B. spirata* occurs in summer. The nutritive value of *B. spirata*, are; Protein-42-68 per cent, Lipid-3-10 per cent and Carbohydrate – 2-6 per cent.

Shells of these snails have been utilized for ornamental purposes. The soft portions of the snails, namely, foot, adductor muscle and mantle are processed and exported to South East Asian countries. The meat is a part of regular diet of the people of some rural coastal villages on the southeast coast of India. It is believed that the opercula of the snail are purchased and are utilized in the industries engaged in the manufacture of perfumes, cosmetics and medicinal drugs.

During 2009, two to three tones per day were exported from Thazhanguda center of Cuddalore coast to Hong Kong, Singapore and Malaysia.

Bivalves

Clams, which play a very important role in coastal rural economy, belong to the group of animal, which is an efficient converter of primary products into food suitable for human consumption. The shell rich in calcium carbonate is used in lime, cement, fertilizer, sugar and shell craft industries and as a feed for prawn and poultry. By temperature manipulation, captive breeding of the great clam (*Meretrix meretrix*) and blood clam (*Anadara granosa*) for production of seed would help in transplanting seed in suitable areas, where a production of 40 tonnes per hectare can be obtained within a period of six months.

Surf Clam

The surf clam, *Mactra violacea*, commonly known as violet trough shell is distributed all along the sandy beaches of north Kerala. It occurs in the depth zone of 75-100 m. It is a large clam upto 80 mm with high meat content and nutritive value, which is collected and consumed by the local people.

Their size ranged from 38-80 mm with modal classes 54-56 mm (12 per cent) and 56-58 mm (11 per cent). The mean sex ratio was 0.7:1 (M:F), with females being dominant during January and July. The mean meat content was 24.2±1.4 per cent and ranged between 22.9 and 26.7 per cent. The meat content increase, when the gonads matured. The surf clam is a continuous breeder with peak spawning season during Feburary to April and another peak spawning during October –November. Clams below 38 mm size were not found in the surf zone. The larger clams possess a large foot and bury deep into the sandy substrate of the surf zone and are able to withstand the frequent disturbances in the sand due to tidal effect.

The meat content of *M. violacea* is very high ranging from 23 to 27 per cent, compared to those reported in *Meretrix meretrix* (7.6-16.1 per cent), *Paphia malabarica* (8.86-20.8 per cent), *Villorita cyprinoids* (6.2-18.76 per cent) and *M. casta* (7.6-16.1 per cent).

Giant Clam

Two species of giant clams of genus *Tridacna* are common in Andaman Sea. Of these, *Tridacna crucea* is the predominant species with a distribution upto 72 nos per square meter on coral rocks and *T. maxima* occurs upto 2 nos per 30 m transact. The juveniles of all the two species have been reared in onshore culture tanks and cages in the intertidal area with 100 per cent and 90 per cent survival respectively over a year.

Oysters

Based on the breeding habits, the mollusks in marine environment are segregated into three major types; (1) year round breeders, (2) winter breeders and (3) summer breeders. The difference in breeding behaviour of bivalves, mollusks exists in different localities of India. *Crassooystrea madrasensis* from Tamil Nadu coast was observed to spawn from April to May and spawning slowly decreased until October, when the salinity was maximum (Paul, 1942), whereas Durve (1965) found *C. gryphoides* to spawn throughout the monsoon period from July to September, when the salinity was lowest at Mumbai coast. *D. cuneatus* from Tamil Nadu coast and those of Palk Bay have only one spawning period of short duration, although at former locality, the breeding is relatively much longer.

C. madrasensis breeds continuously in marine, but discontinuously in estuarine environment (Rao, 1956). Algarswami (1966) showed two prolonged spawning periods of *D. faba* in a year from November to December and May to June, with peak spawning at November, which was probably induced by the lowering of salinity at Tamil Nadu coast. Thus there exists differences in spawning periodicities of bivalves from Indian coasts and the differences may be related with the change in salinity. In pearl oyster (*Pinctada fucata*), the primary breeding takes place during winter followed by a minor secondary breeding in early summer.

Paul reported (1942), that *C. madrasensis* and *Mytilus viridis* attain sexual maturity at the length of 12.5 mm and 15.5 mm respectively. Rao (1951) recorded that in *Katelysia opima*, the sexual maturity was attained when the clams were just 12 mm in length. Abraham (1953) had recorded that *Meretrix casta* attained sexual maturity, when they were 11 mm in length. *C. cucullata* at Ratnagiri attained sexual maturity at 10 to 12 mm. The oysters (*C. cucullata*) spawned at Ratnagiri during October to January, when the salinity rose at moderate temperatures. Spawning reached its peak in November and December, when the salinity was 30 to 31 ppt and the temperature fluctuated from 22 to 27.5°Celsius.In February and March, when the temperature and salinity started increasing, the recovery was observed at a faster rate, during which time unspawned gamets got reabsorbed and gametogenesis with respect to the rapid metabolism was in active phase. The oysters became mature in July. Throughout the monsoon period these mature oysters remained in the same condition due to very low salinity. After monsoon, as the salinity started increasing, the oysters spawned, stimulated by the higher salinity.

Cowry Shells

Of all the living organisms in the oceans, Cowries are probably the most esteemed one. Because of their brilliant colour, uncomparable glossyness, smoothness, polish and attractive shape, they have been attractive to men and women from the time immemorial and their importance is still existing in one way or other. They belong to a large super-family of the class Gastropoda in the phylum Mollusca. Cowries are largely tropical in distribution and common in around coral reefs and islands. They have been utilized in different ways by man from pre-historic period. Ancient feeding grounds, burial grounds, excavations and caverns of Stone Age have shown how primitive men used this soft molluscan flesh as food item, shell for utilitarian purposes in their daily life, for their personal adornment and as a charm against evil eyes and evil spirits. Money cowry has been a familiar form of currency over a large part of the world, since pre-historic period. It has been widely used for decoration purpose by certain tribes in Africa. In its fancied likeness to the organs of reproduction, the cowry has always been consecrated by primitive people to Venus, the Goddess of Fertility, whose spirit was believed to dwell within the shell. Due to this universal belief that cowries will insure fertility, sterile and pregnant women used to wear and perhaps may still be wearing cowries as amulets, hung around the neck, waist and arms. Athenians used cowries as ballot balls in voting. Apart from decorative and ornamental purposes, money and snake head cowries are used in India as an aid in astrological calculations in Kerala, in popular games among youngsters and in the manufacture of porcelene.

Natives of most islands of indo-Pacific regions eat cowry flesh. The Filipinose and Japanese eat tiger cowry (*Cypraea tigris*). Japanese throw live tiger cowry with its shell on to hot coals and roast it; but Filipinose dry it first by hooking the animal out from the shell and fried. Hawiians eat both tiger and hump back (*Mauritia mauritiana*) cowries.

Lakhs of snake head cowries (*Ravitrona caputserpentis*) and crores of money cowry (*Monetaria moneta*) and gold ringers (*Ornamentaria annulus*) along with other commercially and economically less important varieties are available during the lowest low-tide period (September-November), when reefs are exposed during day time. The method of collection is hand-picking.

Phylum – Mollusca; Class – Gastropoda; Order – Archaegastropoda, Super family – Cypraeaceae (Cypracoidea) and Family – Cypraeidae (True cowries)

Sub-family – Pustulariinae – Small globlar shaped, unicoloured or irregularly spotted cowry shells with smooth or granular texture. Extremities of the shell are beaked, margins not calloused and the shell aperture has very fine teeth that do not extend across the rounded shell base. All less than 2.5 cm length.

1. *Pustularia cicercula cicercula; Cypraea cicercula* – Shells are small (about 16 mm) and yellowish white, punctuated with brown. No brownish blotches on dorsum or on the base, but there is a brown groove at the spire of the shell.

2. *Pustularia globules vulavula* – Milky white and globular with finely granulated humped dorsum and with a dorsal groove along with its length. Fine regular teeth across the base.

3. *Pustularia globules brevirostris* – Shell is small and sub-pyriform, with a short dialated posterior extremity, rather curved left ways.

   Sub-family – Staphylaeinae – Pretty small shells, sub-globular, ovate-oblong in shape with produced extremities. The surface is granulose, pustulose, or smooth and regularly spotted. The apertural teeth are strong, varying in the species or races from being produced right across the shell base in a most conspicuous manner. Base of shells are rounded but not spreading and the margins are not calloused.

4. *Staphylaea staphylaea; Cypraea staphylaea* – Numerous small granules and spots, teeth produced right across the base, which is fulvous tinted because of the red lines bordering the teeth. Ground colour greyish, extremities sometimes brown tinted, pustulose. Granules are elevated.

5. *Staphylaea descripta nukulau* – Comparatively smaller and ovate shell with minute granules on the dorsum. Animal is red. Shell 11 to 18 mm in length and teeth extend right across the shell base.

6. *Staphylaea staphylaea consorbina* – Shell smaller and more oblong. Colour light pale yellow. Base has teeth extending across it. Sometimes extremities are orange-red and sides pitted. Upper surface with crowded, irregular sized small elevated pustules. Base having teeth extending across it Length 11 to 24 mm.

7. *Cypraea semiplota; Eustaphylaea semiplota* – Shell comparatively small, oval in shape, margine and below has a narrow aperture. Olive brown above and the upper surface is marked by many small white spots. Light in colour below with a yellowish aperture. Length of the shell is 18 mm.

8. *Nuclearia nucleus sturangi; Cyprea nucleus* – Granules are finer than typical *N. nucleus*, the dorsam is less inflated and the left shell margin is more depressed. Solid whitish shell with brown nodules, the lateral ones uniting together by fine ridges. Extremities of the shell extend as a pair of blunt process. Basal teeth are strong. A prominent sulcus runs along the central dorsal surface. Length is 23 mm.

   Sub-family – Talpariinae – A family of large and prettiest shells. They are cylindrical in shape rather than depressed. Shells are highly polished, with extremities slightly produced, posteriorly they are not umbilicate. They are solid and smooth, and the surfaces of most species are either spotted, speckled or ringed conspicuously, but some shells are practically unicoloured. The apertural teeth are fine and not expanded across the base.

9. *Talparia talpa talpa; Cypraea talpa* – Shell beautifully glossy and smooth, rich creamy yellow fawn on the dorsal surface with extremities, base and lateral areas, a dark gleaming chocolate brown shed. Aperture narrow, teeth linear and numerous; brown in colour with white interstices. Outer lip is elevated. Shells vary 6.5 to 9 cm in length.

10. *Chelycypraea testudinaria testudinaria* – Very large and beautiful shell with more or less cylindrical shape, richly dark chocolate brown, highly polished, smooth shell of great beauty. Even darker coloured blotches of varying size are placed nicely over the surface. Over all a fine peppering of white sand like dots is also there. Where dark colouring is absent, the ground colouring seems to be a pale orange tint. The chalky specks forming the peppering are embedded in the final layers of enamels.

11. *Arestorides argus argus* – Pretty cylindrical, moderately large (6.5-10 cm) shell. Fairly wide aperture. Light brown surface of back side is scattered by number of small brown rings and more sparsely by round brown spots. The base is light fleshy brown with two dark brown blotches on each side of the lip. Teeth brown in colour.

12. *Bacilitrona isabella Isabella; Luria Isabella rumphii; Cypraea Isabella* – Cylindrical shell with somewhat truncated appearance, yellowish fawn on the dorsal surface, with fine interrupted, elongate black lines. Extremities brilliantly marked with orange-red. Smooth, rounded, glossy attractive shells of smaller size are abundant on coral reefs of Andaman Sea.

Sub-family – Erroneinae

Sub-family of smallest shells, 5 cm or less in length, elongate, cylindrical in shape, smooth slightly umbilicate, regularly speckled and sometimes obscurely banded. A dorsal blotch of varying size and density of colour is normally present. Teeth of aperture coarse or fine and not produced across the base. Shell margins weakly to strongly calloused and the base is typically rounded, not spreading.

13. *Ovatipsa caurica caurica; Erronea caurica* – Shell is light brown black due to flecking of brown on light-greenish-yellow. The margins are creamy-orange with large dark brown spots and the bases are yellow and unspotted. Teeth

are not numerous, but ridged across the base and outer lip. Aperture is narrow and the fossula is concave. Shell is cylindrical in shape and bases are highly calloused. Length 4 cm.

14. *Talostolida subteres subteres* – Thin, costly small shells, right side is acutely margined. The steep fossula and columellar sulcus are shallow, posterior callosity of dorsum is obsolete and the dorsum lacks a central blotch. Length 1.75 to 2.5 cm.

15. *Bistolida stolida crossei* – An oblong form with extremities greatly produced and turned upwards and the central blotch of the typical *B. stolida* extended into a more or less and caused by the spots joining together. Teeth are white, columella denticulate within. The shell has the distorted appearance. Length is about 2.6 cm.

Sub-family – Adustinae

A sub-family of inflated, pyriform-shaped, smooth shells with the extremities produced, slightly umbilicate. Teeth are medium size to course structure, not extended across the base, which is rounded and not spreading. Shell margins are not calloused. Shells are either uniformly coloured, banded or spotted. Do not reach more than 3.75 cm in length.

16. *Cribraria cribraria orientalis* – Fairly small shell, surface is oscellated, entirely white throughout the development of whorls and only the last whorl is covered by a light brown layer except along its dorsal margin and on the base. Round white spots of different sizes clearly marked over the back side.

Sub-family – Narinae

Ovate depressed shells, sometimes slightly umbilicate with extremities slightly produced. Shells are smooth, speckled, banded or blotched. Margins are not calloused and the base is flattened, but not spreading round the shell. Teeth are fine to medium and not produced across the base. Length approximately 2 cm or less.

17. *Purpuradusta fimbriata marmorata; Paulonaria fimbriata marmorate* – Smaller and slender shells with numerous teeth, dorsal side is light purple-brown in colour, on which microscopic spots and discontinuous colour bands are marked in dark-brown. It also bears a pair of minute spot, violet in colour on each extremity, which is a distinguishing character. Young shells are dull purple with more prominent bandings. Length 1.25 cm or more.

18. *Paulonaria gracilis japonica; Purpuradusta gracilis japonicum* – A rather pyriform little shell with a grayish back, profuse brown speckling, and a large but distinct dark central blotch. Purplish brown terminal spots protrude into the outlets at the extremities of the shell.

19. *Evenaria asellus kawalawa; Palmadusta asellus; Cypraea asellus* – Shell small elongate pyriform in shape, with three broad chocolate-black bands across the white dorsum, continuing obscurely across the columella into the

aperture. The animal is black with minute processes on its mantle, its tentacles have red tips. Length 11 to 21 mm. Found on rocks inside reefs.

20. *Evenaria punctata punctata; Palmadusta punctata* – A tiny whitish to cream shell, dotted with brown spots, oval in outline. The hind top of the inner lip is callously thickened or swollen, the callosity on the right side of the posterior extremity is less acuminated. A ver fine, short, colourless (or tinted with yellow) teeth on the aperture.

21. *Evenaria punctata atomaria* – Aperture of the shell is more curved behind than in the above species and the hind top of the inner lip is callously thickened or swollen, the callosity on the posterior extremity is much less accentuated.

22. *Evenaria punctata carula* – Smallest of *E. punctata* races of shells. Pyriform shaped and elegant in appearance. White in colour and sparsely dotted with conspicuous brown spots on the umbilical region and one on each side of the anterior extremity. Lateral spots are smaller than the others. The base is white, teeth more distant on the labial margin than on the columellar side, and the former are streaked with yellow. Length 10 mm, breadth 6.5 mm.

23. *Evenaria hirundo neglecta* – Greenish blue shell, small and plump, oblong cylindrical in shape. It is comparatively smaller and having smaller and more numerous teeth to the aperture, the left border of which is angular. The back usually has a brown blotch, dorsal markings are greenish-blue.

24. *Evenaria hirundo camerone* - Shell is cylindrical rather than inflated, creamish with pale greyish blue with faint dorsal bands. The dorsal surface has small scattered brown spots and there is a large blackish brown blotch on each side of the extremities. The base is white, crossed by the teeth, which also extend about half the way over the labial margin. There are a few brown spots on the lateral base. Shell is umbilicate posteriorly. Length 19 mm and breadth 11 mm.

25. *Melicerona feline listeri* – It is narrower, more elongated and darker than *M. feline*. Shell is cylindrical, ovate, depressed, spotted and banded. Colour is greyish-yellow with four brownish-black spots. Length between 125-200 mm.

26. *Melicerona feline fibula* – A short gibbous shell with elevated back. Greenish-blue in colour, freckled with olive, sides and base are yellowish, lateral spots large and black and the teeth are strong.

Sub-family – Mauritiinae

Large ovate to sub-cylindrical shells reaching 10 cm or less in size, extremities not produced and not umbilicate. Smooth, very polished, surface spotted or with Arabic markings, or other patterns. Teeth coarse to medium in strength, not produced across the shell base, margins calloused and base flattened.

27. *Mauritia mauritiana mauritiana; Peribolus mauritiana calxequina; Cypraea mauritiana* – Its size and rich brown, almost blackish colouring with light

spotting on the dorsum, elevated back and the bluish tinge, polish on the dark chocolate coloured sides and base are distinguishable characters. Average size is 10 cm. Teeth are dark brown white between them. In some specimen, the aperture is more wider and labial teeth are less produced than *M. mauritiana calxequina* and fossula and columellar sulcus less concave.

28. *Arabica maculifera; Peribolus maculifera; Cypraca maculifera* – A large solid, pyriform shaped shell, very elevated, margins rounded but very thickened, polished and calloused, extremities blunt, dorsal line broad, slightly curved a little sub-central. Spire concealed by callus. Dorsal surface pattern consists of close set, irregularly sized and shaped whitish spots with longitudinal golden brown, lateral spotted with same colouring which merge on to the same basic colouring, pinky blue with orange tinge, of the base. Teeth on the outer lip are 24 to 34 and inner about 24. Length of shell varies to 90 mm.

29. *Arabica histrio; Cypraea histrio* – This is a smaller (57mm), more pyriform shaped shell than *A. maculifera*. Whitish spots smaller, more crowded and mostly rhomboidal in shape. The dorsal line creamy whitish, much narrower, sometimes one-fourth than that of the former shell. Some race is more elongated and having smaller margins than *A. maculifera*.

30. *Arabica depressa depressa* – This is a very solid, highly polished little shell, like a much smaller *A. histrio*. The surface is reticulated like the above two species, the sides dark purplish blue as in *A. histrio*, but the lateral spots are much smaller, more numerous and extend furtheron the base, which lacks the large basal blotch of *A. maculifera* and is also China white in colouring. Dark reddish, elongated teeth 21 on outer and 18 on inner lip.

31. *Arabica eglantine eglantine* – An oblong ovate shell, rather cylindrical in form, sides more or less callous and swollen instead of being regularly rounded or narrowly margined. The typical species is greenish grey in colour, with whitish patches, but at times has a yellowish misty tinge over the surface and has fine lines. Length 68 mm and breadth is 36 mm.

32. *Arabica scurra scurra; Arabica scurra; Cypraea scurra* – An oblong-ovate, sky blue, obscurely banded shell, covered with a delicate, open, net-work pattern, golden brown in colour. Sides of the shell are brownish, sprinkled with small crowded, smoky black spots. Aperture narrow, teeth fine. Length 3 to 5 mm.

Sub-family – Erosariinae

Pyriform to ovate shaped shells with extremities slightly produced or not at all. Shells are not umbilicate, marginally blotched, and are usually either spotted, obscurely banded or peripherally ringed. They are smooth or tuberculate, with coarse teeth not produced across the base, which is flattened and spreading, and the margins are calloused. Shells are 3.75 cm or little less in length.

33. *Erosaria erosa erosa; Cypraea erosa* – Oval in outline. Very common form. Shell is pale bluish to yellowish, thickly covered with white spots and

sprinkled with ocellated brownish ones. The sides are white, stained in the middle with a squarish brown-black spot. Extremities are ridged with chestnut brown. Teeth are strong, the outer ones extending across the lip and frequently over the sides. The base is white, sometimes spotted and streaked with chestnut brown. Length varies from 2.5 to 5 cm.

34. *Erosaria nebrites ceylonensis* – Shell is rather humped and deltoidal in shape and has accentuated fossular denticle, In Lakshadweep race any kind of blackish blotch is not seen anywhere. General colour is yellow brown on the back, base pale flesh tinted crossed by brown dots and lines almost to the teeth. Base is more convex and aperture more curved behind than the typical *E. nebrites*.

35. *Ravitrona caputserpentis caputserpentis; Cypraea caputserpentis* – Very shiny, solid, dark shell, reddish to blackish-brown sometimes almost black-margined with dorsal spotting or maculations of white. Extremities have a white blotch. Dark unicolour margins are wide and smooth like rest of the shell. Base is greyish white with white teeth.

36. *Ravitrona caputserpentis reticulatum* – Shells of this race are broader and smaller than the above race, with more thickened sides and dorsal spots more crowded. The posterior blotch is often orange tinted rather than greyish white.

37. *Ravitrona labrolineata labrolineata; Erosaria labrolineata* – A small greyish yellow shell with numerous small dorsal spots and occasional brown "eyes" dotted over the dorsal surface. The marginal spots are obsolete and the under surface of both the extremities are stained orange. Length 1.25 to 2.5 cm found among the crevices of coral rocks of shallow waters.

38. *Ravitrona labrolineata maccullochi* – A small beautiful shell. Dorsum pale greenish-grey, marked with large and small distinct chalk white spots, the dorsal line broad and well defined. Terminal spots ill-marked, pale but more definite anteriorly; sides with few brown spots and the base is shining like China white. Teeth are strong and stout, the posterior ones produced, the labial ones in particular, are thick and pronounced. Teeth number 13 on both the margins. The fossula recedes and the sulcus is obsolete. Length 12 mm and breadth 7 mm. Found in barrier reefs.

39. *Ravitrona helvola helvola; Erosaria helvola; Cypraea helvola* – Broad shell with blunt lilac extremity, ferruginous base, brown dorsal spots are so crowded that they occupy as much as that devoted to the white specks that characterize its surface. Normally bluish grey above mottled with brown spots speckled with much smaller white dots and the base is reddish brown.

40. *Ravitrona helvola callista* – The brown spots in this shell are less numerous than the above race, so that the white specks appear more numerous. The base is lighter and the extremities tend to be whitish rather than the more usual lilac-tint. Also the dorsum of the shell and the lateral bands are inclined to become lilac-pink instead of brown shades.

41. *Ravitrona poraria poraria; Erosaria poraria; Cypraea poraria* – A small solid ovate, pinkish mauve shell with numerous brown ocellated spots on the

surface and sides, base pinkish mauve, aperture whitish. Length about 2 cm.

42. *Ravitrona poraria scarabeaeus* – Shell richly tinted, dorsum brownish with white spots enclosed in violet colour. Base of shell is purple and concave. Shell is rather deltoidal in shape.

43. *Monetaria moneta rhomboids* – Shell is flattened and somewhat triangular in outline. The upper surface is smooth but uneven due to a few humps in the back. Surface layer is deep violet in colour, but is hardly visible under the yellow or light yellow surface callus. Teeth on both lip margins are very small and short.

44. *Monetaria moneta monetoides* – This is a small shell with four prominent dorsal nodules, the base is strongly nodulose, the aperture is open and colour yellowish-green showing fainter, but darker transverse bands. Labial and columellar teeth number 10. Length 22 mm and breadth 16 mm. Some specimen is yellowish white above.

45. *Monetaria moneta isomers* – Smaller and narrower shells than *M. rhomboids*. Pale greenish in colour, especially in younger stage. Juvenile forms showing three narrow transverse zonal bands, characteristic of *Monetaria* races. Dorsal nodulation is obsolete and the aperture is not so narrowed as in larger species. Teeth are coarser and the labial ones produced a little, but are not nodulose as in *M. monetoides*.

46. *Ornamentaria annulus annulus; Cypraea annulus* – the dorsal ring is distinct and the teeth are fine. Bluish-grey all over the surface. Central part of this cowry is particularly expanded on both sides. Animal is greenish white to orange, black tinged, usually cluster in numbers under and on coral boulders and rocks in shallow water of reefs and lagoons.

47. *Ornamentaria annulus dranga* – A Samoan race. Here the raised area of the dorsum does not envelop the annulus like ring. It is a medium sized shell. Length 20 mm and breadth 15 mm with eight columellar and 11 labial teeth. Some specimen is smaller, length 15 mm and breadth 10 mm.

48. *Ornamentaria annulus scutellum* – The sides of the shell appear slightly less depressed than the Pacific forms, but they are more callous, the inner lip also exhibiting a slight central callous deposit. Teeth are coarse and long, dorsum is bluish, dorsal ring distinct.

Sub-family – Cypraiinae

Contains some of the largest and most beautiful cowries including brilliantly spotted and polished tiger cowry and much admired orange cowry. Shells mostly reach about 7.5 to 10 cm in length. Ovate in shape, swollen and not umbilicate. Surface smooth and well-polished and if not unicoloured like orange cowry. The teeth are coarse and not produced across the base.

49. *Cypraea tigris* – Large shell, dark, smoothly polished and brilliantly spotted. Base is white, with large strong, sometimes bifurcate teeth. Length of the

shell varies from 6.5 cm to 10 cm. Young shells are chestnut to whitish with interrupted bands of zigzag, rusty brown flashes over the surface.

50. *Lyncina lynx lynx; Cypraea lynx* – Bluish white to brownish shells, profusely spotted and coloured with fulvous-brown and blue, interspersed with dark, blackish-brown spots irregularly scattered over the surface. Sides, base and teeth whitish, space between the teeth blood red. Young shells are whitish, faintly banded, and profusely banded with light brown. Lenth varies from 2.5 to 6.5 cm.

51. *Lyncina lynx williamsi* – Shells have rather blunt extremities, the posterior extremity is short, sides are rounded, the aperture is wide, fossula flattened and the base is more liable to be tinged with flesh colour. In some forms, the callosity of the sides extends far above and covers about half the dorsal region and the shell is more thicker and solid than the rest of the member of the genus.

52. *Mystaponda vitellus vitellus; Cypraea vitellus* – Commonly found on coral and rocky reefs of the tropical seas. Shells are ash to fulvous-bay coloured, profusely sprinkled with milky white spots of various sizes. The sides are olive-brown and striated, base and teeth white. Sand like striations on the sides, extending from the base up towards the dorsum.

53. *Mystaponda camelopardalis; Cypraea camelopardalis* – A Red Sea form. It is distinguished from *M. vitellus* by constricted and margined extremities, the inner lip is much produced and bent behind and the sand like striae on the sides are replaced by white spots. The space between the teeth are blackish. The white dorsal spots are more scattered and smaller than *M. vitellus*. Lateral zones are not darker than dorsum. Length between 8 to 10 cm.

54. *Ponda carneola carneola; Cypraea carneola* – A pale flesh coloured shell with 4 and sometimes 5 transverse bands of deeper tint. Surface smooth, highly polished and softly tinted. Base and sides are fulvous pink, teeth and interstices deep purple. Length 2.5 to 7.5 cm.

List of Gastropods found in Andaman Sea

(1) *Nautilus pompilus* (Linnaeus), (2) *Cassias rufa* (Linnaeus), (3) *Turbo marmoratus* (Linnaeus), (4) *Chicoreus ramosus* (Linnaeus), (5) *Cassis crnuta* (Linnaeus), (6) *Trochus niloticus* (Linnaeus), (7) *Conus litteratus* (Linnaeus), (8) *Turbo petholatus* (Linnaeus), (9) *Vasum cornigerum* (Linnaeus), (10) *Cerithium nodulosum* (Bruguiex), (11) *Haliotes diversicolour* Reeve), (12) *Turbo cochlys* (Linnaeus), (13) *Trochus pyramis* (Born), (14) *Nerita semirugosa* (Linnaeus), (15) *Strombus gibberulus* (Roeding).

Fifty one species of *Conus* have been reported from Andaman and Nicobar waters as mentioned below.

(1) *Conus achatinus* (Hwass in Bruguiere), (2) *C. adansoni* (Lamarck), (3) *C. amadis* (Gmelin), (4) *C. andamanensis* (Smith), (5) *C. araneosus* (Solander); (6) *C. arenatus* (Hwass in Bruguiere); (7) *C. aulicus* (Linnaeus); (8) *C. bandanus* (Hwass in Bruguiere); (9) *C. canonicus* (Hwass in Bruguiere); (10) *C. chaldeeus* (Roeding); (11) *C. capitaneus*

(Linnaeus); (12) *C. caracteristicus* (Fischer); (13) *C. catus* (Hwass in Bruguiere); (14) *C. cevlanensis* (Hwass in Bruguiere); (15) *C. coronatus* (Gmelin); (16) *C. distans* (Hwass in Bruguiere); (17) *C. ebraeus* (Linnaeus); (18) *C. eburneus* (Hwass in Bruguiere); (19) *C. edwardi* (Preston); (20) *C. emaciates* (Reeve); (21) *C. episcopus* (Hwass in Bruguiere); (22) *C. figulinus* (Linnaeus); (23) *C. flavidus* (Lamarck); (24) *C. generalis* (Linnaeus); (25) *C. geographus* (Linnaeus); (26) *C. insculptus* (Keiner); (27) *C. janus* (Hwass in Bruguiere); (28) *C. litteratus* (Linnaeus); (29) *C. lividus* (Hwass in Bruguiere); (30) *C. masoni* (G and H Nevill); (31) *C. miles* (Linnaeus); (32) *C. miliaris* (Hwass in Bruguiere); (33) *C. millepunctatus* (Lamarck); (34) *C. mitratus* (Hwass in Bruguiere); (35) *C. monile* (Hwass in Bruguiere); (36) *C. mutabilis* (Reev); (37) *C. nicobaricus* Hwass in Bruguiere); (38) *C. nobilis* (Linnaeus); (39) *C. nussatella* (Linnaeus); (40) *C. pennaceus* (Born); (41) *C. piperatus* (Dillwyn); (42) *C. pretiosus* (G and H Nevill); (43) *c. scabriusculus* (Chemnitz); (44) *C. straturatus* (Sowerby); (45) *C. striatus* (Linnaeus); (46) *C. terebra* (Born); (47) *C. tessulatus* (Born); (48) *C. textile* (Linnaeus); (49) *C. tulipa* (Linnaeus); (50) *C. virgo* (Linnaeus); (51) *c. zonatus* (Hwass in Bruguiere).

Bivalves

A variety of bivalve resources, such as, edible oysters. Pearl oysters and mussels are distributed along the coastal areas of Indian Ocean, including Andaman Sea. The ecological set up of bivalve habitat is intricately connected with a variety of factors or combination of factors.

The marine ecosystem of Andaman and Nicobar Islands, with numerous creeks and protected bays offer some of the best sites for pearl culture operations. The main species of pearl oyster, which occurs at several regions of Andaman Sea is the black lip pearl oyster, *Pinctada margaritifera*. The other three species available in the region are *P. fucata*, *P. sugillata* and *P. anomioides*. *P. margaritifera* generally occurs in the intertidal ref flats to depths of 10 meters. The reef flat is coralline interspersed with hard sandy bottom. The oysters are found attached to live or dead corals. Their population density is low on the intertidal reef flats. The ecosystem of Andaman Sea is suitable for the culture of black lip oyster.

The Indian pearl oyster, *Pinctada fucata* settles down in large numbers on the rocky substratum of calcium grains formed from the comminuted remains of shells and corals, thus giving thr true character of limestone, and attains harvestable size at the end of the third year of growth, when pearl formation also reaches exploitable proportions. The rock bed supports a variety of marine plant and animal assemblages, which are characteristic of the region constituting well recognizable and defined pearl bank biota.

The size of granules, kind and amount of organic matter associated with the substrate, and mobility of substrate, degree of hardness of solid substrate and the total area of a given type play a key role in the prosperity of pearl oysters in the natural beds. The importance of water movement over the pearl banks is important as an ecological factor. There is a general drift of water over pearl banks from south to north between April to September and from north to south during height of northeast monsoon with intermittent periods of calm and variable movement from February to April. Population replenishment might depend upon conditions of water movement.

It is a recognized fact that throughout euphotic and littoral regions, the various modalities of light modified by the time of the day, year and latitude, the presence or absence of water movement and the depths and clarity of water exert effects upon functions and structure of marine invertebrates. The periodicity and the width of the shell opening is said to be influenced in natural beds by alternating light and darkness. The width is narrower in light, but broader in darkness, due to the effect of light on adductor muscle. So the rate of filtration of water is likely to be affected during day time in areas, where light penetration is greater. This is more frequent in depths beyond 20 m and less, so in shoreward banks during the months from December to April.

The pearl oyster larvae are photo-positive in the Veliger stage and metamorphosis proceeds normally under favourable light. There is a view that turbidity during metamorphosis is desirable shielding the larvae against ultraviolet radiation in shaded areas. Conditions in nature, over the pearl banks in May-July satisfy the requirement. Flood discharge during northeast monsoon in October-November carry with it considerable silt, which creates great turbidity over the pearl beds, particularly over the shore lying banks, causing the problem of growth and survival of oyster population met within 12-15 m depth range.

Some correlation exists between temperature and the breeding behaviour of pearl oyster. The breeding season is more restricted in higher latitudes and occurs during warmer months. In lower latitudes, there is less restricted breeding season. The major spawning occurs outside warmer months. (*Pinctada fucata* 7-10°N; *P. margaritifera* 17°S. Orton, (1920) states that in those parts of the sea where temperature conditions are constant or nearly so, and where biological conditions do not vary much, animals breed continuously. There is a particular threshold temperature which, when exceeded all the year round, permits continuous breeding and when exceeded only in summer, restricts breeding intensity to summer months. The Indian pearl oyster fertilization was noticed to take place in the 27-28°Celsius range. The lower the temperature, the greater the time taken for completing metamorphosis and settlement.

The average value of salinities range between 33.5 to 31.26 ppt. Seasonal fluctuations of salinity in conjunction with other environmental factors are known to exert influence on the physiological functions and reproductive activity of oysters. By and large, the pearl oyster being truly marine form in its entire life cycle is not known to tolerate great variations in salinity. In salinities of 13.9 ppt and 50.96 ppt mortality was total after 48 hours, whereas 3 out of 5 oysters died in the salinity of 19.9 ppt in the same period.

Dissolved oxygen ranging from 6.84 ml/l in October to 3.4 ml/l in September appear to be common in pearl oyster bed.

Pearl Banks

The very fact that the fauna and flora of the pearl banks comprise the whole assemblage of more than 2700 species of animals and 200 species of plants, small and large, makes the study of inter-relationship among them very complicated, although it is well recognized that the nature and density of such animate surroundings have a profound effect on the well being of the stock of oysters in the beds.

Characteristic of the area is the dense growth of sponges, *Aulospongus tubulatus, Phakellia donnani, Siponochalina communis, Itrochota* spp., *Clathria procera, C. indica, Mycle grandis, Zygomycale parishii, Phyllospongia* spp., *Spongionella* spp., and *Suberites* spp. are abundant. Dense forest-like growth of gorgonid *Juncella juncea* and *J. gemmacea is also noticed.*

The growth of the coral *Heteropsammia* sp. is characteristic of the inner series. *Montipora* sp. and *Echinopora* sp. and the other corals in addition to *Porites* sp.

The molluscan fauna is mostly represented by myriad numbers of *Modiolus* spp. spreading like mattress on the bottom. Large *Pinna* spp. are found in good numbers rooted in this layer of sand covering the rock in many places. *Cypraea tigrinus* are seen in rocky pits. *Oliva* spp., *Conus* spp., *Nassa* sp. and *Bulla amoulla* are other common shells.

Among the echinoderms, *Lamprometra palmate palmate* and *Comanthus (Comanthussis timorens)* are the most common under rocky crevices and over the gorgonids and sponges. *Holothuria edulis, Protoreaster lincki* and tests of *Clyspeaster humilis* are the other common species.

The fish fauna is fairly rich and consists of *Scolopsis bimaculatus, S. vosmeri, Abalistes stellaris, Upeneoides* spp., *Chaetodon* spp., *Pomacanthodes annularis* and *Lutjanus lineolatus.* Large fishes like, *Gaterin* spp., *Ennaeacentrus miniatus, Epinephelus* spp., *Lethrinus* spp., and *Siganus* spp. are abundantly seen in the inner series.

The flora is poor, *Gracilaria* spp., *Hypnea* spp. and *Sargassum* spp. are common.

Compared to the inner series, the outer series is richer in fauna and flora qualitatively and quantitatively. The formation of outer series, generally between 15-25 m depth range are fairly extensive stretches of rock, whose outcrops differ greatly from tubular fragments, rock of a meter or two across, to great areas of a km in extent. Fine grained sand covers the rock filling up the hollows and crevices occasionally cutting off the continuity of paars to give the impression of sandy bottom. The hard core of the bottom can be easily detected by removing the engulfing sand of 5-10 cm thickness.

Live corals are seen as a low fringe running along the 18-19. depth. Broken and worn out fragments of pearl oyster shells, cockles, *Pecten* spp., *Conus* spp. etc are scattered about in great profusion. Balls of *Porolithon* sp. from the size of a nut to that of a lime are seen on the edges of the rocky expanse. All through the length and breadth of the paar are a number of pits ranging from 0.5 to 1 m diameter and of equal depth. Such pits are inhabited by a number of small and large fishes, eels and lobsters. The general set-up of the area is ideal for the settlement of oysters as the horizontal clarity at he bottom exceeds 15 m on most days and because of variety of fauna and flora inhabiting the area.

The concentration of sponges are very high. The predominant species are, *Petrosia testudinaria, P. similes, Aulospongus tubulatus, Axinella donnani, A. symmetrica, Spirastrella inconstans, Suberites* spp., *Cliona vostifica, Clathria indica, C. procera, Mycale grandis, Raspailia hornelli, Myxilla arenaria, Iotrochota purpurea, Pachychalina subcylindrica,* and

Phakellia donnani. There are other species of *Auletta, Spongionella, Hippospongia, Phyllospongia* and *Hircinia* met in the 25 m depth line also.

The area is rich in coelenterates with a conspicuous growth of anemones, alcyonarians, and gorgonids. Some of the fleshy alcyonarians that are common are, *Sarcophytus* spp., *Lobophytum* spp. and *Sclerophytum* spp. *Spongodes rosea, Nepthya* sp., *Solenocaulon tortuosum, Suberogorgia* sp., *Acanthogorgia* sp., *Lopohogorgia* sp. and the gorgonids, *Juncella juncea* and *J. gemmacea* harbouring many commensals are noticed commonly.

Octopus (*Polypus* spp.) are common in pits and holes. Great numbers of dead, empty broken shells are found in crevices and faults in the rocks haunted by the octopus. Pearl oysters are particularly preyed upon by them, thus posing the question as to whether they are the chief enemies of pearl oysters. On many occasions, the octopus have been noticed to open the shell valves of the oysters and eat the flesh.

The echinoderm fauna is found to be lacking in abundance as a whole. By far, the crinoids are the most abundant, found attached to the gorgonids, under coral blocks or on sponges. *Lamprometra palmate palmate* and *Comanthus annularis* are the most common. Among holothurians, *Holothuria edulis* is the most common. The synaptid, *Chondrocloea striata* is common in deeper waters. Of the sea stars, *Protoreaster lincki* is the most abundant, although *P. affinis* and *P. uustralis* are also seen rarely. The other sea stars are *Pentaceraster multispiralis, Linckia laevigaeta* and occasional specimens of *Culcita schmideliana, Protoreaster nodosus, Astropecten indicus* and *A. monocanthus* are also seen.

Of the cake urchins, *Clypeaster humilis, Echinodiscus auritus* and *Laganum depressum* are common. Of the heart urchins, *Echinolampus ovata* and *E. alexandri* appear here and there. Among sea urchins, *Salmacis bicolour, Salmaciella dussumieri* occur where ever dead coral blocks are covered by coarse sand. In the crevices of the coral stones and under the boulders live many numbers of ophiuroids of which *Astrob clavata, ophiocnida echinata, Ophioenemis marmorata, O. cataphracta* and *Ophiocnermis dubia* are more common. All over the rocky bottom, fishes are found abundantly. Numerically, *Abalistes stellaris, Sufflamen capistratus, Odonus niger, Scolopsis bimaculatus* and *S. vosmeri* are the most abundant. But where ever the area is rugged with boulders and pits fishes like, *Gaterin* spp., *Lethrinus* spp., *Enneacentrus* sp., *Epinephelus* spp., *Pomacanthodes annularis, Lutjanus sebae, Pterois milis, Chaetodon* spp., *Zanclus cornatus* and *Henio acuminatus* live in large numbers.

Throughout the rocky expanse, the density of algal vegetation seems to be moderate, especially on the edge between 17-25 m line. The flora seem to be luxuriant with *Sargassum* spp. dominating in most of the areas. Among the red algae, *Gracillaria edulis* and *Hypnea valentiae are common. The other common species in the pearl banks of the outer series* are, *Caulerpa* spp., *Codium* sp., *Halimeda* spp., *Dictyota, Padina* spp., *Porolithon* sp. and *Spathoglossum* sp. The algal flora of the pearl beds is most of the types found in coral beds or rocky region of Indian coasts irrespective of depths. In other words, there appears to be no selectivity for algae with regard to depth.

Important enemies of the pearl oysters are; *Modiolus* spp. (Weaving mussel) by competion and smothering oyster spat settlement. *Octopi*, predation by killing and preying flesh. Fishes – Predation by *Balistes,* Sreeanids, Rays and Skates by crunching the shells and eating the meat. Boring polychaetes – By drilling the shell valves after setting on them (*Polydora* sp.). Boring sponge – By riddling the shell valves with minute holes for gaining substratum for living. Sea stars – By tearing the shell valves apart and feeding on oyster meat. Boring gastropods – By gyrating holes on shell valves and feeding on the meat (*Nassa* spp., *Sistrum* spp., *Murex* spp. and *Cymatium* spp.). Crabs and lobsters - By destroying byssal threads and killing the oysters.

The black-lip pearl oysters, *Pinctada margaritifera* are found in inter-tidal and sub-tidal regions of Andaman Sea. For nearly a century, the shells of this oyster have been extensively used for production of exquisite shell-craft products. The lustrous black perals produced by the black-lip pearl oysters, *Pinctada margaritifera* in the inner surface of its two shells have become one of the most valuable pearls on earth.

In the natural spawning of *P. margaritifera*, the fertilized eggs pass through different larval stages like the veliger, umbo, pediveliger and finally by thirty first the spat settled. The settled spats measured 300-400 micron. The larvae feed on *Chaetoceros* sp., *Isochrysis galbana, Pavlova lutheri* and *Nannochloropsis* spp.

Mabe Pearl

" Mabes " are dome-shaped pearls and can be either half pearls or pearls depecting an image. These are generally produced in winged pearl oyster of the genus *Pteria* available in the Andaman Sea. These oysters have an elongated hinge and the nacre (pearl coating) colour is multi-hued and brilliant.

Though there are several species of winged oysters, two species, available in Andaman Sea, are commercially important, namely. *Pteria penguin* and *Pteria sterna*, which are used for commercial scale culture in Southeast Asia, Australia, Pacific Island nations, Gulf of California and Mexico.

Using the same technique, but with different anaesthetizing procedure, Maybe pearls were produced in the black-lip pearl oyster, *P. margaritifera* and the winged pearl oyster, *Pteria penguin* available in Andaman Sea.

The base images, which are used for Mabe pearl production are produced using finely sieved shell powder and resins. They can also be curved from shells. The size of the base images is very important and range from a small 1 cm square to a large 3 cm square. The carefully prepared fine base images are placed in specific locations on the inner valve of the oysters with the use of specific equipments after anaesthetizing the pearl oysters.

The implanted pearl oysters are stocked in small lantern cages and suspended from rafts moored in sites, which have clear oceanic waters. Within 60 to 90 days, the base images will be fully coated with fine pearly nacre. The colour of the nacre depends on the location of base image on the shell. The Mabe along with the shell can later be mounted on wooden stands or can be converted to jewellery pieces. The Mabe pearl production is simple and can be easily adopted by islanders of Andaman and Nicobar.

Mussels

Mussels in Andaman Sea were found attached to concrete structures of culvert, sluice gate and rocky substratum in Sippighat and on concrete structures, stones and hard muddy substratum in other areas. Sippighat is a narrow tidal creek, 30 to 120 m wide near Port Blair. The well pronounced tidal amplitude of 2 m high neutralize the effect of fresh water influx of run-off water during months. Tidal water flows with rich nutrients and plankton during high tide and slowly recedes during low tide through the sluice gates. The water level in the mussel bed was about 20 cm or sometimes exposed during low tide and during high tide reached to 1.5 to 2 m.

Ecology of mussels in Mithaghari, Hathitope, Kadakkachan Sluice, Minnie Bay and Rangat are different from Sippighat creek. Mussels settled on the piers, just below the jetty in Minnie Bay, and on the hard rock substratum or on the hard muddy bottom in Mithaghari jetty. Mussels were found in the sub-tidal region. The sea water is oceanic in character and changes in hydrological parameters during the monsoon are also negligible in these areas.

Mussels are found in thin population, along with pearl oysters *Pinctada margaritifera* and *Pertia penguin*. Mussels observed on their natural habitat have shown hard and thick shells with heavy settlement of fouling organisms, such as, barnacles, bryozoans, serpullds, sponges, corals, ascidians and hydroids, which are sedentary in nature and transit forms like, polychaetes, crabs, fishes, carideans etc. Turbidity is comparatively less in this area, compared to Sippighat.

Mussels were in thick population in the form of mat underneath the bridge at Sippighat. The total area of the mussel bed was 250 square meter. The population of mussels per square meter was 158. The size ranged between 35 and 118 mm with mean size of 72.3 mm and the average weight was 35.1 gram. The total biomass of mussel population in the area was estimated as 1386.5 kg. The percentage edibility was 30.2 per cent. The sex ratio showed that females were dominant (31 : 69) in the population.

Mussels were sparsely distributed in the hard muddy bottom and on the granite stones of Bimbleton creek of Andaman Sea. The total mussel bed area was approximately 200 square meter and the density of population was 61.3 per square meter. The size ranged between 47 and 136 mm with a mean size of 96.7 mm and weight 76.3 gram. The total biomass, which could be realized from the bed was 935.4 kg. The percentage of edibility was 29.9 per cent. Females outnumbered males (35:65) in the population.

The population of mussels in the black rocky area of Kalapathar was observed in small patches at the rate of 47.4 mussels per square meter. The size of mussels ranged between 62 and 132 mm. The mean size and weight were 98.2 mm and 80.1 gram respectively. The total biomass which could be realized from the bed was 1139 kg in a total area of 300 square meter. Females were found to be dominant in the population (37:63) and the condition index was 30.8 per cent.

The density of mussel population was thick underneath Garacharma Sluice gate bridge, where the depth of water during low tide was 50 cm and the turbidity was high. On either side of the creek, fringing mangrove vegetation is present. They occur in moderately small patches in outer area of the sluice. The average number of mussels per square meter in this area, was 76.2 and the sizes were between 74 qnd 121 mm with an average weight of 48.7 gram. The total biomass from 300 square meter mussel bed could be realized at 1113.3 kg and meat weight obtained was 37.7 per cent. The sex ratio was observed in equal proportions.

In all the vertical pillars of Mithaghari jetty have good settlement of edible oysters *Saccostrea cucullata, Crassostrea rivularis* and pearl oysters *Pteria penguin* and *Pinctada margaritifera*. The mussels were thickly populated and settled in clusters along with the oysters. The water depth is 2 to 5 m. The bottom is muddy with granite stones, which are scattered on either side of the jetty. The mussels were found attached to the granite structures and sometimes partly buried in the hard muddy bottom. The mussels fished for market were comparatively larger in size and all the mussels were thickly deposited by the fouling organisms. The density of population per square meter was 64 and the mean weight was 181.6 gram. The total mussel biomass could be realized to 1743.4 kg in 150 square meter area of bed.

The creek with rocky structures on either side of the bank and the jetty at Hathitope provide a suitable substratum for settlement of mussels on the pillars. Mussels were observed in patches on rocks and thickly populated on piers. The density of mussel per square meter was 84 and the size was ranging between87 and 130 mm with a mean size of 117.7 mm. The total biomass of mussels was arrived at 2215.9 kg in an area of 200 square meter area of mussel bed. The sex ratio in the population was 30:70 and females outnumbered males. The percentage edibility was 29.7 per cent.

Moderate settlement of mussels have been observed on the concrete structures of Kadakkchan Sluice area. The maximum water depth near the sluice gate was 1.5 m. The average number of mussels per square meter was 48 and the size ranged between 79 mm and 109 mm with a mean weight of 78.6 gram. The estimated biomass was 754.6 kg.

The mussels in Minnie Bay (0.8 square km roughly, deeply curved and surrounded by isolated patches of mangroves on all the three sides. The eastern entrance and the western side harboured coral stones and shallow areas get exposed during low tide. The water is turbid and maximum depth of water in the bay is 4 m) were larger in size and the age group may be 2 to 4 years. Their distribution was scattered with a minimum density of 10 and a maximum of 26 per square meter with an average of 17.8 mussels on these pillars. Their size ranged between 99 and 201 mm with a mean size and weight of 149.1 mm and 272.9 gram respectively. The total biomass estimated was at 3886.1 kg in 80 square meter area. Females were dominant in the population (46:54).

In Rangat, mussels were found attached to submerged rocks in a small creek confluent with Andaman Sea. The bottom was slashy with rocks. The depth of water in the mussel bed was about 1 meter and it got exposed during the receding tide. The mussels were in dense patches at three places and in small patches on several rocks submerged in the creek. The mussel bed area was estimated at 300 square meter. The average number was 73 per square meter with a mean size of 78.2 mm and weight of 47.8 gram. A total biomass of 1046.8 kg was estimated in Rangat area. The percentage edibility was 25.2 per cent. Females were dominant in the population (40:60).

Green Mussel

Specimens of *Perna viridis* ranging from 65 to 230 mm in total length and weighing 21 to 298 gram (total weight) and 10.5 to 84.2 gram (mean weight) were collected from Sippighat area near Port Blair during 1978.

The species is characterized by the presence of pitted resdial ridge, the anterior portion of the foot retractor, and the absence of anterior adductor. The shell is highly arched in the middle, posterior margin broadly rounded. Shell is covered with thick green periostracum. Anterior adductor absent. Posterior adductor and retractor impression widely apart, the latter almost in the middle position. The mantle edges are whitish and fimbriating without any papilla.

The species is widely distributed along the west coast of India. On the east coast of India it is recorded from Pamban, Porto Novo, Pundicherry, Tamil Nadu, Kakinada, Visakhapatnam and Chilka lake.

The meat weight percentage in relation to total weight of Andaman specimens ranged from 28.4 to 50, higher in small specimen and low in large specimens. Though the Andaman specimens attain large size, corresponding increase in the meat weight is not attained.

Rope culture of *Perna viridis* has been demonstrated in the tidal creek, with a production upto 26 kg per meter of rope length.

Edible Oysters

Four species of edible oysters occur in Andaman coast, namely, *Crassostrea gryphoides, C. madrasensis, Saccustrea cucullata* (marine), and *C. rivularis* (brackish water). The concentrations of *C. gryphoides* and *C. madrasensis* (ideal species in the bay) is 25 per square meter and 56 per square meter respectively. *S. cuculata* is distributed upto 60 numbers per square meter and *C. rivedaris* upto 85 per square meter. The natural oyster resources of Andaman Sea justifies oyster farming in bay waters.

Chapter 23

Crustaceans

Shrimp, prawn and crabs are the most evolved species among the crustaceans. Decapod crustaceans, including shrimps, prawns and lobsters are found throughout the world from frigid to tropical zones in both fresh and salt waters, shallow and deep. The density of their distribution is greater in waters lying in the tropical, subtropical and temperate zones in latitudes 40°N to 40°S. The majority of crustaceans are marine. They are benthic living on the sea bed even upto a depth of 8000 meters.

They usually possess 20 segments and 19 pairs of appendages. The head and thorax, fused to form what is known as the *Cephalothorax* is enclosed in a calcareous shell, the carapace. The abdomen is visibly segmented, each segment being dorsally protected by a calcareous surface, like that of *Cephalothorax*. The appendages, the feelers, legs, swimmerets etc are modified to suit the various functions to which they are adapted. Respiration is carried on by the branchia or gills which are enclosed in the carapace. Their sexes are separate. Prawns carrying fertilized eggs are sometimes referred as to "Berried" prawns. Between the egg and adult stages, there is frequently a striking, varied and interesting metamorphosis.

All prawns have hard shells on the outside, which affords protection to the animal and supports the internal organs. The shell also provides points of attachment for the muscles with which the animal moves. It plays, therefore, the part of a skeleton, but unlike vertebrates, the exoskeleton of the prawn is outside the body. As it does not increase in size after it has been formed and is inflexible, the animal has to cast its shell periodically as it grows. During moulting, the shell splits and is cast away leaving the animal with only a thin flexible membrane. This, however, rapidly grows by the absorption of water and gradually hardens by the deposition of calcium salts. The process goes on over and over again. Some of these animals have the power of voluntarily throwing off their limbs and regenerate them later.

All species of prawn are edible. They have the property of storing great quantities of glycogen or animal starch and of fat, which renders them nutritious. They are rapid growing and prolific animals, capable of extensive fishing.

Prawns are generall of a very timid and retiring disposition, hiding themselves in holes and crevices of the surface on which they live. At the slightest approach of danger, they hurry away to their burrows or lie quiescent on the substratum, where in the midst of weeds etc they gradually escape unnoticed. Prawn is produced in the tropics for consumption across the world. Prawn is not a staple protein, but a luxurious item and will not be bought when the family is on a budget.

The marine prawn fishery, which supports the export industry is generally confined to shallow coastal areas within 40 meter depth and comprised of the species like, *Penaeus indicus, P. merguiensis, P. monodon, P. semisulcatus, Metapenaeus dobsoni, M. affinis, M. monoceros, M. brevicornis* and *Parapenaeopsis stylifera*. Though different species show variations in the breeding season, most of them spawn throughout the year, having two peak periods, November to December and February to April, the former being most productive.

The body of a crustacean is segmented and covered by a chitinous integument, the exoskeleton. Typically each segment has ring of exoskeleton covering, the adjacent rings being connected by thinner cuticle, the arthroidal membrane, making feasible the movement. The dorsal region of the ring is called tergite, and the ventral, the sternite. A pair of appendages are found on the lateral sides of the sternite portion, when tergite overhang freely over the sternites, they are called pleurons.

The outer integument of a crustacean is made up of several layers. On the posterior part of the cephalic region, a dorsal shield or carapace arises as an integumental fold, which may take the form of a bivalve shell, a shield or mantle in the lower crustaceans. In the prawn family, the carapace fuses with some or all the tergites of other thoracic segments, the posterior extremity projecting freely at the sides to form branchial chambers. The anterior region may be produced into a toothed prolongation, the rostrum.

The appendages or limbs of prawns are of three types, the uniramous, biramous and multiramous. A limb has a basal portion, which is attached to the body, the protopodite consisting of two segments, the proximal coax and the distal basis. A precoxa also may be present proximal to the coax. In a uniramous limb only one ramus develops on the protopodite, as in the antennules of nauplius larvae and also of several adult crustaceans. If two rami develop on the protopodite the limb is biramous, the inner one is the endopodite and the outer one the exopodite. The endopodite is usually made up of five segments or podomeres. From the proximal to the distal extremity, the podomeres are the ischium, merus, carpus, propodus and dactylus respectively. Most of the appendages other than antennae in the higher crustaceans are biramous, but due to suppression of one ramus, some limbs may appear as uniramous.

On the protopodite leaf-like processes may develop on the outer margin, called epipodites or exites, and on the inner margin, the endites. A foliaceous limb may have a protopodite or axis attached to the body, which in some cases is made up of a few segments or podomeres. On the inner margin of the axis, there are generally six endites, the proximal one directed inwards in the axial direction of the body. On the outer margin, there are two exites, the proximal one being called bract and the distal one, flabellum. The structures are almost flattened and leaf-like. Due to the action of the appendages food particles are directed forward to mouth along the mid-ventral groove of body, the proximal endites helping in masticating the food and act as gnathobases. All or some of the epipodites may be respiratory.

The important characters considered in the identification of prawns are the carapace and its spines, the rostrum and its dorsal and ventral teeth, the ridges or carinae, the groves or sulci, abdominal carination, telson, appendages and their segments, petasma and appendix masculine in the male and the thelycum in the female.

In order of commercial importance, a list of different species of prawns found in Indian Seas are given below.

Genus – *Penaeus*

Penaeus indicus

Available in all coastal waters, estuaries and backwaters. Maximum length 20-23 cm, gastro-orbital carina of *Penaeus indicus* occupy the posterior two third distance between hepatic spine and orbital angle. Rostral crest may be elevated, but triangular in profile. The species is distributed in India, Sri Lanka to the west through Gulf of Aden and Madagascar and east coast of Africa and to the east through Malaysia and Indonesia to Philippines, New Guinea and northern Australia. In India, the species occur in all coastal waters of east and west. Young migrate to estuaries. In estuaries there is a fishery of young ones of 12-14 cm size. The adults form one of the most important commercial species.

Penaeus monodon

Tiger prawn. Occur more on east coast, especially in Orissa and Bengal. On the west coast, more on the northern sector. Largest marine prawn. Attain a maximum size of 30-32 cm. *P. monodon* is distributed from South Africa to Southern Japan. Juveniles enter estuaries. It is an important commercial species and form substantial fisheries in Bengal and Orissa. In west coast the species forms a fishery, though, not as dominant as east, but large ones are caught in northern areas of west coast. Young ones are caught from estuaries. *Hepatic carina* of the species is horizontally straight and its fifth pereiopod is without exopodite.

Penaeus merguiensis

Occur in the middle regions of east and west coasts. Sparingly in other regions. Maximum length observed 24 cm. *P. merguiensis*, characterized by triangular rostral crest, adrostral carina not reaching as far as epigastric tooth and dactylus of third maxilliped of adult male is half the size of propodus, is distributed from West Pakistan eastwards to New Caledonia, penetrating southwards to Australia. In India, the species is available mostly in the middle of east and west coasts. In other areas they are found in small numbers. Juveniles enter into the estuaries. They form small fishery in the middle region of east and west coasts. Juveniles are fished from estuaries.

Penaeus semisulcatus

More common on east coast. Attains a maximum length of 23-25 cm. *P. semisulcatus* is widely distributed in Indo-Pacific, from Durban Bay through Red Sea, India, Malaysia, Indonesia to northern and north-eastern Australia through New Guinea, Philippine Islands to Southern Japan, mostly in tropical habitats. In India, the species is more common on the east coast, and form a small fishery. *Hepatic carina* in the species is inclined at an angle of 20°anteroventrally. Fifth pereiopods are provided with small exopodite.

Penaeus penicillatus

Occur in coastal waters of north Bombay and Orissa coasts. Attain a maximum length of 21 cm. dactylus of third maxilliped of adult *P. penicillatus* is much longer than propodus. *Adrostrul carina* of the species reaching just beyond epigastric tooth. Its rostral crest is markedly elevated. It is distributed from Karachi coast in West Pakistan through Malaysian waters to Taiwan.

Penaeus japonicus

Sparingly occurs in Tamil Nadu (Pulicat lake) and Bombay coasts. Attains a maximum length of 27 cm. *P. japonicus* is widely distributed throughout the greater part of Indo-Pacific region, from Africa to Fiji. In India, it occur along Tamil Nadu coast and on the west coast especially in Bombay. It form

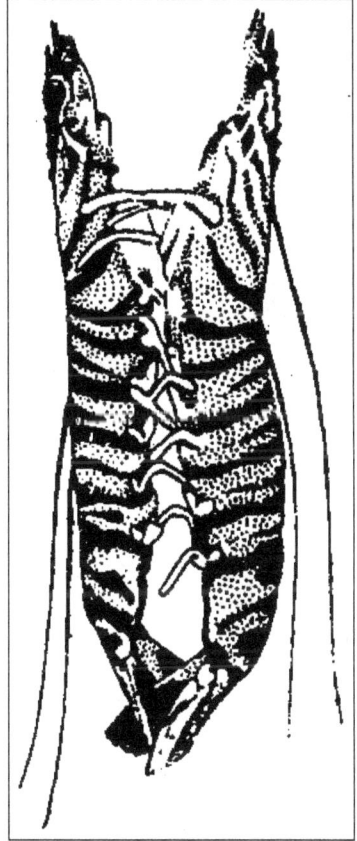

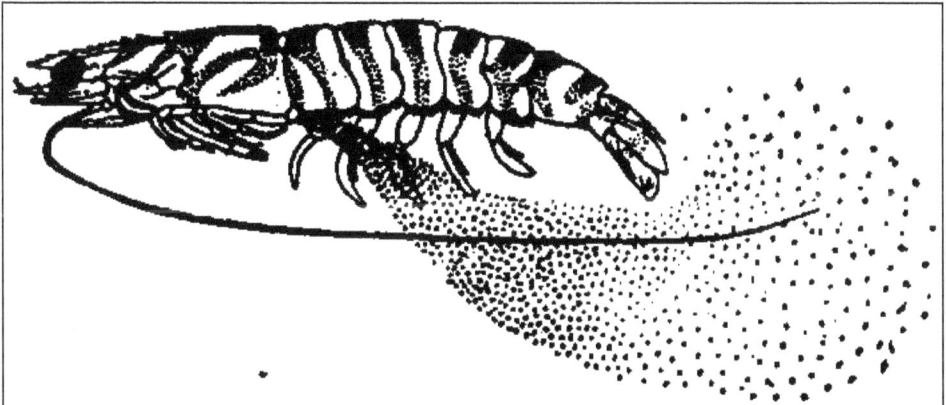

Figure 23.1: Copulation and Spawning, *Penaeus japonicus*

Figure 23.2: Deep Sea Shrimp, *Gnatheuphausia*

a small fishery in Tamil Nadu area, especially in Pulicat lake in post monsoon months. In Bombay area they are found in small numbers. *Adrostral sulcus* of the species is narrower than post-rostral carina. Anterior plate of thelycum is rounded at the apex.

Penaeus canaliculatus

Very sparingly occur on south west coast. Grows to a maximum length of 15 cm. Telson of *P. canaliculatus* is unarmed and rostrum with one ventral tooth. The species is found in South Africa, Mauritius and Red Sea through India to East Indies and Fiji Islands. In India they are available in west coast, where they are caught in small numbers.

Penaeus latisulcatus

Only one or two specimens obtained from south west coast. In *P. latisulcatus*, adrostral sulcus is as wide as post-rostral carina. Anterior plate of thelycum is bifid at the apex. Scattered distribution of the species have been recorded from Red Sea through Malaysia and Malaccas to Korea and Japan. It attains a maximum size of 16 cm.

Genus – *Metapenaeus*

Metapenaeus dobsoni

Largest single species landed in India. Found in all coastal waters. More common on south west coast. Attains a maximum length of 12.5 cm. The posterior part of rostrum of *M. dobsoni* is without distinctly elevated crest. Basal spine on male third pereiopod is long and barbed. Apical petasmal filaments are not readily visible. Anterior thelycal plate is tongue-like. The species is distributed in India through

Malaysia and Indonesia to Philippine Islands. They are found in estuaries also. In India, the species is more along south west coast upto 40 meter depth. The species form one of the most important commercial fisheries, a major fishery along south west coast. Juveniles of 70 to 75 mm size are fished from estuaries and river mouths.

Metapenaeus affinis

Occur all along Indian coast. Attain a maximum length of 18 cm. *M. affinis*, characterized by distinct branchicardiac carina, extending from posterior margin of carapace almost to hepatic spine, has anterior thelycal plate longitudinally grooved, wider posteriorly than anteriorly. Distomedian petasmal projections are cresent shaped. The species is distributed in Indian Seas through Malaysia and parts of Indonesia to Hong Kong and Japan. In India, it is found along the coasts and form a very important fishery along the coasts, majority are caught from inshore waters upto 45 to 50 meter depth.

Juveniles enter the estuaries and are fished from there.

Metapenacus monoceros

Available along entire coastline, estuaries and backwaters. Largest size of 18 cm was recorded. Lateral thelycal plates of *M. monoceros* are with salient and parallel ear-shaped lateral ridges. Distomedian petasmal projections are hood-like. The species is distributed to South Africa through Mediterranean and Indian Seas to Malaysia. Eastern limit of the species is Malacca Strait. In India, the species is available along the entire coastline. Adults are found in the deep sea at 50 to 60 meter depth and juveniles in the estuaries.

Metapenaeus brevicornis

Available only in the northern region of both east and west coasts including estuaries. Attain a maximum length of 12.5 cm. *M. brevicornis* form a good fishery in northern regions of south and east coasts. Juveniles enter into the estuaries from where they are caught. The species is distributed in West Pakistan through India, Malaysia, Thailand and Indonesia to about East Borneo. In India, the species is available in the northern region of both coasts. Juveniles are found in estuaries. Posterior part of rostrum is with distinctly elevated crest. Basal spine is simple found on third pereiopod of the male. Apical petasmal filaments are slender and slightly converging. Thelycum is with a large anterior and small lateral plates.

Metapenaeus ensis

Reported only from east coast off Waltair. Maximum size attained 17 cm. *M. ensis*, caught on the east coast of India, where the species form a small fishery along with *M. monoceros*. They are distributed in Indian waters, Sri Lanka through Malacca Strait and Indonesia to New Guinea, south east China to Japan, South western Australia, Queensland and New South Wales. In India, the species is available on east coast of Waltair and a little south. Distomedian petasmal projections of the species are directed anteriorly. Lateral thelycal plates are with raised lateral ridges, each with a posterior inwardly curved triangular plate.

Metapenaeus lysianassa

The species occur mostly on east coasr off river Hoogly, Orissa coast and Gulf of Mannar. Rarely found on south west coast. Maximum length attained is 9 cm. The species is found in India, Sri Lanka to North Borneo. In India, it is only available in stray catches. Rostrum of the species is wide and short, not reaching to distal end of basal antennular segment. Thelycum is with ovoid anterior and lateral plates of sub-equal size. Conjoined pads are usually set askew. Apical filaments of petasma are vestigial represented by a pair of rounded processes.

Metapenaeus burkenroadi

Recently reported in Indian inshore waters and estuary in Cochin, Attains a maximum length of 10 cm. The species is distributed north of equator from Japan through Hong Kong ans Malaysia to south India and Sri Lanka. Though the species attain a maximum size of 10 cm, only a few have been recorded recently. Branchiocardiac carina of the species is feeble or ill-defined. Anterior end not exceeding posterior one-third of carapace. Distal margin of anterior thelycal plate is convex to indistinctly triangular. Petasma is found with laminose and strongly diverging distomedian projections.

Metapenaeus stebbingi

Recently reported in India from Gulf of Kutch. Largest size of 12 cm was recorded. The species is available in Eastern Indian Ocean region, western South Africa through Suez, Red Sea to West Pakistan and north west coast of India. The species is found in stray catches only. Branchiocardiac sulcus of the species is almost completely absent. Distomedian petasmal projections, are anteriorly filiform each with a serrated ventral margin.

Metapenaeus kutchensis

The species form a good fishery in Gulf of Kutch area. Attain a maximum length of 13-14 cm. The species, recognized by posterior extension of the anterior median thelycal plate not bound laterally by oval plate on either side and distomedian petasmal projections not overlying lateral processes.

Metapenaeus alcocki

Recently reported from Gulf of Kutch area. Grows to a maximum length of 9.7 cm. Posterior extension of the anterior median thelycal plate of the species is bound laterally by an oval plate on each side. Disto median petasmal projections are overlying lateral projections and distally trilobed. They are caught in small numbers.

Genus – Parapenaeopsis

Parapenaeopsis stylifera

Entirely marine species. Available all along the coastline, especially west and south east coasts. Attain a maximum length of 14 cm.

Telson of *P. stylifera* is with a pair of fixed sub-apical spines. At least distal half free portion of rostrum is unarmed. The species is available in India and Sri Lanka through Malaysia to Indonesia and Borneo. There is no estuarine phase in its life history. It is considered a very important commercial species along the entire coast of India with maximum abundance along the Kerala coast.

Parapenaeopsis hardwickii

Forms fairly a good fishery in north west coast, Bombay waters and off Godavari estuary. Maximum length recorded is 12 cm.

Antennular flagella of *P. hardwickii* are 0.7 length of carapace or longer. Thelycum is without a median tuft of setae on posterior plate. The species is available in Indian waters through Malaysia to South China. It form a fairly good fishery in Godavary and more in Bombay coast.

Parapenaeopsis sculptilis

The species occur in northern regions of both east and west coasts only. The largest size is 16.5 cm. The species form a small fishery in northern east and west coasts of India. The species is distributed in India through Malaysia, Indonesia to Hong Kong and Australia and New Guinea. Antennular flagella of the species are 0.5-0.6 length of carapace. Its thelycum is with median tuft of setae on posterior plate.

Parapenaeopsis uncta

The species was recorded from Orissa and south west coasts. Attain a maximum length of 10 cm. *P. uncta* occurs in India, Sri Lanka and Malayan waters. It is found as stray catches from Orissa and south west coast. The second pereiopods of the species is without basal spines.

Parapenaeopsis nana

Recorded only from east coast, off Orissa and Tamil Nadu. Maximum length observed 5.5 cm. Post-rostral carina of the species reaching three fourth carapace. Petasma with a pair of distolateral projections directed laterally. Cup-like distal projections are absent. The species is found in Indian and Sri Lankan waters. In India, it is recorded only from east coast off Orissa and Chennai coasts, where it is obtained in stray catches only.

Parapenaeopsis cornuta maxillipedo

Available in Bombay, Kerala and Tamil Nadu coasts. Not significant. Attain a maximum length of 12.5 cm. In the species petasma is with a pair of long slender caliper-like distolateral projections directed forwards. Thelycum is with median tuft of long setae behind posterior edge of last thoracic sternite. Third pereiopod of female is provided with basal spine. The species is distributed over equatorial region, spread over from west coast of India and Sri Lanka through Malaysia to Philippines and New Guinea. It occurs in small numbers.

Parapenaeopsis acclivirostris

Occur both in east and west (Bombay) coasts. Maximum size available is 5 cm. The species is distributed in South Africa, Persian Gulf and Indian Seas. In India they are available in east and west coasts, and found in small numbers with other commercial species. Anetrior plate of thelycum is with a more or less straight transverse posterior edge. No accessory ridges are found on anterior edge of posterior plate. Rostrum is inclined upwards at angle to carapace for whole of its length.

Parapenaeopsis tinella

Onlly recently collected from east coast, Gulf of Mannar and Palk Bay. Maximum size is 5 cm. Anterior plate of thelycum of *P. tinella* is with V-shaped posterior edge; and two accessory ridges on anterior edge of posterior plate. Rostrum of the species is with proximal one-third rising from carapace. The remainder is more or less horizontal. The species occur in east coast of India and Myanmar through Malaysia to Northern China, Southern Japan and Northern Australia. In India the species is recorded from Palk Bay and Gulf of Mannar only in small numbers.

Genus – *Macrobrachium*

Macrobrachium rosenbergii

Available in most of the lakes and estuaries along the coastline of India. They grow to a maximum length of 30-32 cm. Carpus of the second pereiopod in adult male of *M. rosenbergii* is slightly longer than half as long as chela besides the absence of branchiostegal spine. Hepatic spine is present, dactylus of last three legs are simple. Fingers of that leg (2nd pereiopod) are of the same length as the palm. The species is widely distributed in the Indo-Pacific region, The western limit being the Indus delta area and extending upto Indo-China in the Asian mainland. They are found in fresh and brackish water. The species used to form a very good fishery in the monsoon and post-monsoon months, but the fishery is now diminished due to indiscriminate fishing.

Macrobrachium malcomsonii

Most common in peninsular rivers draining into Bay of Bengal. Also available in river Indus. They attain a maximum length of 20-23 cm. *M. malcomsonii* have distinctly elevated basal crest and provided with five to nine teeth. In younger specimen, the second leg has swollen palm and the fingers longer than the palm. Carpus of second leg in adult male is shorter than chela. The species is available in India and Mayanmar. The species migrates into brackish waters during breeding season and forms a fairly good fishery in north east coast in monsoon months. The males and females of the species attains a maximum size of 23 and 20 cm respectively.

Macrobrachium villosimanus

Available in Kolkata and Chittagong region. Maximum size attained is 15 cm. The species has a very limited distribution and occur in small numbers in India and Bangladesh. They also occur in Rangoon of Mayanmar. Carpus of second pereiopod in adult male of the species is as long as, slightly longer or slightly shorter than chela. Fingers of that leg is a little less than half as long as the palm.

Macrobrachium lamarrei

Found in northern region of east coast, Chilka and Bengal. The species form a small fishery in India and Pakistan in fresh and brackish waters. They grow to a maximum length of 13 cm and mostly found in northeast coast of India, Chilka lake and West Bengal. The basal crest in the species is not much elevated and provided with five to nine teeth. Palm of the second leg is not swollen and fingers are shorter than palm.

Macrobrachium rude

Found in south west region of west coast and throughout the east coast, especially in deltaic Bengal, Orissa and Andhra coasts. They grow to a maximum size of 12-13 cm. The species form a good seasonal fishery in Bengal and Orissa. It is available in East Africa, Madagascar, India and Sri Lanka. Larger chela of second leg of adult male of the species are with tubercles at both sides of the cutting edges. Carpus of second leg in adult male is shorter than chela. All joints of second leg in adult male is pubescent.

Macrobrachium idella

More common in east coast. Attain a maximum length of 10-11 cm. Rostrum of the species is with nine to eleven teeth dorsally, three of which are generally placed behind the orbit. The carpus of second leg in adult male is larger than chela. The species form a small fishery in East Africa and Madagascar through Indian coasts to Java, Sumatra and Malayan Archipelago. In India they are present in estuaries and rivers in south west region and more common in the east coast. From sea stray records are available.

Macrobrachium equidens

Found only in Kerala in small numbers. Maximum length is 10 cm. The species is distributed from Africa to south west New Guinea. In India, it is only found in small numbers in fresh and brackish waters of Kerala. Their rostrum is curved upward and lower margin with five to seven teeth, Their fingers are covered with stiff or velvety hairs on the entire surface or in the proximal part.

Macrobrachium mirabile

Available mostly in Gangetic delta area. Grows to a maximum size of 6.5 cm. *M. mirabile* have conspicuously, about one and one third, longer fifth legs than the fourth leg. Rostrum is short and high with many dorsal teeth. Second leg of adult male is smooth. The species is distributed in India through Myanmar to Malayan Archipelago and Borneo. In India, this estuarine species are mostly found in Gangetic Delta, where they form a small fishery.

Macrobrachium javanicum

Occurrence restricted to deltaic Bengal. Attain a maximum length of 10 cm. The species form a small fishery in India through Myanmar to Malayan Archipelago and Borneo. In India they are available in freshwaters and estuaries in deltaic Mengal.

The fingers of second legs of adult male are with one or two fairly large teeth. Smaller teeth are present between the first tooth and the base of the fingers. The anterior tooth of the dactylus is placed in or slightly before the middle of the finger.

Macrobrachium sabriculum

Occurs in deltaic Bengal, Chilka lake, Kerala and Tamil Nadu. Maximum size is 10 cm. The species inhabits around Indian Ocean, extending from Africa through south west and eastern coasts of India to Malaya Archipelago. It forms a small fishery in deltaic Bengal, Chilka lake, south west coast of Kerala and south east coast of Tamil Nadu. The species is recognized by fingers of second leg of adult male with more than four teeth placed at regular intervals, sometimes restricted to the proximal part. Teeth are generally of equal size, but one of the proximal teeth may sometimes be larger. Fingers are with a velvety pubescence in their basal portion. Dorsal teeth of the rostrum beginning in the distal third of the carapace.

Besides 24 other species of this genus have been reported to occur in India, but they are only of minor importance.

Genus – *Palaemon*

Palaemon tenuipes

Mostly found in the northern area both on east and west coasts andform a good fishery. They attain a maximum size of 8 cm. Dactylus of last three pereiopods of *P. tenuipes* are very long and slender. The fourth and fifth pairs are excessively long, flagelliform with dactylus much longer than carapace. Pleopods are very long, first pair much longer than carapace. Carpus of second pereiopod much more than half as long as palm. Basal crest of rostrum with almost seven teeth. The species occur in India through Malaysia to New Zealand coastal waters upto 20 meters depth, and in estuaries. In India, the species are found in northern regions of both coasts, and form one of the most important fisheries in Bombay and Gangetic Delta.

Palaemon styliferus

Occur in the northern area both on east and west coasts. Maximum size attained 9 cm. Dactylus of last three pereiopods of the species are not abnormal in length, that of third scarcely one and half length of propodus and that of fifth at most one third length of propodus. Pleopods are normal in length. One or more subapical dorsal teeth on rostrum. Last four abdominal somites bluntly carinate dorsally. The species is distributed in West Pakistan and India to Malayan Archipelago in shallow coastal waters, brackish and sometimes in fresh water areas. It form one of the most important fisheries in Gangetic Delta.

Genus – *Acetes*

Acetes indicus

The species is most common in Bombay waters, where they form an important fishery. They are also available on entire east coast, both in sea and brackish waters. Attains a maximum size of 4 cm. *Acetes indicus* is distributed in India through Mergui

Archipelago and Gulf of Siam to Malaya and East Indies. In India, it is common in northern area in west and in the east throughout the coastal and estuarine regions. It form good fishery in north west coast and east coast. In Bombay the species contribute 20 per cent of prawn fishery. The trochanter (basis) of the third pereiopod is with tooth on inner free margin. Petasma is without membraneous coupling folds.

Acetes erythraeus

Fairly good quantities are available in Bengal, Orissa, Tamil Nadu and Trivandrum coasts. They grow to a size of 2 to 3 cm. *A. erythraeus* trochanter (basis) of third pereiopod is without tooth on inner free margin. Petasma with a pair of folded coupling membranes armed with hooks. It is available in Red Sea through Bay of Bengal and Gulf of Siam to Malaya Archipelago. In India the species is found in east and south west coasts. It form a fairly good fishery in Bengal, Orissa and Tamil Nadu. In Trivandrum coast, the species enter in large quantities from December to April.

Acetes sibogae

Only available off Quilon. They grow to 3.5 cm in length. The species is distributed in India, Bay of Bima and Java Sea, Malaya and New South Wales. In India, stray catches are available off Quilon in south west coast. Apex of telson is triangular in the species. External antennular flagellum in male is with single clasping spine.

Acetes serrulatus

Available in Travancore coast. Attain 2 cm in length. Apex of telson is truncated in *Acetes serrulatus* with a tooth in each corner. Segment preceding the one bearing the clasping spines is with angular process pointing backwards. The species is available in Indo-China Sea, Borneo and Singapore. In India, only the variety, *A. serrulatus johni* was recorded from south west coast. It grows to a maximum size of 2.3 cm and form a fairly good fishery in coastal waters of Travancore (South Kerala) from December to April.

Acetes japonicus

Available in west and south west coasts. Attain a length of 2.6 cm. The species occur in India, Gulf of Siam, Java, Korea and Japan. In India, the species is available in west, south east coasts and lower parts of Bay of Bengal. It occurs in small numbers off Trivandrum coast, large numbers are found in July. Ciliated and non-ciliated portions of external border of exopod of uropod is not separated by a tooth in the species. Distal portions of pars externa are with tubercles.

Acetes cochinensis

Reported recently from Cochin waters. Attain 2 cm in length. They grow to a maximum length of 3 cm and occurs in plankton of sea and back waters in May and June. In the species, ciliated and non-ciliated portions of external border of exopod of uropod is separated by a tooth. Distal portion of pars externa are with tubercles.

Family – *Pandalidae*

Prawns of this family (pink in colour) have been found recently in deep water trawling operations off southern Kerala coast of 150 to 200 fathoms depth.

Parapandalus spinipes

Found only in Arabian Sea off south west coast. They attain a size of 13 cm. Pereiopods of the species are without epipods. Upper margin of rostrum is finely and evenly serrated along its whole length. Carpus of fifth leg is shorter than propodus. Sixth abdominal somite bear minimum thickness, when looked at dorsally two fifth length of this somite. Telson is almost one and half as long as fifth somite. The species occur at Zanzibar through Red Sea, Gulf of Aden and Arabian Sea to Malaya Archipelago and New Guinea. In India, the species is recorded in the Arabian Sea off the south west coast off Tuticorin. The species is considered as potential commercial importance. Obtained in fairly good quantities in deep water trawling from 150 to 200 fathoms depth off Kerala coast and at 250 to 400 fathoms off Tuticorin coast.

Plesionika martia

Available in Arabian Sea and Bay of Bengal. Largest size attained is 12.5 cm. The species is widely distributed in Pacific and Atlantic regions, Eastern Atlantic and Mediterranean through Indian Seas to Japan, Australia and Hawaiian Islands. Its posterior border of third abdominal tergum though convex is not acutely produced. Its rostrum is 45 to 67 per cent of the length of the body from orbit to tip of the telson. The species is considered to have potential commercial importance. Caught in good numbers from deep water trawling off Kerala coast at 150 to 200 fathoms depth.

Plesionika ensis

Found only in Arabian Sea off south coast. Attain a length of 12.5 cm. Atleast the first two pereiopods of *Plesionika ensis* are with epipods. Posterior lobe of scaphognathite is truncate. Stylocerite pointed anteriorly. Posterior border of third abdominal tergum acutely produced into a sharp tooth that overlaps the next tergum. The species has a world-wide distribution, Pacific, Atlantic and Indian Oceans. The species occurs in Barbados, Martinque and Granada of West Indies, Barra Granada in Brazilian coast, Rio Muni in Gulf of Guinea, Hawaiian islands and Australian Sea. In India, from Arabian Sea the species is available in deep water areas off south west coast. Specimens are obtained from deep sea trawling off Kerala coast.

Genus – *Heterocarpus*

Heterocarpus gibbosus

The species isavailable in Arabiam Sea, Bay of Bengal and Andaman Sea. Grows to a length of 14 cm. the species is characterized by carapace with longitudinal carinae on the lateral surfaces. Its integument is very firm and pereoipods of second pair is very unequal. Abdominal terga, though carinated, never produced posteriorly into overhanging spines. Post-ocular carinae is present. Proper upper margin of the rostrum is armed with two or three teeth. The species is distributed off Tablas Island,

Indian Seas, Bali Sea and Kei Islands. They are available in small numbers from deep water trawling off Kerala coast.

Heterocarpus woodmasoni

Available in Andaman Sea and Arabian Sea off Kerala coast. Attain a length of 13 cm. The species occurring in East African coast through Indian Seas to Kei Islands. The species is considered as potential commercial importance. In deep water trawling, this species is predominant in the catch. The third abdominal tergum of the species is armed with an acute spine arising from the anterior half. Post-ocular carina is completely wanting.

Genus – *Hippolysmata*

Hippolysmata ensirostris

The species forms a fairly good fishery in Bombay coast and Godavary estuary. Present in most regions in small numbers. Attain a length of 8 cm. The species is characterized by longer rostrum than carapace, with an elevated dentate basal crest with acute and unarmed apex and concave lateral margin of telson, is distributed in India, Sri Lanka, Myanmar and Sumatra. In India, the species is along the coasts and form a fairly good fishery in Bombay and Godavary estuary. It also form small fishery in other areas.

Hippolysmata vittata

Present in small numbers in Indian Seas. Attain a length of 4 cm. The species occur in South Africa through Red Sea, Persian Gulf and India to East Indies and Japan. The species is recognized by shorter rostrum than carapace and without elevated basal crest. Lateral margin of its telson is convex. The apex is blunt with a pair of spines. The species is found only in small numbers.

Genus – *Trachypenaeus*

Trachypenaeus pescadorensis

Rcently reported from south east and south west coasts. Attain a length of 9 cm. In the species, epidodites are absent on first and second pereiopods. Distolateral projections of petasma are with sharp tips reaching coxae of fourth pereiopods. Anterolaterally with large wing-like flaps on outer curvature. Besides India, the species is available in eastern Malaya and northern Australia. Stray catches were only obtained from south east and south west coasts of India recently.

Trachypenaeus sedili

Recently reported from Visakhapatnam coast. Attain 6 cm in length.The species was recorded from Malaya and Sri Lankan waters. The species has the plates of thelycum with raised anterior and lateral margins.

Trachypenaeus curvirostris

Available both in east and west coasts in small numbers. Attain 9.5 cm in length. The species is distributed in Eastern Africa through India, Sri Lanka and Malayasian waters to Japan and Australia. They are found at 10 to 30 fathoms in both east and west coasts of India. They do not occur in large numbers to contribute to a fishery. Third pereiopod is with epipodite. The anterior plate of the thelycum may have a raised anterior margin, but laterally the margins are not raised. An excavation present between the anterior plate and the transverse sternal ridge.

Genus – *Atypopenaeus*

Atypopenaeus stenodactylus

Available both in east and west coatsa of Tamil Nadu and Bombay. Attain a length of 5 cm. The species is found in Indian Seas through Malaysia and Hong Kong to Japan. The species form a fishery in Bombay and are caught in bag nets(Dol nets) from 6 to 15 fathoms in large numbers throughout the year. Small numbers are available in east coast. The carapace of the species is without longitudinal sutures. Ischial spine is present on second pereiopod. Hepatic spine is present. Petasma is not constricted distally. The anterior plate of thelycum is rounded posteriorly.

Genus – *Metapenaeopsis*

Metapenaeopsis stridulans

Available in Bombay as well as northern region of east coast. Attain a length of 10 cm. The species is distributed in Indian, Sri Lankan through Malaysian waters to eastern New Guinea. In India, they are available in Bombay and northern area of east coast, at 5 to 30 fathoms depth. They form a fishery in Bombay, caught in "Dol" nets in fairly large numbers, especially in October and November. Dorsal carina of third pleonic somite of the species is sulcate. It's stridulating organ is almost straight. The anterior edge of thelycal plate is entire. Its left petasmal lobe is sharply pointed and triangular.

Metapenaeopsis mogiensis

Occur off Malabar coast and Andaman Islands. Attain a maximum size of 9 cm. The species is characterized by the presence of a pair of tooth like platelets behind thelycal plate and lacking of posterior tubercles, grow to a maximum size of 9 cm and obtained in stray catches only in India to South China Sea and Japan in north east and tropical Queensland and Sri Lanka in south east. Their petasma is asymmetrical and third maxilliped is with basal spine.

Metapenaeopsis andamanensis

Occur in deep waters of south west coast and Andamans. Recently caught at 150-200 fathoms off southern Kerala. Attain a maximum size of 13.5 cm. The species, is distributed in Indian seas through Malaysian waters to Kei Islands and Japan. The species is considered as potential commercial importance. Fairly good numbers are obtained from deep sea trawling off south west coast of India from 150 to 200 fathoms, off Tuticorin at 250-400 m depth. Petasma of the species is asymmetrical.

Third maxilliped is with basal spine. Posterior extension of thelycal plate with indistinct median sulcus and angular postero-lateral corners.

Metapenaeopsis philippii

Recently caught from deep waters of south west coast. Attain a maximum length of 13 cm. The species have distinct median sulcus in the posterior extension of thelycal plate and evenly rounded posterolateral corners. Their petasma is asymmetrical and third maxilliped is with basal spine. They are distributed in Zanzibar through Indian Seas to Philippine islands. In India they are caught from south west coast. The species is considered as potential commercial importance. They are caught in deep sea trawling off Kerala coasts in good numbers.

Genus – Parapenaeus

Parapenaeus longipes

Found in sea off Mangalore, Cochin, Ganjam, Visakhapatnam and mouth of Hoogly river. Grows to a maximum size of 8 cm. Branchiostegal spine in the species is absent. Fifth pereoipods of the species exceeds antennal scale by dactylus. The species is distributed in East Africa through India and Indonesia to New Guinea and Japan. They are caught in small numbers.

Parapenaeus fissures

Only available in east coast off Ganjam and Andamans. Attain 12 cm in size.

Branchiostegal spine of the species is on anterior margin of carapace. Sixth abdominal somite is less than twice the length of fifth. Process "a" of petasma bifurcate and directed laterally. Thelycum is with anterior, intermediate and posterior plates. The species is distributed in East Africa through Indian Seas, Malaysia and Indonesia to Philippines, South China Sea and Japan. It forms stray catches only.

Parapenaeus investigatoris

Occur in Gulf of Mannar, Pulicut lake, Andaman Sea and off Cochin. Attain a maximum size of 8 cm. The species is distributed in East Africa through India and Indonesia to Japan. They are caught off Cochin in deep sea trawling and occur in small numbers. The branchio-stegal spine of the species is situated a little behind anterior margin of carapace. Sixth abdominal somite is more than twice the length of the fifth. Rostrum is reaching distal end of first segment of antenular peduncle.

Genus – Penaeopsis

Penaeopsis rectacuta

Occur in seas off Tamilnadu, Andamans and deeper waters off Kerala coast. Attain a maximum length of 13 cm the species occur in Gulf of Aden through India and Malaysia to Philippines, South China Sea and Japan. The species is considered as potential commercial importance. It is obtained in appreciable numbers off Kerala coast in deep sea trawling at 100 to 200 fathoms depth. Carapace of the species is without longitudinal sutures. Branchiostegal spine is present. Telson with three pairs of movable marginal spines in addition to the fixed pair.

Genus – *Solenocera*

Solenocera indica

Available both in east and west coasts. Attain a length of 11.4 cm. The species is distributed from Bangladesh along Indian coasts and Sri Lanka to Malaysia, Borneo and Hong Kong. In India, the species is available along east and west coasts at 40 meter depth or less. It attains a maximum size of 13 cm and forms a small fishery in Bombay. Telson of the species is simple and devoid of any spine on lateral margin.

Solenocera pectinata

Occur off south west coast and are too small. The species has externo-distal margin of the exopod of the uropod without spine. Post-rostral carina is not extending beyond cervical grove. The species is distributed in Arafura Sea, Flores Sea, Ceram Sea off Owase, Japan, Arabian Sea Tennesseram coast, Mayanmar and South China Sea. In India, the species is available off south west coast at 25 to 60 fathoms. They grow to a maximum length of 75 mm and only found in small numbers. Therir size is small and commercially is not attractive.

Solenocera hextii

Available along the entire west coast and Bay of Bengal. In *Solenocrea hextii* spine in cervical grove ventral to posterior most spine of the rostral series is present. "L" shaped grove on either branchiostegal region is also present. The species inhabit in northern Indian Ocean regions. In India, it is available along entire west coast and Bay of Bengal at 65 to 276 fathoms depth. It grows to a maximum size of 14 cm. During deep water explorations at 150 to 200 fathoms depth, the species was caught in varying numbers, but never in large quantities. The large sized specimens is attractive to commerce.

Solenocera choprai

Occur only in Arabian Sea. Stray catches of the species occur from Arabian Sea from a depth of 56 to58 fathoms. The maximum size attained by the species is 13 cm. Spine on the cervical groove ventral to the posterior most spine of the rostral series are absent. "L" shaped groove on either branchio-stegal region is absent.

Solenocera koelbeli

Available only in south west coast.

Solenocera melantho

Recently reported from Godavari estuarine system.

Genus- *Hymenopenaeus*

Hymenopenaeus acqualis

Available off south west coast and Andamans. Antennular flagella of the species is cylindrical or sub-cylindrical. Its rostrum is straight, inclined upwards at an angle of 20°with 7-8+2 spines dorsally. The species is available in east coast of Africa along Indian Seas to Japan. In India, it is available on the south west coast and Andaman Sea from beyond 150 fathoms. The species grow to a maximum size of 9 cm and only stray catches are available.

Genus – *Sicyonia*

Sicyonia lancifer

Post-rostral carina of the species is armed with five teeth. Abdominal pleura of first and second segments are unispinose and third, fourth and fifth are with three spines. The species is available in Japan, Penang, Gulf of Mannar, Sri Lanka and India. In India, they occur on the south west coast at 12 to 17 fathom depth. It grows to a maximum length of 8 cm and only a very small numbers are caught from Arabian Sea.

Genus – *Aristeus*

Aristeus semidentatus

Recently caught at 150-200 fathoms depth off Cochin and Alleppey. Largest prawns were caught in these operations. Considered to be potentially important in commercial fisheries, this species appear in Kermadec Islands, Kei Islands and Arabian Sea. In India, the species is recently discovered off Cochin and Alleppy, south-west coast. They grow to a maximum size of 15 cm and obtained in fairly large numbers in deep water trawling at 150 to 200 fathoms off Kerala coast. They can be identified with the presence of pleurobranchiae on ten to thirteen segments on distinct filaments provided with pinnules.

Aristeus alcocki

Found in Bay of Bengal, Arabian Sea, Laccadives and Cape Comorin. The species can be identified with three-toothed rostrum dorsally. Hepatic spine is absent. Pleurobranchiae on segments ten to thirteen is reduced to mere papillae. Occur in Gulf of Aden, Bay of Bengal, Arabian Sea near Laccadives and Cape Comorin. It grows to 15 cm of maximum length. A few numbers were caught from exploratory trawling off south-west coast of India.

Aristeus virilis

Found in Andaman Sea. The species is characterized by the presence of pubescent integument, occur only in stray catches from Andaman Sea through East Indian Archipelago to Japan. In India, the species is available from Andaman Sea at 188 to 405 fathoms. The species attain a maximum length of 15 cm.

Aristaemorpha woodmasoni

Available in Bay of Bengal, Andaman Sea and Arabian Sea. The species occurs only in small numbers in Bay of Bengal, Andaman Sea and Arabian Sea. The species has also been recorded from south east Australia. In India, the species is obtained from deep waters of 180 to 271 fathoms. Attained a maximum length of 12 cm. The rostrum of the species bear many teeth on upper border. Hepatic spine is present. The length of pterygostomian region is more than two and half times its greatest breadth.

Besides, the deep sea prawn fishery of Kerala coast also include, "Gondwana shrimp" and a deep red coloured prawn.

Penaeopsis jerryi

Known as "Gondwana shrimp" is a widely distributed small penaeid represented the landings from the depth of 251-300 meter. Peak period of abundance was during January to March. Size range of 5.8 cm to 10.8 cm in males and 6.7 cm to 11.4 cm in females contributed the fishery. The modal sizes between 7.6 cm to 8.9 cm dominated in the catches in both sexes.

Acanthephyra sanbuinea

A deep red coloured prawn, often mistaken as *A. alcocki* by the fishermen is landed in minor quantities and represented the fishery in April at 320 meter depth off Cochin. Size range of males was 8.8 to 10.0 cm and 10.2 to 11.8 in females. The impregnated females outnumbered the males in the total landings.

Decapod Crustaceans in Andaman Sea

Eighteen species of decapod crustaceans were collected from Andaman Sea. Out of these, six penaeid prawns, namely, *Penaeus canaliculatus, Penaeus merguiensis, Metapenaeus dobsoni, Metapenaeus affinis, Metapenaeus burkenroadi* and *Parapenaeopsis cornuta* are considered to be potential resources of penaeid prawns in Andaman Sea, which has been reported to be as 800 tonnes. Decapod crustaceans new to Andaman and Nicobar Islands consists of;

Penaeus canaliculatus, collected in trawl net at a depth between 2-12 meters of Andaman Sea, bears lateral spines in the telson, which are the distinguishing characters, by which the species can be separated from *P. japonicus*, which also has more or less similar colouration while alive. The rostrum is with only one lower tooth as in *P. japonicus*. The thelycum also is characteristic in shape and structure of the anterior plate.

Penaeus merguiensis is collected and reported from Andaman Sea for the first time in 1977. This species can be easily distinguished from the other two allied species, namely, *P. indicus* and *P. penicillatus* by the presence of the characteristic deltoid crest at the base of the rostrum, which is usually reddish in colour with darker margins in the adult specimens. The gastro-orbital carina is clearly defined in adults.

Metapenaeus dobsoni, collected from Andaman Sea has free filaments of the disto-median projections of petasma on the dorsal aspect are well developed in the adult specimens. The impregnated females have conjoined white pads on the thelycum as in *M. lysianassa*.

Metapenaeus burkenroadi, collected from Andaman Sea were 11.1 cm in length for the male and 13.8 cm for the female. The distance between the rostral teeth are variable in specimens of Andaman Sea. Similarly, the pubescence on the dorsal durface of the carapace is more prominent in females than males. Females have first four and the last two abdominal segments less glabrous than those of the males, although in both the sexes the fifth and sixth segments are more pubescent than the preceeding segments.

The carapace of *Calcinus lateens,* with short rostrum, which is pointed and triangular, more prominent than the antennal angles. Eye stalks more than one-third the anterior border of carapace, longer than antennular and antennal peduncles. Ophthalmic scales simple and pointed anteriorly. Antennular peduncle longer than the antennal peduncles. Antennal acicle with serrated outer margin and provided with few spines in inner aspect, reaching nearly one-fourth the terminal joint of antennal peduncle.

Left cheliped larger than right, upper border of merus faintly serrated. Upper border of carapace with a spinule at the far end and bears a conspicuous tubercle near the middle, at the proximal end. Merus has a pair of spinules at the far end near distal extremity. Setae are spareely arranged near the inner aspect of fingers. Fingers meeting at tip with obscure teeth on inner aspect. Right chelipeds with large terminal tooth, distally on the inner side of the carpus. Superior surface of palm with a crest of five stout, drak tipped spines extending the entire length of palm, each spine with setae at base. Inner border of free finger is also toothed. Tips of finger typically spooned. Lower inner aspect of merus with two sharp and one blunt spine. Setae long and sparse. Second and third legs with distal tooth on carpus. Long thin tufts of setae on the distal and of propodus and dactylus, rest with sparsely arranged small setae.

In preserved specimens, the colour of the bases of dactyli of second and third legs dark purple, rest of the body being yellowish and light red in colour.

The carapace of *Pagurus janitor* collected from Andaman Sea is elongate, rostrum broadly triangular with pointed tip, which is more prominent than the antennal angles, eye stalk short, less than anterior width of carapace. Cornea dialated and reniform. Ophthalmic scales broad and separated at bases, pointed anteriorly and ending in a sharp spine. The narrow anterior portion has a groove on the dorsal surface. Antennular peduncle shorter than antennal peduncle and longer than ophthalmic stalks. Antennal acicle elongated, slightly curved outwards and extending well beyond the base of the last antennal segment.

Chelipeds unequal, right vastly larger and thickly covered with setae, which partly conceal the spines and granules; scattered spines and granules present on the dorsal surface of the carpus and chela and their margins. Cutting edges of fingers close without leaving any space. Lower side of cheliped less setose. Second and third legs with long setae on superior surface, although, some scattered setae are present on the lower margin also.

Colour of preserved specimens cream with light brown patches, eye stalks dark brown, antennal flagella pale yellow.

Other varieties of prawns recorded from Andaman Sea are; *Microbrachium lar* (Fabr), *Heterocarpus gibbosus* (Bate), *Pandalus martius* (A.M.Edw.), *Hoplophorus gracilirostris* (A.M.Edw.), *Nematocarcinus paucidentatus, Cardinilla* sp., *Metapenaeus mogiensis* (Rath), *Metapenaeus coniger andamanensis* (Wood-Masson), *Palaemon concinnus* (Dana).

In Andaman Sea, *Penaeus monodon* brooders are almost exclusively caught from the eastern coast of the Andaman Sea in North Andaman district. The fishing area of tiger prawn brooders are restricted only to about 2-3 km from north to south and about half a km towards the sea (about 8-10 m depth). In the northern edge a local stream empties copious quantity of fresh water into the sea, which constantly dilute the sea water in the area. The whole area is quite shallow and with muddy bottom. The tiger prawn brooders migrate to this area for spawning, due to ideal environmental conditions and bottom topography. The brooders are available from June to November, with peak season August to September. Hard shell female tiger prawns at advanced stages of maturation weighing around 120 gram and males weighing 70-90 grams each is preferred for inducing spawning in hatcheries.

Marine Ornamental Shrimps

Marine ornamental shrimps, belong to the families, Stenopodidae (Claus), 1872; Alpheidae (Rafinesque), 1815; Gnathophyllidae (Dana), 1852; Hippolytidae (Dana), 1852; Hymenoceridae (Ortmann), 1890; Palaemonidae (Rafinesque), 1815; and Rhynchocinetidae (Ortmann), 1890. They have the widest distribution throughout the Indo-Pacific region.

Marine ornamental shrimps available in the aquarium trade are mainly collected from the Indo-Pacific, as well as, Carribean region. Among the ornamental shrimps, *Lysmata* spp. (Hippolytidae) and *Stenopus* spp. (Stenopodidae) are the most commonly collected marine ornamental invertebrates and Indonesia, Philippines and Sri Lanka are the main exporting countries. The others are anemone shrimp (*Oericlimenes brevicarpalis*, Schenkel,1902); harlequin shrimp (*Hymenocera elegans*, Heller, 1861); caridean shrimp (*Pycnocaris chagoae*, Bruce, 1972).

The real number of marine ornamental shrimps collected from the coral reefs, each year, world wide, may be upto ten times higher than the level currently reported. The income generated from one kg of animals for the same species collected for human consumption increases over hundred folds, when sold for aquarium purposes.

There are several advantages of shrimps as exhibit species for aquarium keeping as many of them appear in bright colour pattern. They feed on detritus and faecal matters of other associated animals, which could also contribute to maintaining the tank condition also. The shrimp body is generally transparent or translucent and this adds to the attractive appearance of the tank. The movement of multiple appendages of shrimps adds to the attraction. They display one or more of certain unique features, such as, dazzling colouration, delicate external morphology, unusual reproductive traits, symbiotic behaviour, cleaning and caring of associated animals, control of inhabitants of nuisance in an aquarium and also keeping reefs same.

Because of the increasing demand and growing popularity, it is expected that an increase in the collection of marine ornamental shrimps, as well as, diversification in the commonly exported species material would occur in the coming years.

Pairs of cleaner shrimp, anemone shrimp associated with sea anemone, *Heteractis magnifica* and Harlequin shrimp exclusively feeding on star fish, *Linckia* spp. In captivity, they are fed with chopped clam and fish meat thrice.

Lobsters

Spiny rock lobsters, *Panulirus* of family Palinuridae are common throughout tropical or sub-tropical seas. Ecologically, palinurids are important links in marine food webs. A nocturnal species, the lobster is an omnivorous scavenger with a preference for mussels, while juveniles feed mostly on barnacles. Palinurids are major predators of various benthic species, namely, snails, clams, urchins, and important prey for larger predators, like sharks and some fin fishes.

The spiny lobsters have a particular preference for corals and rocky reefs that are washed by sand surfs. They inhabit mostly clear waters with low runoff on islands or continents. They are among the largest crustaceans, the total body length sometimes attaining a length of 60 cm. The body parts are divided into cephalothorax, which

Figure 23.3: Squat Lobster, *Galathea strigosa*

Figure 23.4: Rock Lobster,
***Panulirus* spp.**

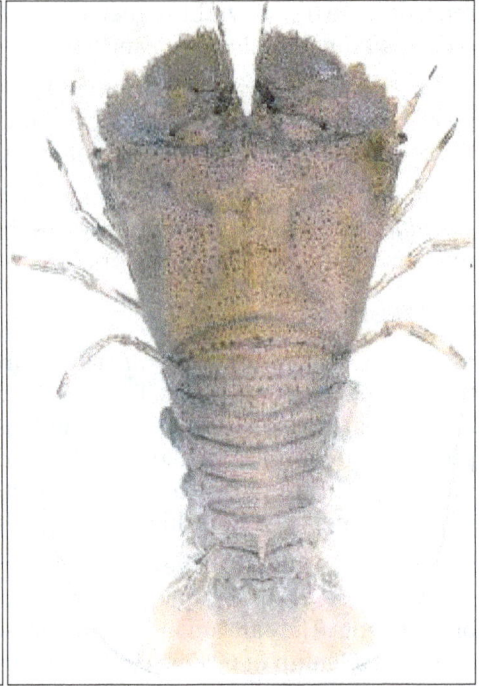

Figure 23.5: Sand Lobster,
Thenus orientalis

consists of the fused head and thorax, and an abdomen with their respective appendages. Appendages on cephalothorax include the eyes, the antennae and antennules, which provide protection, mechano-reception and chemoreception; the mouth parts, which include mandibles, maxillae and maxillipeds and five pairs of walking legs. The ventral portion of cephalothorax froms a sternum, which bears the gonopores at the bases of third pair of pereiopods in females and fifth pair in males.

In the abdomen of males, the first two pairs of pleopods are often formed into copulatory organs, whereas in mature females the pleopods become setose to enclose the external egg mass. Posteriorly, the sixth abdominal somite forms the tail fan consisting of uropods and telson, which enables the swift backward escape response.

The spiny lobsters carry eggs externally and clutch varies according to the size of the lobster. Gardener and Frusher (2000) reported that the size of the egg clutch is also influenced by the size of sperm packet that the female receive at mating. In general, it ranges between 35000 to over 750000 eggs. Incubation period for the eggs is relatively short, being about three to five weeks, but the larval period, however, is lasting about 11 months.

Fertilization is external, the male deposits spermatophoric mass on the female's sternum. Fertilization occurs when the eggs are extruded on to the abdomen and pleopods. The female carries the eggs on the abdomen until hatching. Unlike other crustaceans, eggs of spiny lobsters pass through a naupliar phase before hatching

Figure 23.6: Spiny Rock Lobster

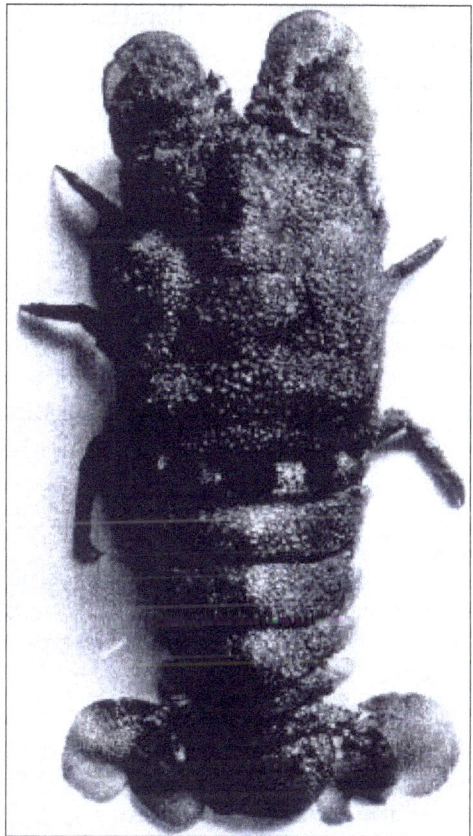

Figure 23.7: Dorsal View of
Scyllarides tridacnophaga

into the distinctive phyllosoma larvae, which are transparent and measure about 2 mm in length.

The spiny lobsters have long pelagic larval phases, during which time, the larvae can be transported long distances by oceanic currents. It has been reported that the larvae are transported over thousands of miles. The early pelagic larvae are known as phyllosoma stage that have an extended oceanic larval period of upto 11 months before returning to inshore waters as pueruli, where they settle into small holes and crevices in the reefs and sea bed. The number of pueruli settling on the fishing grounds every year may depend on the success of the previous year's migration and breeding.

Spiny lobsters exhibit five major phases within their life-cycle; egg, phyllosoma larva, puerulus (post-larval stage), juveniles and adult. During growth, lobsters like other crustaceans, undergo moult (shedding of hard outer shell). Shortly before reaching adulthood, many spiny lobsters undergo movements (migrations) from the nursery habitat to the characteristic deeper reef habitats, where reproduction occurs.

Lobsters are distributed throughout the Indian coast line and in the Andaman Sea.

Occurrence of the scyllarid lobster, *Scyllarides tridacnophaga* was reported for the first time in trawl landings at Sakthikulangara, Kollam on the south west coast of India. One male specimen (71 mm carapace length) was reported from Gulf of Mannar. The species is of limited economic interest due to rare occurrence of the species. The species has a depth of distribution range of 5 to 12 meters and the collected specimen was fished from 70 m off Kollam. The known geographical distribution range of the species is Indo-West Pacific region comprising of Red Sea, East Africa, Gulf of Aden, Pakistan and west coast of Thailand.

Among six species of spiny lobsters of genus *Panulirus*, the juveniles of green lobster, *P. versicolour* are abundant in near shore waters of Andaman Sea. Juveniles of 23 mm size to adults of 180 mm size have been reared 3 to 6 numbers per square meter in onshore sea water tanks. 100 per cent survival without any cannibalism was found to be feseable on moderate feeding with fish pieces and mussel meal. A net production of 150-250 gram per square meter could be obtained in a farming period of 6-8 months.

Crabs

Crabs belong to the class crustacean of phylum Arthropoda. The Arthropoda contain many species and aquatic arthropods are mostly crustaceans. Among these, crabs, spiny lobsters and tiger prawn are numbered among the important aquatic resources.

Most species of crabs live in the sea; however, a small number of species also live in fresh water, in the brackish water of estuaries and on the coastal land. Crabs are widely distributed from the tropical to frigid zones, and they show great variety in terms of morphology and ecology.

Distribution

The high latitude zone in the northern hemisphere has vast resources. The low latitude zone still remain unsurveyed in many areas. In many districts of Japan and South east Asia, various types of local fishery are carried out. In the coastal areas of Indian Ocean and north west region of Africa, a small numbers of crabs are caught along with the shrimps, but the catchable amount of crabs is still unknown.

The warm current dwellers are Gazami crab (*Portunus trituberculatus*); Swimming crab (*Portunus pelagicus*); Blue crab (*Callinectes sapidus*); Dana swim crab (*Callinectes danae*); Mud crab (*Scylla serrata*); Black stone crab (*Menippe mercenaria*).

In the Indian Ocean, only Swimming crab and Mud crab are distributed.

Reproduction, Development and Morphology

Crabs are dioecious, and the sex can be easily determined by the external shape. As to reproduction, ectosomatic fertilization takes place by copulation and the female carries eggs in the abdomen for protection. Crabs grow through various larval stages with body shapes quite different from those of mature adult. The first larval stage, *i.e.*, nauplius stage, occurs within the egg, and the hatching occurs at the zoea stage. During the planktonic life stage, the zoea metamorphoses into the megalopa, a more advanced stage of larva. The megalopa grows further into the adult crab.

Figure 23.8: Hermit Crab, *Pagurus arrosor*

Figure 23.9: Blue Crab

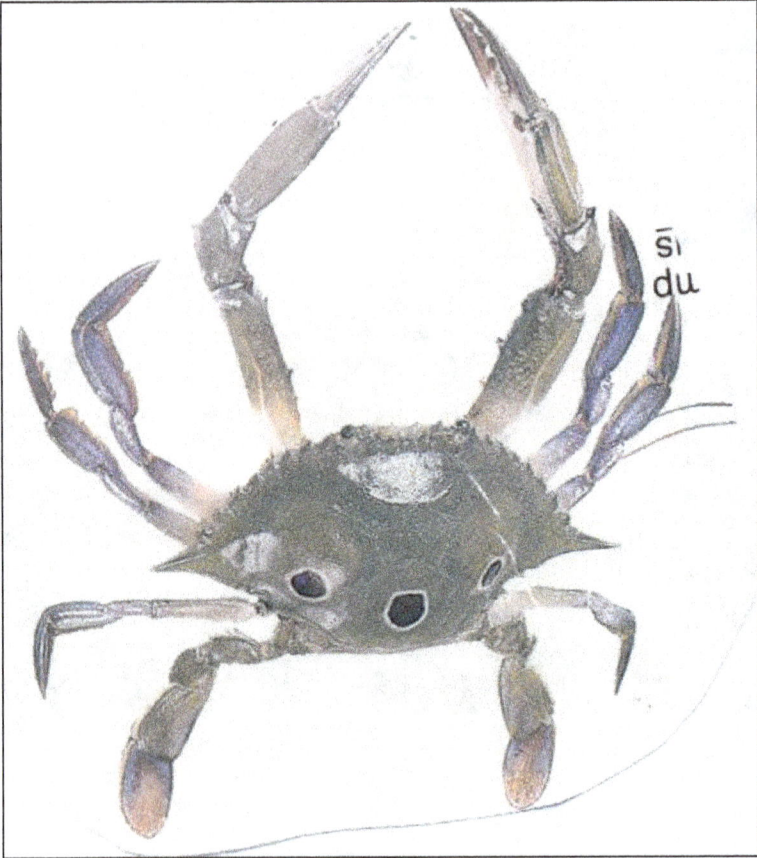

Figure 23.10: Sea Crab, *Portunus sanguinolentus*

Figure 23.11: Sea Crab, *Charybdis* spp.

Figure 23.12: Sea Crab, *Portunus pelagicus*

Figure 23.13: Mud Crab, *Scylla serrata*

Figure 23.14: Berried Female of *Portunus sanguinolentus*

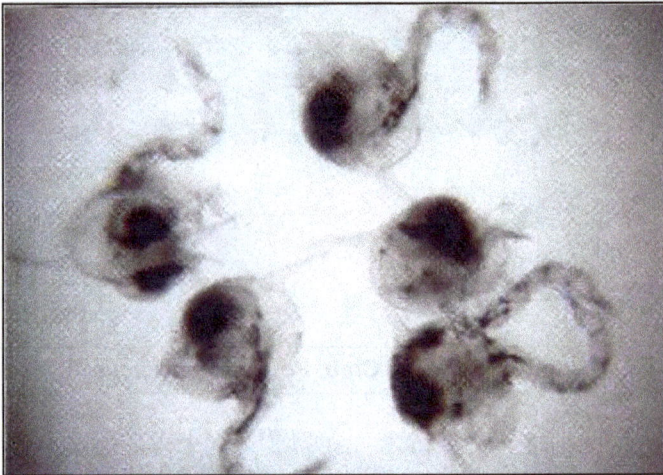

Figure 23.15: Zoe of *Portunus sanguinolentus*

The body and appendages of the arthropods are covered with exoskeleton, containg chitinous substance. Ecdysis occurs several times during growth, because the exoskeleton does not grow at the same rate as the inner body. The body consisting of twenty segments is divided into head, thorax and abdomen. Each body segment has paired appendages, and of these five pairs function as walking legs. Other appendages are transformed into antennas, mouth parts and a part of the copulatory organ. Out of the five pair of walking legs, the most anterior pair is transformed into the cheliped, and for the species which have a swimming capacity, the last pair is transformed into pleopod.

Ecology and Life-cycle

The ecology and life cycles of crabs vary greatly with latitude and living environment.

The carapace of the cephalothorax of Blue crab (*Portunus trituberculatus*) is elongated from right to left, and both edges project like a spine. This species lives in the sandy mud zone in shallow waters.

The crab stays in the mud in the day time, and it moves about seeking food at night. The end segment of the fifth leg is shaped like a flat plate, being adapted for active swimming. Spawning season is from April to October (spring-summer-autumn) and the same crab spawns several times in one season. The spawning ground is in the submarine forest of shallowwaters.

Blue crab have strong swimming ability, and migrate over a large area. In winter, they move to the deeper bottom or into the open sea and when the temperature drops to 14-15°Celsius and below, they burrow into the sandy mud and enter a period of hibernation.

In spring, when the temperature rises over 17°Celsius, they begin to move about, migrate to shallow waters or bays and gather in shallow places. After 12-13 ecdysis, young crabs reach the adult stage (the carapace is about 130 mm in width). The larvae hatch from May to August and grow into adult crabs, which have a carapace width of about 130-170 mm by September or October, and even for the crabs with slow growth, maturity is reached by October or November.

King crab (*Paralithodes camtschaticus*), a large one, living in the North Pacific Ocean, is characterized by the fact that the fifth pair of appendages are atrophied and concealed under the body and there are many spines on the surface of the body and appendages.

It inhabits the continental shelf at depths upto 250 meters and its optimum temperature is minus 1 to minus 12°C.

The female reaches sexual maturity at 5 to 6 years of age. Both males and females become fertile, when they reach a size of about 90 mm in carapace length. The breeding season of the species is during the spring, and these crabs come to the shallow coastal waters in groups to copulate and spawn. Fertilized eggs are brooded by the female until the next spring and hatch from March to June entering into the planktonic stage of their life cycle. The planktonic life stages is about two months, and after that benthic life begins. Young crabs live gregariously in shallow waters between 20 to 60 meters, but with growth they gradually migrate to the deeper bottom in what becomes a seasonal depth migration.

In the waters around the northern districts of Japan Islands, they generally live in the sandy mud zone between 30 to 300 meter in depth. The life span is over 15 years. The growth is slow and reproduction rate is low.

The species, *Chionecetes opilio* (Tanner crab) lives on the continental shelf from Alaska and Kamchatka to the Japanese islands. There are many small protuberances on the dorsal surface of the carapace. The legs are slender, and all leg segments are flat. The body colour is pale testaceous. The male is larger than the female and it has a carapace of more than 140 mm. The female's carapace is 75-85 mm long. They live in the sandy mud bottom, 200-400 meter deep, but they move to comparatively shallow waters in winter.

It is believed that tanner crabs reach the adult stage after 9th instar, and they obtain complete copulatory ability after 10th instar. They reach this stage about eight years after hatching.

For crabs, like other crustaceans, there are no traits by which their age can be determined easily. Therefore, studies of the age of crabs have been carried out using the characteristics of the growth by ecdysis. The young crab, which has just finished its first metamorphosis and has begun benthic life is called the 1st instar. After that ecdysis occurs at regular intervals and one instar is added for each ecdysis that the crab passes through.

Deep Sea Crab

The deep sea crab, *Thalamita crenata* mostly inhabit only deeper waters (more than 100 m) and occur in deep sea gill net operations along with fishes.

However, *T. crenata* was noticed to occur at 20 m depth, north of Tuticorin. The species were caught in huge numbers. The size of these crabs ranged from 50-52 mm for males and 45-47 mm for females weighing 10-20 gram per crab. The migration of these crabs from deeper waters to inshore waters might have taken place due to the sudden changes in the environment, or may be for their searching better feeding ground.

Sand Crab

The sand crabs, *Emerita asiatica* and *Albunea symnista* belonging to the family Hippidae are abundant in the intertidal sany beaches of Tamil Nadu coast, India. Prior to tsunami (26.12.2004), the fisherwomen used to collect this sand crabs by hand picking during the low tide period from the intertidal region and were sold at low prices, since the sand crabs are very small containing very little flesh. The catching of sand crabs had become a source of daily earnings after post-tsunami period (January-March, 2005) due to destruction of fishing crafts and gear along Chennai coast.

A total of 4970112 weighing 3984 kg were hand picked from an inter-tidal area of 50000 sq km approximately. In January 2005, when the exploitation started, the total catch amounted to 1625 kg, which increased to 1855 kg in February, since more fishermen took part in hand picking. But in March the catch was reduced to 504 kg due to intensive fishing in previous months.

Among the two species, *E. asiatica* dominated the catch forming 81 per cent. Its total length ranged from 36-40 mm (males) and 40-50 mm (females). The average weights for male and female *E. asiatica* were 6.5 g and 10.5 g respectively. Among females, 50 per cent were found berried. The number of eggs found attached ranged from 4950 to 6200.

The total length of *Albunea symnista* ranged from 32 to 46 mm (males) and 30-48 mm (females), while the average weight was 4.98-6.34 g (males) and 5.03-6.13 g (females).

Spider Crab

Generally the spider crabs, *Doclea ovis* (Fabricius, 1787) are known to carry some objects and animals on their backs using their hooked setae, just to camouflage or to get mutual benefit. Earlier investigations reveal carrying sea anemones on their carapace. Large animals, especially females carry 2 to 4 anemones of varying sizes on their carapace. The anemones attach themselves by their sole to the dense velvet like setae on the crab's carapace. During the day, the crabs themselves hide in the sand or mud and only the anemones are obvious. When one physically detaches the anemones from the crab and leaves them next to it, the crab "plants" the anemones back. The recent observations reveal the berried female with its orange coloured developing eggs carrying the sea anemone on the carapace.

Poisonous Crabs

Among a total of 990 species of marine brachyuran crabs belonging to 281 genera and 36 families, reported from Indian waters, more than a dozen of species are poisonous. There are two categories of poisonous crabs; the first one include highly toxic species of Xanthid crabs, such as, *Lophozozymus pictor, Demania* spp., *Zosimus aeneus, Platypodia granulose,* and *Atergatis dloridus* and the other one is mildly poisonous or occasionally poisonous Xanthid crabs (*Atergatis integerrimus, Carpilius* spp., *Eriphia* spp., *Etisus* spp.) and Land crabs (*Cardisoma* spp.). It is believed that the poisonous crabs have also been found associated with red tide algal or diniflagellate blooms. In the second category, the crabs are not always poisonous, but their toxicity varies from place to place and is also seasonal.

It is believed that the primary source of toxins in the crabs is from marine red macroalgae, species of the genus *Jania* from southern Japan.

Crab toxins are heat resistant and are found in crab's viscera, and hence they can not be destroyed by cooking and also no known cure is available for these toxins.

Poisonous crabs belong to the families Dairidae, Dromiidae, Eriphiidae, Majidae, Parthenopidae, Pilumnidae, Portunidae and Xanthidae. Majority of the species, however, belong to the family Xanthidae. The toxins are in the flesh and can not be transmitted, if crab is bitten/pinched by its claws. Most of the poisonous crabs have a very bright warning colour, which is a good indicator to avoid consuming them.

Dromia dormia (Family-Dromiidae) is known as sponge crab. Its carapace, convex in shape is 200 mm width, wider than length and is covered with dense hairs. Five anterolateral teeth, with the first to third of them larger, rostrum is tridentate, there is no spine on the outer margin of dactylus of last walking leg, body colour is dark brown with pink fingers. It is commonly found in rocky and coral reef areas of Andaman Sea. The species is discarded after capture as the Indian fishermen know its poinonous nature.

Eriphia sebana (Family-Eriphiidae), known as "Red-eyed rock crab" has carapace width 76 mm and is oval shaped and smooth. Chelipeds are also smooth, covered with polished round tubercles, free edge of frontal lobes are cut into blunt teeth, anterolateral border with six or more spines. Chelipeds are unequal in size, legs stout and smooth, walking legs have stiff hairs on their upper and lower border. It is found among coral reefs and rocky boulders of Andaman Sea and is aggressive and nocturnal habits.

Eriphia smithii (Family-Eriphiidae) has oval shaped carapace with a width of 57 mm. Anterior regions of carapace and chelipeds are covered with small pearly tubercles. Posterior regions of carapace are almost smooth, frontal lobes bluntly spinate. Chelipeds are unequal in size. Lives among coral and rocky boulders of Andaman Sea.

Micippa phylira (Family-Majidae) has 16 mm carapace width and is subquadrate. Spines are present in both antero- and postero-lateral borders of the carapace. Chelipeds are smooth and stout in male and slender and smooth in female. Scattered tubercles are present on carapace. First pair of legs is longer than the chelipeds. Body and legs are covered with dense hairs, each segment being indistinctly granulated and the fingers widely gaping near the base. This crab is an inhabitant of coral reef and sea weed beds of Andaman Sea.

Schizophryx aspera (Family- Majidae) is known as spider crab. Its carapace width is 45 mm. It is piriform (pear shaped), circular to sub-ovate and anterior half to one-third of it is usually distinctly narrower than posterior part. The front part has two long horn-like projections. Anterolateral margins of carapace is armed with six well developed spines. Legs are with spines and stiff hairs. The species inhabit coral reef and rocky areas of Andaman Sea.

Daldorfia horrid (Family- Parthenopidae), known as "Rubble crab" has its carapace somewhat pentagonal in shape and is 40 mm in width. Its surface is deeply eroded and sharply tubercular and posterolateral angle is produced much outwards. Anterolateral border is coarsely spinate and posterolateral border granular. Chelipeds are massive and unequal in sizes. Chelipeds and legs are with granular spines. It lives under rocks and remain buried in sandy mud of Andaman Sea.

Parthenope longimanus (Family- Parthenopidae), known as long pincer legged crab, has its carapace width 40 mm. It is broader than long and covered with sharp tubercles. Anterolateral border has 10 irregular teeth and posterior border has hooked spines. Chelipeds are longer, more than four times as long as the carapace length and there are sharp spines on both anterior and posterior borders. Legs are slender. It inhabits sandy and muddy areas with shell fragments of Andaman Sea.

Pilumnus vespertilio (Family- Pilumnidae), known as hairy coral crab has transversely oval carapace of 40 mm width. Large tubercles are present on carapace and chelipeds. Chelipeds are unequal in size. Anterolateral border is shorter than posterolateral and it is cut into three sharp teeth. Body and legs are usually covered with thick hairs.Inhabits in rocky shores and coral reefs, especially in areas with dense sea weed, *Sargassum* spp. of Andaman Sea.

Thalamita prymna (Family- Portunidae), known as red swimming crab has its carapace 72 mm in width. Carapace is broader than long, less convex and smooth with distinct transverse ridges, front ones are straight with 6 short and truncated lobes. Five anterolateral teeth, first three larger, fourth rudimentary and fifth smaller than third. Chelipeds are unequal in size. It is a inhabitant of inshore seas, sandy, muddy or rocky bottom of Andaman Sea.

Actaeodes tomentosus (Family- Xanthidae), known as coral crab, has its carapace width 30 mm. Carapace is broader than long, with five lobes on its anterolateral border. Round tubercles are present on carapace. Chelipeds and legs, fingers of chelipeds are stout and somewhat pointed. It inhabits among coral reefs and boulders of Andaman Sea.

Atergatis floridus (Family – Xanthidae), known as floral egg crab has 51 mm carapace width. Carapace is convex with a lace work colour pattern on its smooth surface. Angle it ends anterolateral in a blunt tooth. Chelipeds are equal in size with black tips. It inhabits coral and rocky area of Andaman Sea. Source of toxin is presumed due to the grazing habit of the species on red calcareous alga *Jania* species.

Atergatis integerrimus (Family- Xanthidae), known as red egg crab has its carapace width 79 mm. It is a mildly toxic crab with a broad oval carapace, brownish-yellow in colour. The surface of its carapace is irregular and rather distantly pitted, especially near the front and anterolateral borders.

Edges of anterolateral borders being sharp and crest like. Chelipeds are equal in size and movable and fixed fingers is black or dark brown in colour. Walking legs hardly pitted with fine hairs. It is a rocky and coral area dweller, occurring between 10 to 30 meter deep zone of Andaman Sea.

Carpilius maculates (Family- Xanthidae), known as nine-spot crab has its carapace width 122 mm. It is a coral reef dweller. The carapace is oval-shaped. It has characteristic three dark brown spots in the middle of the carapace. Anterolateral border ends in a blunt tooth. Chelipeds are unequal in size and are stout. It is found in Andaman Sea.

Demaniabaccalipes (Family- Xanthidae), known as "demon crab" has its carapace width 45 mm. Carapace is wider than long and there are 3 to 4 teeths on its anterolateral border. Round tubercles are present on carapace and chelipeds. A triangular blunt tubercle is present on the internal angle of wrist (carpus) of cheliped. Upper and lower borders of walking legs are sharp. The species live in coral and rocky areas of Andaman Sea.

Demania cultripes (Family- Xanthidae), known as "demon crab" has its carapace width 50 mm. Carapace is broad and its median regions are relatively smooth. Its anterolateral margin is cut into three blunt and distinct triangular teeth. Inner angle of wrist (carpus) of cheliped has a sharp tooth or spine. Its habitat is coral reef and rocky substratum of Andaman Sea. Palytoxin was found in all tissues with high concentrations in gills, viscera and eggs of the species.

Demania scaberrima (Family- Xanthidae), known as "demon crab" has its carapace width 57 mm. Dorsal margin of carpus and propodus of ambulatory leg is with prominent well-defined sharp spines or teeth. Inner angle of wrist (carpus) of cheliped is with sharp or distinctly acute tooth or spine. Dorsal surface of carapace and chelipeds is covered with numerous large rounded or sharp granules. Frontal margin is lobulated and trapezoidal projecting beyond carapace. The species is found in rocky and coral reefs of Andaman Sea.

Euxanthus exculptus (Family- Xanthidae), known as "coral crab" has a carapace width of 28 mm. The carapace is smooth with five lobes on anterolateral border.Chelipeds and legs are with large granules. Chelipeds are equal in size, their cutting edges are strongly toothed. Tips of fingers are with hollow spoon. It is an inhabitant of coral and rocky boulders of Andaman Sea.

Pilodius areolatus (Family- Xanthidae) have regions of carapace well defined and separated by deep grooves. There are small granules on carapace, chelipeds and legs. Legs are thickly fringed with plumose setae. It inhabits in coral reefs of Andaman Sea.

Zozymus aeneus (Family- Xanthidae), known as "jewel crab" has its carapace width 100 mm. Coral reef and rocky boulders of Andaman Sea are its habitat. Hides under coral reefs during the day and crawls out at night. Its body has eye-like chocolate brown spots with smooth and well defined lobules. Anterolateral border is divided into four lobes. The first three are rounded and the fourth is tooth-like. Chelipeds are rough. Walking legs are with hairs. Toxic symptoms of this crab range from numbness of the tongue to eventual paralysis of the extremeties, respiratory failures leads to death, mortality rate is at 60 per cent. Source of toxin is due to grazing of the crab on red calcareous alga, *Jania* sp.

Ecology

Crabs are distributed throughout the world with freshwater, brackishwater, seawater and semiterrestrial waters as habitats.

Most of the crabs live in burrows, while some are free swimming. Only a few crabs, mostly inhabiting estuarine and marine waters are preferred as food commodity due to their palatability, preferably meat/body ratio and nutritive value with relatively high protein content (upto 73 per cent dry weight) and their having essential vitamins, amino-acids and minerals.

Large diversity of crabs are seen in the estuarine habitats, out of which four species of crabs of the genus *Scylla* (*S. serrata, S. paramamosain, S. oceanica, S. tranquibarica*), the estuarine crab, *Thalamita crenata* and semi-terrestrial burrowing crab, *Cardisoma carnifex* are preferred as food crabs.

Amongst the marine crabs, *Portunus pelagicus, P. sanguinolentus* and *Charybdis cruciata* are commercially exploitable species in India. *Menippe rumphii*, an intertidal form is distributed along the rocky shore on the east and west coasts of India, is also preferred as food by many sects.

Good demand exists for estuarine and marine crabs. The mud crabs are generally marketed as a live commodity, throughout the world, especially in Asian markets. The marine crabs are frozen after removing carapace and visceral organs for export. There is also good demand for crab meat extracted from the three species of marine crabs mentioned above. The female red crabs, *Scylla serrata* with ripening gonads in the is preferred throughout the world as a delicacy. Such of these crabs in the weight range of 180-250 g fetch money in the domestic as well as in the export market. There is also a good demand for soft crabs (post-ecdysial crabs). Soft mud crabs in the weight range of 100-180 g are used for this purpose.

Crabs landing by trawlers in Mumbai consists of three major species, namely, *Charybdis fereatus, Portunus sanguinolentus* and *P. pelagicus*. Of these, *C. feriatus* has traditionally been the dominant species at all landing centers forming approximately 60-70 per cent of the landings, while *P. sanguinolentus* has stood at second place (15-20 per cent) of the landings. The landings of *P. sanguinolentus* showed an unusual upward trend in September, 2004. Though this trend was not sustained for following two months, the landings again showed a marked increase in late December and early January, 2005. There was a corresponding decline in the landings of *C. feriatus*.Landings of *C. feriatus* revived in June, 2005 once again. No change in fishing grounds, gear or efforts were noticed.

Marine Decapods of Southwest Coast, off Karnataka

A total of 33 species of prawns, belonging to 8 families were collected from coastal and marine zones of Karnataka. Penaeidae was the largest family with 16 species and all the species belonging to the family were collectedfrom within 50 m depth zone and these are commercially very important. *Metapenaeus dobsoni, M. monoceros, M. affinis, Fenneropenaeus (Penaeus) indicus, F. merguiensis, Penaeus monodon, P. canaliculates, P. semisulcatus, Parapenaeopsis stylifera, Trachysalambria (Trachypenaeus) curvirostris* and *T. sedelli* form regular part of prawn fishery, whereas others were found seasonally and sometimes in stray numbers. *Parapenaeus fissuroides indicus* is the first record from Indian coast. Solenocera family along Karnataka coast is represented by species from mid-shell (*Solenocera choprai* and *S. pectinata*) and from deep sea (*S. hextii*). *S. choprai* is a commercially very important species, which was caught in huge numbers from a depth of 60 to 100 m off Mangalore and Malpe. Other deep-sea varieties belong to Aristidae, Pandalidae and Sicyonidae families. Out of these, Aristidae is commercially the most important family. *Aristeus alcocki*, known as "red-rings' and belonging to Aristidae were caught in good numbers from a depth of 150 to 500 m off Mangalore-Kundapur. From the same fishing ground, six species belonging to pandalidae family were also caught, of which *Heterocarpus gibbosus* was found in good numbers and all others represented in stray numbers. "Jawala" shrimps (*Acetes* spp.) belonging to Sergestidae family were found distributed along the coast and they seasonally formed a good fishery and most important, that these shrimps form major food items of most of the carnivorous fishes and other marine fauna. *Rhynchocinetes durbanensis* shrimps, belonging to Rhynchocinetidae family, were collected from Netrani Island and these were the first record of the species from the coast. Commercially, this very important ornamental species is having heavy demand all over the world.

Among decapod macro-fauna, brachyuran crabs or true crabs are the most abundant in terms of species diversity. 105 species of true crabs belonging to 18 families were collected from Karnataka coast. Among others collected were, those of Portunidae and Xanthidae, the biggest families with 19 and 17 species respectively. From a commercial angle, Portunidae is the most important family for the reason that it mostly consists of edible species. Portunidae is represented by 19 species, out of which *Portunus pelagicus, P. sanguinolentus* and *Charybdis feriatus* are the major species froming a commercial fishery. *C. lucifera* is also considered as edible species. *Scylla serrata* and *S. tranquebaruica*, which belong to the estuarine ecosystem also occur in coastal waters of Mangalore. *C. smithi*, which is caught from deeper waters also is an edible species, landed occasionally. *P. gracillimanus* was identified in the catches from west coast of India and this as the first record of its occurrence in the west coast. Eventhough many of the crabs are non-edible, these crabs are caught as by-catch by the trawlers and are used for making fish meal. From catches of commercial trawlers, 9 species of coral reef related crabs (7 species from Xanthidae family and 2 from Carpiliidae family) were collected. This is indicative of the presence of a submerged coral reef off Karnataka coast. As far as zonal distribution is concerned, Portunid crabs, *Portunus pelagicus, P. sanguinolentus, Charybdis lucifera, C. feriatus, C. annualta, Scylla serrata, S. tranqebariica,* Grapsid crabs, *Grapsus albolineatus* and *G. tenvicrustatus,* Parthenopid crab *Ashtoret(Matuta) lunaris,* Ocypod crab *Uca(Celuca) annulipes* and Xanthid crab, *Leptodius exartus* has been found to have a wide distribution throughout the coast.

Nine species of lobsters belonging to three families were collected from Karnataka coast. While coastal species, like, *Panulirus homarus, P. polyphagus, P. versicolour* and *Thenus orientalis* could be collected in few numbers, deep sea species, *Puerulus sewelli* and *Nephrosis stewartii,* caught from deep waters (beyond 150 m) off Mangalore coast were found in good numbers. *P. polyphagus* and *P. versicolour* were found on the coral reef around Netrani Island and other coastal species were collected from trawl collections. Until late 1970s,while there were reports that good number of *P. homarus* were caught from Mangalore coast, during the period of survey in 2011, the catches of this species were in few numbers. In the case of *T. orientalis* also there was alarming reduction in the distribution and now the species is recorded as a rare occurrence.

Fourteen species of hermit crabs were collected from the marine regime of Karnataka waters. Out of these, 13 belonged to Diogenidae family and one belonged to Paguridae family. Four more species of these crabs also were collected from inter-tidal area and one (*Clibenarius aeqabilis*) was collected from Devgadh Island.

List of Decapod Crustaceans Collected from Marine Zones along Karnataka Coast

Prawns (8 families and 33 species)

☆ Family – Alpheidae – *Alphius malabaricus, Alphius palaudicola.*

☆ Family - Aristeidae – *Aristeus alcocki*

☆ Family - Pandalidae – *Heterocarpoides levicarina, Heterocarpus gibbosus, Heterocarpus sibogae, Heterocarpus woodmasoni, Plesionika spinipes, Olesionika martia.*

☆ Family - Penaeidae – *Fenneropenaeus indicus, Fenneropenaeus merguiensis, Metapenaeopsis andamanesis, Metapenaeus affinis, Metapenaeus dobsoni, Metapenaeus monoceros, Metapenaeus moyebi, Metapenaeus brevicornis, Parapenaeus fissuroides, Parapenaeus longipes, Parapenaeopsis stylifera, Penaeus canaliculatus, Penaeus monodon, Penaeus semisulcatus, Trachysalambria curvirostris, Trachysalambria sedilli*

☆ Family – Rhynchocinetidae – *Rhynchocinetes durbanensis*

☆ Family – Sergestidae – *Acetes indicus, Acetes erythraeus, Acetes johni*

☆ Family – Sicyoniidae – *Sicyonia lancifera*

☆ Family – Solenoceridae –*Solenocera choprai, Solenocera hextii, Solenocera pectinata*

True Crabs (18 families and 105 species)

☆ Family – Eriphiidae – *Myomenippe hardwickii*

☆ Family – Calappidae – *Calappa philargius, Calappa lophos, Calappa gallus*

☆ Family – Carpiliidae – *Carpilius convexus, Carpilius maculates*

☆ Family – Dorippidae – *Dorippe frascone*

☆ Family – Dromiidae – *Cryptodromia hilgendorfi, Dromia dehaani*

☆ Family – Grapsidae – *Sesama tetragonum, Metopograpsus maculates, Varuna litterata, Pseudograpsus elongates, Pseudograpsus intermedius, Sesarma quadratum, Sesarma edwarsi, Sesarma lanatum, Percnon planissimum, Grapsus tenuicrustatus, Grapsus albolinealus, Metopograpsus messor, Metaplax indica, Metaplax distincta*

☆ Family – Hymenosomatidae – *Neorhynchoplax demeloi, Elamena cristatipes*

☆ Family – Leucosiidae – *Leucosia sima, Leucosia pubescens, Myra fugax, Leucosia anatum, Ebalia malefactrix, Philyra scabriuscula, Leucisca squalina, Eucrate dentate, Philyra globosa, Philyra corallicola*

☆ Family – Majidae – *Ophthalmias cervicomis, Naxioides hirta, Schizophrys aspera, Aethra scruposa, Doclea ovis, Achaeus lacertosus, Dehaanius limbatus, Menaethius monoceros, Doclea hybrida, Menaethius monoceros, Doclea gracilipes*

☆ Family – Menippidae – *Menippe rumphil*

☆ Family – Ocypodidae – *Uca annulepis, Ocypode ceratophthalma, Ocypode cordimana, Macrophthalmus parvimanus, Macrophthalmus brevis, Scopimera proxima, Dotilla myctiroides, Dotilla malabarica, Macrophthalmus latreille, Macrophthalmus depressus, Macrophthalmus crinitus*

☆ Family – Parthenopidae – *Matuta planipes, Galene bispinosa, Cryptopodia angulata, Ashtoret lunaris, Pinnotheres placunae*

☆ Family – Pinnotheridae – *Pinnotheres placunae, Pinnotheres gracilis*

☆ Family – Plagusiidae – *Plagusia tuberculata*

☆ Family – Portunidae – *Scylla serrata, Scylla tranquebarica, Thalamita crenata, Charybdis japonica, Charybdis annulata, Charybdis natator, Charybdis smithi, Charybdis helleri, Charybdis hoplites, Podophthalmus vigil, Portunus gracilimanus, Thalamita danae, Thalamita integra, Thalamita prymna, Charybdis riversandersoni, Charybdis feriatus, Charybdis lucifera, Portunus pelagicus, Portunus sanguinolentus*

☆ Family – Raninidae – *Ranina ranina*

☆ Family – Trapiziidae – *Tetralia cavimana*

☆ Family – Xanthidae – *Etisus levimanus, Leptodius exaratus, Zozymus aeneus, Pilodius areolatus, Phymodius monticulosus, Phymodius ungulatus, Phymodius nitidus, Chlorodiella nigra, Pseudoliomera speciosa, Atergatis subdentatus, Atergatis integerrimus, Menippe rumphii, Ozius rugulosus, Ozius tuberculosus, Epixanthus frontalis, Pilumnus vespertilio, Eurycarcinus orientalis*

Hermit Crabs (2 families and 14 species)

☆ Family – Diogenidae – *Paguristes incomitatus, Clibanarius infraspinatus, Clibanarius padavensis, Clibanarius aequabilis, Clibanarius arethusa, Diogenes Diogenes, Diogenus affinis, Diogenes planimanus, Diogenes viloaceus, Diogenes miles, Diogenus avarus, Dardanes setifer, Troglopagurus manaarensis*

☆ Family – Paguridae – *Pagurus kulkarnii*

Lobsters (3 families and 9 species)

☆ Family – Nephropidae – *Nephropsis stewartii*

☆ Family – Palinuridae – *Panulirus homarus, Panulirus polyphagus, Panulirus versicolour, Panulirus ornatus, Panulirus penicillatus, Panulirus longipes, Puerulus sewelli*

☆ Family – Scyllaridae – *Thenus orientalis*

List of Decapod Crustaceans Collected from Andaman and Nicobar Islands

Prawns

Macrobrachium lar (Fabr), *Heterocarpus gibbosus* (Bate), *Pandalus martius* (A.M.Edw), *Hoplophorus gracilirostris* (A.M.Edw), *Nematocarcinus paucidentatus, Cardinilla* sp., *Metapenaeus mogiensis* (Rath), *M. coniger, V. andamenensis* (Wood Masson), *Palaemon concinnus* (Dana)

True Crabs

Scylla serrata (Forsk de Haan), *Uca (Deltuca) dussumieri* (M.rd), *Uca (Celuca)luctea annulipes, Thalamita crenata* (Latr), *Uca vocens* (Linn), *Sesarma bidens* (Haan), *Maluta victor* (Fabr), *Thalamita exitastica* (Alcock), *Uca annulipes* (Mrdw), *Grapsus strigosus* (Herbst), *Thalamite prymna* (Herbst), *Uca (Thalassuca) vocaus vocaus* (Linn), *Lophactaea cristata* (A.M.Edw), *Eriphia laevimana* (Latr), *Mictgris longicarpus* (Latr), *Etisus laevimanus* (Randell), *Uca (Thalassuca) tetragonon, Calappa hepatica* (Linn), *Leptodius sanguineus* (M.Edw)

Hermit Crabs

Clibanarius longitarsus (de Haan), *C. striolatus* (Dana), *C. olivaceus* (Henderson), *C. arethusa* (de Man), *C. merguiensis* (de Man), *C. corallinus* (H.M.Edw), *C. humilis* (Dana), *Calcinus herbstii* (de Man), *C. lateens* (Randall), *C. gaimardii* (H.M.Edw), *Dardanus megistos* (Herbst), *D. vulnerans* (Thallwitz), *D. guttatus* (Oliver), *D. deformis* (H.M.Edw), *D. varipes* (Heller), *Aniculus aniculus* (Herbst), *Coenobita clypeata* (Latreille), *C. rugosa*

(H.M.Edw), *C. perlata* (H.M.Edw), *C. cavipes* (Stimpson), *C. violascens* (Heller), *Diogenes miles* (Dana), *D. avarus* (Heller), *D. custos* (Fabricius), *Pagurus puncttulatus* (Oliver), *P. zebra* (Alcock), *P. pergranulatus* (Henderson), *Paguristes ciliatus* (Heller), *Aniculus strigatus* (Herbst), *Birgus latro* (Linnaeus)

Isopod

The deep sea isopod, *Bathynomus giganteus*, belonging to the order Isopoda, Sub-order Flabellifera and family Cirulanidae is distributed in Bay of Bengal, Arabian Sea, Gulf of Mexico and the south western Atlantic, off Brazil. It's habitat is the sea bottom at a depth ranging 1200 to 2000 feet. It lives in an area, where the light penetration is very less and the temperature is extremely cold. In India, it was first reported from the coastal waters of bay of Bengal at a depth of 300 feet. The specimen was having a total length of 200 mm and maximum breadth of 95 mm. the specimen from west coast of India was bigger (280mm) than the one reported from east coast. It was reported that *B. giganteus* grows upto a length of 420 mm (18 inch) and attains a weight of 1360 grams.

The species is not a fast swimmer and does not migrate to a long distance in search of food or for the purpose of spawning. They are carnivorous and feed on fish, sponges, small crustaceans, nematode worms and protozoans. They prey on diseased or injured fish and also attack fish that have been caught in the net. *B. giganteus* is an egg laying animal. The females are provided with a brood pouch in which the eggs are brooded and the young ones are released.

Sea Spider

Pycnogonids, popularly known as sea spiders are marine arthropods, belonging to the phylum Arthropoda and class Pycnogonida are more related to spiders, but have a skinny body. They are rarely observed as they are small, cryptic or hidden amongst other organisms, moving slowly over sea weeds, corals, sponges, hydroids. They are usually white and are camouflaged beneath the rocks and among the algae, that are found along the shore lines. Their distribution is worldwide and they occur commonly in shallow waters, but can be found in water as deep as 7000 meters.

These marine arthropods are long legged and their body is so reduced that reproductive organs are located in the legs. While most species have 4 pairs of legs, there are some species have 5 to 6 pairs of legs. Proboscis is the most prominent external feature of pycnogonids, which is a movable organ and shows wide variation in shape and size among families. The shape and internal structure has been related to specialized feeding habits, sometimes specific to a particular host among parasitic species.

Pycnogonids are carnivorous, preying on bryozoans, hydroids and sedentary polychaetes. Adult pycnogonids, either suck the juices from the soft bodied invertebrates or browse on hydroids and bryozoans through their long proboscis. They have taste preferences for particular type of prey that develop depending on what they are fed on, while they were juveniles.

Figure 23.16: Sea Spider (*Endeis mollis*)

Seasonality in the occurrence of pycnogonids has been observed. They generally occur during March to August and mostly on April.

There are more than 1300 species of pycnogonids described and it is believed that many more species are yet to be discovered mainly from remote deep sea habitats.

In Andaman Sea, two pycnogonid species, namely, *Eurycyde flagella* and *Pallenopsis ovalis* have been reported. These animals are not necessarily rare in the wild, but since they are cryptic and small, they are easily been overlooked.

Sea spider predation or parasitism is a potential threat to all corals, soft corals and sea anemones.

Chapter 24

Syngnathids

The sea horses, pipe fishes, pipe horses and sea dragons are members of the family Syngnathidae. There are about 215 species classified in 52 genera in the Syngnathidae family. They can be distinguished from other bony fishes, by the presence of fused jaws, pore-like opening of the gill chamber, bony plates encasing the whole body and absence of teeth and pelvic fins.

Pipe fishes, derived their name from the long, slender and angular body. Sea horses have a head and neck, which resemble that of a horse, body like caterpillar, tail like monkey, brood pouch like a kangaroo and independently moving eyes like a chameleon. Pipe horses are intermediate in form between the sea horses and pipe fishes with an elongated body, grasping tail and head slightly angled to trunk. Sea dragons have elaborate and permanent leaf like appendages on their body.

There are 34 species of sea horses throughout the world belonging to the only genus *Hippocampus*. In India, nine species of *Hippocampus* was reported. Among 175 species of pipe fishes belonging to 47 genera in the Indo-Pacific region; Day (1958) recorded 12 species from India. About ten species of pipe horses, belonging to the genera *Accntronura*, *Apterygocampus* and *Stigmatopora* described in Indo-Pacific region, only *Accntronura gracilissima* was reported from Indian waters.

Sea horses are exclusively marine, distributed from 50°N to 50°S latitudes. Pipe fishes are almost cosmopolitan in distribution between 71°N and 56°S latitudes.

The bulk of syngnathids are near shore, shallow water species living with benthic invertebrates, algae and rocks. Sea horses, pipe horses and few pipe fishes require suitable anchorages due to their prehensile tail. Dead corals and sea grasses are the preferred habitats for *Hippocampus fuscus*, whereas, *H. kuda* is found among sponges also. *H. trimaculatus* and *H. spinosissimus* are common among sea grasses and sea weeds. In addition, the latter showed preference to sponges only.

Syngnathoides biaculeatus is the most abundant and common pipe fish and it inhabits sea grasses habitats in near shore areas, although up to 5 m depth. *H. cyanospilos* and *H. splcifer* are also found in the sea weed, sea grass and mangrove areas in fewer numbers. *H. grayi* and *Trachyrhamphus* spp. are observed at depths above 20 m.

Syngnathids have a small mouth located at the tip of a long tubular snout. They are an ambushing kind of predators and they employ sit and wait atrategy. They wait for the prey to come nearer. When the prey is in close reach, they suck the prey with a rapid intake of water. They lack teeth and a well distinguished stomach. Syngnathids are diurnal predators and consume only live and moving prey, whereas feeding of sea horse with frozen food is gaining momentum due to aquarist demand. The major food items of *H. kelloggi* are copepods, mysids, amphipods, tanaeids, *Acetes* spp. and shrimp larvae. The gut contents of *H. kuda* mainly consisted of mysids, *Acetes*, amphipods, copepods polychaetes and larval forms. *H. trimaculatus* mainly consumed amphipods, *Acetes*, shrimp post-larvae and copepods. Female sea horses consume more quantity of feed compared to males. Brood males consume more food than non-brooder males. Bulk of the diet of pipe fishes consists of copepods, amphipods, mysids etc. Since syngnathids have limited mouth opening, their food preference is limited by the prey size. There is also an ontogenetic shift in prey items in syngnathids, with juveniles consuming smaller prey, such as, copepods and adults preferring larger prey items like *Acetes* and juvenile shrimps. Diet of syngnathids also vary with sex and reproductive condition.

The young sea horses show inflection points in growth because of their habit of switching over among prey types. *H. kuda* males attain a growth of 116.02 mm by the first year, 162.67 mm by second year, 197.34 mm by third year and 223.08 mm by the fourth year, whereas females were found to grow to a length of 112.84 mm in the first year, 159.47 mm in the second year, 191.09 mm in the third year and 214.90 mm in the fourth year. *H. kelloggi* attains an average growth of 140.2 mm in the first year, 201 mm in the second year, 246.8 mm in the third year and 281.1mm in the fourth year. *H.trimaculatus* attains a growth of 91.49 mm in the first year, 119.99 mm in the second year and 141.98 mm in the third year. The life span of all the three species is found to be over three years.

Syngnathids are fishes, and show extreme parental care in which male fishes carry the developing embryos in a brood pouch until birth. Brood pouches in males differ within the family from simple gluing of eggs in few pipe fish genera to completely enclosed pouches in sea horses. During mating, the female deposit her egg clitch into brood pouch of male, where the eggs are subsequently fertilized and further development occurs. The brood pouch offer protection, aeration, osmoregulation and nourishment to the developing embryos.

All the sea horses with one exception accept eggs only from a single female showing monogamous mating behaviour. The exception is *H. abdominalis* which is found to mate with several females in a single breeding season. The pair bond remains faithful and is reinforced by daily greetings. Pipe fishes present examples of sex role reversal in which females compete more intensively than males for access to mate.

However, monogamous mating behaviours are reported in other pipe fishes like, *Corythoichthys* spp. and *Hippichthys penicillus*.

Size at first maturity of sea horses varies within species depending on the maximum size attained by each species. Smaller species, like *H. zosterae* attain maturity in three months, while many other species start breeding at six months to one year of age after birth. Indian sea horses are found to mature at or after first year of birth. Males and females of *H. kuda* attain first sexual maturity at 110 mm size. In *H. kelloggi,* the males mature at 150 mm and females at 165 mm. The size of maturity of males of *H. trimaculatus* is 70 mm and for females it is 75 mm.

Sea horses exhibit continuous spawning throughout the year. The duration of breeding period in sea horses is longer in tropical species. Syngnathid eggs are oval or pear-shaped, semi-transparent and orange in colour. The eggs attain maturity continuously in their ovaries and hydrate the eggs before transferring into male's brood pouch. The eggs transferred in a single mating in sea horses is found to range from 5 in *H. zostarae* to over 1000 eggs in *H. erectus* and *H. reidi*. In *H. kuda*, the fecundity is found to vary from 2451 to 2793. In *H. kelloggi,* this varies from 80 to 1895, and in *H. trimaculatus* from 34 to 1530.

New born syngnathids resemble the adult in body form and are free living in nature. The size of young ones ranged from 2 to 20 mm. Gestation period varies within the species (20-25 days in *H. kuda* and 13 to 14 days in *H. trimaculatus*.

Syngnathids are popular among aquaria and hobbyist and fetching high prices for size and body colour. Except sea dragon, the syngnathids are used as ingredients in Chinese traditional medicines.

Captive Breeding

Sea horses are a fascinating and a remarkable group of fishes with their unusual body shape. A surprising aspect of their biology is that the males incubate the fertilized eggs in a brood pouch. They inhabit many ecologically sensitive aquatic habitats including coral reefs, sea grasses, mangroves and estuaries with most species in the Indo-Pacific and western Atlantic regions. It is their popularity that places them in danger as they are sought in large numbers for use in traditional medicines, as aquarium fish and curios (souvenirs). It has been estimated that atleast 20 million sea horses are being caught annually for the traditional medicine market. In addition, more than one million live sea horses are caught for aquarium trade, mostly destined for sale in North America. India was contributing about 30 per cent of global sea horse trade until 2001, and now all species of sea horses had been brought under the schedule I of the Wild Life Act, 1972 to prohibit their exploitation. The value of sea horses are quite high, the price of dried sea horse in the Hong Kong market ranges from Rs 11500/- to Rs 50400/- per kg depending on the species, quality and size. About 50 countries are involved in sea horse exploitation and trade. Dried sea horses are used as medicine to arrest whooping cough in children.

In response to a significant increase in international demand and the relaxation for captive breeding of sea horse for stock enhancement program, National Institute of Oceanography, Goa have succeeded in breeding of yellow or spotted sea horse (*Hippocampus kuda*) under captivity.

Sea horse collected from wild were grown to maturity. There was pair bonding, mating, complete gestation and spawning under captive condition through manipulation of feed, feeding and environmental conditions.

Two male sea horses delivered 320 new born juveniles. The 300-odd babies with a length of barely 8 mm have survived the more critical pelagic stage. When they grow beyond this stage, they will be able to attach to any objects instead of drifting in the waters.

Unlike other fishes, a sea horse deliver anywhere between a few hundreds to a thousand babies. It takes about a year for them to grow up into adults. It is the father, who incubates the babies in his pouch and the gestation period is about 12 to 15 days. The size of the adult is around 15 to 20 mm.

Endangered Species of Sea Horse

Sea horses are highly modified pipe fish. The earliest known sea horse fossils are dated back to about 13 million years. Strong mate and site fidelity, combined with essential parental care and limited mobility are believed to have made sea horse populations vulnerable to disruption. The smallest sea horse is the dwarf sea horse (an inch long). Large sea horses are found along the Pacific coast of Central America (the foot long *Hippocampus ingem*). The young ones are seen in sea grass meadows, mangrove stands and coral reefs, where they are camouflaged by murky brown and grey patterns that blend into the sea grass back ground. During their social moments or in unusual surroundings, sea horses turn to be bright in sea weed beds in the sea. They can change their body colour, according to the surroundings. The scientific and common names of important sea horses are;

Hippocampus abdominalis (Pot bellied sea horse), *H. barbouri* (Barbours sea horse), *H. bargibanti* (Pygmy sea horse), *H. breviceps* (Reunion sea horse), *H. camelopardalis* (Short snouted sea horse), *H. capensis* (Giraffe sea horse), *H. colemani* (Knysna sea horse), *H. comes* (Colemans sea horse), *H. coronatus* (Tiger tail sea horse), *H. denise* (Crowned sea horse), *H. erectus* (Denises pygmy), *H. fisheri* (Lined sea horse), *H. fuscus* (Fishers sea horse), *H. gulloelatus* (Sea pony), *H. hippocampus* (Speckled sea horse), *H. histrix* (European sea horse), *H. ingeus* (Thorny sea horse), *H. jayakari* (Pacific sea horse), *H. kelloggi* (Jayakars sea horse), *H. kuda* (Indian sea horse), *H. lichtensteinil* (Spotted sea horse), *H. minofauri* (Lichtensteins sea horse), *H. mohnikii* (Bull neck sea horse)

The head of the sea horse is located at right angles of the body. It has a fully prehensile tail, which wraps around sea grass system and corals, sticks or any other suitable natural or artificial object. Sea horse has a thin layer of skin, stretched over a series of bony plates. They do not have any scales on their body. There are, however, ring like formations around its trunk. These rings are useful for identifying the species. Some of the species have bony bumps or skin filaments protruding from these bony rings. Sea horses are master of camouflage, changing colour and growing skin filaments to blend with their surroundings. They have short time colour changes, which are for courtship display. Males can be identified from females from the presence of a brood pouch on the front of the abdomen. Sea horses do not have either stomach or teeth. They feed on prey through a tubular snout and pass food through an efficient deigestive system. They breath through gills, extracting oxygen from the water that passes over them unlike other fishes. However, the gills are small and compact, almost a "grape-like" structure. Sea horses can swim using the propulsive of a quickly oscillating dorsal fin. They also use the pectoral fins on either side of the body for steering and stability. They are adapted to maneuverability than speed. Therefore, they rely primarily on camouflage to avoid detection from predators. Luminous sea horses swim among corals in the dark waters.

The female sea horse releases its eggs into the male pouch, where they hatch out. The days of pregnancy are about 40 to 50 days. Sea horse is the only animal in which the father is pregnant. Sexual maturity in males is usually determined by the presence of a brood pouch. Male sea horses can become pregnant any time during the breeding season. The breeding season varies among the species and this is related to environmental factors. Temperature is one of the key factors for the onset of breeding season. Most of the species of sea horses are monogamous. These monogamous species form into pairs and bond up during the entire breeding season. Although some species may not be pair bonded, they are all marked by daily greetings between them. During breeding, males and females may change colour and promenade and pirouette together. This dance lasts several minutes, and there after they separate for the rest of the day. The aforesaid greetings take place through the male pregnancy. Further, the male and female are ready again to mate on the same day.

The female inserts her ovipositor into the brood pouch of the male, where she deposits her eggs. The fertilized eggs are then embedded in the pouch wall, and they become enveloped with tissues. The pouch acts like the womb of a female mammal, complete with a placental fluid that baths the eggs, and provides nutrients and oxygen to developing embryos, while removing waste products. The pouch fluid is altered during pregnancy from being similar to body fluids to being more like the surrounding sea water. This helps in reducing the stress on the offspring at birth. The pregnancy period is about four weeks. With the increasing temperature at the end of gestation period, the male releases its brood.

Sea horses are pre-trained by the keepers to feed on frozen shrimps (mysis), 2-5 frozen mysis shrimp enhanced per sea horse per feeding as they grow.

While sea horses are often found in sea water less than one meter deep, most species of sea horses have been observed to live at depths between 1 to 15 meters and as deep as 77 meters, when they move out into deeper waters during winter. *H. hippocampus* is found in waters below 5 meters, whereas, *H. guitulatus* is found in shallower depths. They occupy only certain parts of seemingly suitable habitats, for example, sticking to the edges of sea grass, since sea horses find more food in areas of good water exchange. Habitat and substratum preference may be seasonal and related to seasonal migration. Adults may dispense during short range migrations, but most of the movement to new areas happens when adults are cast adrift by storms or carried away while grasping floating debris. Sea horses are globally distributed occurring in the marine waters of atleast of hundred countries. In India, sea horses are diversely distributed around the coast of Ratnagiri, Maharashtra, Gulf of Mannar, Thondi, Thoothukudi (Tamil Nadu), Kollam and Sakthikulangara (Kerala) and Andaman Nicobar Islands.

Chapter 25

Fish Fauna in Indian Seas

Fish Fauna in Near Shore Waters

Puri, one of the southern districts of Orissa state, in east coast of India has 155 km of coast line. Marine fishes, crustaceans and cephalopods, which congregate, within 10 to 22 km from the shore are fished out by the 9765 full time active fishermen of 35 fishing villages by trawl nets, gill nets, drift nets, hooks and lines and shore seines. The catch of these fishing gears, operating at different coastal depth zones (Seine net in near shore surface areas, gill nets in water column and trawl nets in ground bottom) represents a cross section of fish fauna congregations in the coast.

Cat fishes, croakers, carangids, clupeids, ribbonfishes, sardines, prawns, seer fishes,eels and pomfrets are the main migrants and habitants along Puri coast in the order of abundance. More than 60 fin fish species and nearly 20 crustaceans including 15 prawns were found to occur in Puri coast during 2006.

Catfishes, eels, sharks, perches, carangids, seer fishes and tunnies were the main groups of fishes, caught in the hooks and lines. Prawns, big jawed jumper, ribbon fishes, croakers, clupeids and cat fishes are caught in good numbers by bottom trawling. Wolf herring, anchovies, other sardines, other clupeids, horse mackerel, pomfrets, seerfishes, mullets and crabs are maximum in drift and gill net catches.

The details of the fin fish and shell fish species, appeared in this coast during 2006 is given below.

- ☆ Elasmobranchs – Sharks – *Carcharhinus sorrah, Sphyrna lewini.*
- ☆ Rays – *Himantura bleekeri, H. uarnak.*
- ☆ Eels – *Anguilla bicolour bicolour, Muraenesox cinereus.*
- ☆ Cat fishes – *Arius tenuispinis.*

☆ Clupeids – Wolf herring – *Chirocentrus nudus.*

        Oil sardine – *Sardinella longiceps.*

        Other sardines – *Sardinella fimbriata, S. gibbosa.*

        Hilsa shad – *Tenualosa ilisha.*

        Other shads – *Tenualosa toli.*

☆ Anchovies – Setipinna – *Setipinna taty.*

        Stolephorus – *Stolephorus devisi.*

        Thryssa – *Thryssa mystax.*

        Other clupeids – *Ilisha elongate, I. melastoma, Raconda russeliana, Anodontostoma chacunda.*

☆ Bombay Duck – *Harpodon nehereus.*

☆ Half beak and Full beaks – *Strongylura strongylura, Hyporhamphus limbatus.*

☆ Perches – Rock cods – *Epinephelus chlorostigma.*

        Snappers – *Lutjanus johni.*

        Threadfin breams – *Nemipterus japonicus.*

        Other perches – *Pomadasys hasta, P. maculatum, Pricanthus hamrur, Drepane punctata, Kurtus indicus, Sillago sihama, Terapon jarbua.*

☆ Goat fishes – *Upeneus vittatus.*

☆ Threadfins – *Eleutheronema tetradactylum, Polynemus indicus.*

☆ Croakers – *Johnius carutta, J. dussumieri, Kathala axillaries, Nibea maculate, Otolithes rubber.*

☆ Ribbon fishes – *Trichiurus lepturus.*

☆ Carangids – Horse mackerel – *Megalaspis cordyla.*

        Scads – *Decapterus russelli.*

        Leather-jackets – *Scomberoides commersonianus.*

        Other carangids – *Rachycentron canadum, Carangoides armatus, Caranx ignobilis, Selar crumenophthalmus, Coryphaena hippurus.*

☆ Silverbellies – *Leiognathus bindus, Secutor insidiator.*

☆ Big-jawed jumper – *Lactarius lactarius.*

☆ Pomfrets – Black pomfret – *Parastromateus niger.*

        Silver pomfret – *Pampus argenteus.*

        Chinese pomfret – *Pampus chinensis.*

☆ Mackerels – Indian mackerel – *Rastrelliger kanagurta.*

☆ Seerfishes – Narrow-barred Spanish mackerel – *Scomberomorus commerson.*

        Indo-Pacific Spanish mackerel – *Scomberomorus guttatus.*

☆ Tunnies – Little tunas – *Euthynnus affinis.*

☆ Billfishes – *Istiophorus platypterus*.

☆ Barracudas – *Sphyraena barracuda*.

☆ Mullets – *Mugil cephalus*.

☆ Flatfishes – Soles – *Cynoglossus bilineatus*.

☆ Crustaceans – Penaeid prawns – *Solenocera crassicornis, Metapenaeus affinis, M. dobsoni, M. monoceros, M. lysianassa, Metapenaeopsis stridulans, Parapenaeopsis hardwickii, P. stylifera, Parapenaeus longipes, Penaeus indicus, P. japonicus, P. Penicillatus, P. merguiensis, P. monodon*.

Non-penaeid prawns – *Acetes indicus, Nematopalaemon tenuipes*.

☆ Crabs – *Portunus sanguinolentus*.

☆ Stomatopods – *Oratosquilla neap*.

☆ Molluscs – Cephalopods – *Sepia aculeate*.

☆ Miscellaneous – *Antennarius hispidus*.

P. merguiensis and *P. penicillatus* were mainly caught by bottom set gill nets, whereas, all other varieties of prawns were the main stay of trawl nets.

Cat fishes, eels, sharks, perches, carangids, seer fishes and tunnies were the main resources found in the hooks and lines catches. Prawns, big jawed jumper, ribbon fishes, croakers, clupeids and cat fishes were found in good numbers in trawl catches. Wolf herring, anchovies, other sardines, other clupeids, horse mackerel, pomfrets, seerfishes, mullets and crabs are maximum in drift and gill net catches.

Appearance of Fishes in Near Shore Waters during Monsoon at Karwar, Karnataka

In the near shore areas, within 10 m depth during monsoon of 2008 (July and August) at Karwar, the burrowing goby, *Trupuachen vagina* was found to form 62.5 per cent of 190 kg followed by the tail-eyed goby, *Parachaeturichthys polynema* forming 12 per cent of 36.2 kg, shrimps (10.81 per cent) with a CPUE of 32.7 kg, *Ambassis* sp. (7.74 per cent) with a CPUE of 23.43 kg, portunid crabs (4.48 per cent) with CPUE of 13.5 kg and species like, *Lactarius lactarius, Platycephalus crocodiles*, Eels, *Scomberomorus commerson* (juveniles), *Lagocephalus inermis, Johnius* spp., *Cynoglossus macrostomus, Solea elongate, Leiognathus* spp., *Thryssa* spp.

T. vagina form the dominant species with their proportion increasing in August. Size range vary from 95 to 200 mm in total length. *P. polynema* of 84 to 104 mm appear in near shore area with two species of Sciaenids (*Johnius belanguri* and *J. carutta*) and six species of Leiognathids, namely *Leiognathus bindus, L. blochii, L. brevirostris, L. splendens, Secutor insidator* and *S. ruconis*. Of these *L. blochii* and *S. ruconis* were dominant. The other species observed to appear are *Pisodonophis canerivorous* among eel, *Thryssa malabaricus, T. setirostris* and *T. vittirostris* among Engraulids; *Penaeus merguiensis* and *Fenneropenaeus indicus* among penaeid prawns and *Portunus sanguinolentus* and *P. pelagicus* among Portunid crabs.

Juveniles of *Scomberomorus commerson, Leiognathus* spp., *Alectis ciliaris, A. indicus, Gnathonodon speciosus, Trachinotus blochii, Lutjanus russelli, L. johni* and *Sardinella* spp were observed regularly occurring in the area.

The annual occurrence of *T. vagina* and *P. polynema* is a purely monsoon phenomena, and is absent during other seasons of the year. They get transported to near shore seas by local currents during monsoon, a southerly drifts of current along the north Kanara coast. Due to lowering of surface salinity (6-15.5 ppt) in monsoon months, important commercial species like *Sardinella longiceps* and *Rastrelliger kanagurta* fail to appear.

Fish Fauna in Coastal and Off-Shore Waters of India

West Bengal, the northern most maritime state in the east coast of India, situated between 20°38 minute to 27°10 minute N latitude and 85°38 minute to 89°38 minute E longitude with a coastal length of 158 km. The fish fauna occurring in a 70 m depth zone during 2007 consisted mainly of hilsa shad (22 per cent), bombayduck (15 per cent), other clupeids (10 per cent), croakers (9 per cent), non-penaeid prawns (7 per cent), pomfrets (6 per cent) and catfishes, anchovies, and penaeid prawns (5 per cent each).

Main species, which contributed to each category were as follows;

☆ Hilsa shad – *Tenualosa ilisha*.

☆ Bombayduck – *Harpodon nehereus*.

☆ Croakers – *Otolithes* sp., *Johnius* spp., *Protonibea diacanthus, Otolithoides biauritus, Johnieops* spp. and *Nibea maculate*.

☆ Non-penaeid prawns – *Acetes* spp., *Plesionika* spp.

☆ Pomfrets – *Pampus argenteus, P. chinensis* and *Parastromateus niger*.

☆ Catfishes – *Arius thalassinus, A. tenuispinis, Osteogeneiosus militaris, Pangasius pangasius* and *Plotosus limbatus*.

☆ Anchovies – *Coilia dussumieri* and *Setipinna taty*.

☆ Penaeid prawns – *Solenocera hextii, Metapenaeopsis stridulans, Metapenaeus affinis, M. brevicornis, M. monoceros, M. dobsoni, Parapenaeopsis hardwickii, P. stylifera, Parapenaeus* spp., *Penaeus canaliculatus, P. indicus* and *P. monodon*.

☆ Clupeids – *Amblygaster leiogaster, Chirocentrus nudas, Stolephorus commersonii, S. indicus, Thryssa mystax, T. dussumieri, Ilisha megaloptera, Opisthopterus tardoore, Raconda russeliana, Escualosa thoracata, Dussumieria acuta,* and *Anodontostoma chacunda*.

☆ Ribbonfishes – *Lepturacanthus savala* and *Trichiurus lepturus*.

☆ Seerfishes – *Scomberomorus guttatus*.

Orissa, located between 17°48 minute to 22°34 minute N latitude, and 81°24 minute to 87°29 minute E longitude has a coast line of 480 km. Fish fauna occurring within 40 km from the shore, mainly consists of croakers (14 per cent), penaeid prawns (11 per cent), ribbonfishes (9 per cent), carangids (9 per cent), sardines (8 per cent), anchovies (6 per cent), catfishes (4 per cent), pomfrets (4 per cent), silverbellies (3 per cent), other clupeids (7 per cent) and non-penaeid prawns (3 per cent).

Different species which contribute to the commercially important fish fauna are;

☆ Croakers – *Pennahia macrophthalmus, Johnius carcutta, J. dussumieri, Otolithes cuvieri, O. rubber, Nibea maculate, Otolithoides biauritus* and *Protonibea diacanthus.*

☆ Penaeid prawns – *Solenocera crassicornis, Metapenaeus dobsoni, M. affinis, M. monoceros, Metapenaeopsis stridulans, Parapenaeopsis hardwickii, P. maxillipedo, P. sculptilis, P. stylifera, Penaeus indicus, P. merguiensis, P. monodon* and *P. penicillatus.*

☆ Ribbonfishes – *Trichiurus lepturus* and *Lepturacanthus savala.*

☆ Carangids- *Megalaspis cordyla, Selar crumenophthalmus, Scomberoides commersonianus, S. lysan.*

☆ Catfishes – *Arius dussumieri, A. caelatus, A. jella, A. platystomus, A. tenuispinis, A. thalassinus* and *Osteogeneiosus militaris.*

☆ Pomfrets – *Parastromateus niger, Pampus argenteus,* and *Pampus chinensis.*

☆ Silverbellies – *Leiognathus bindus, L. equulus, L. fasciatus, L. splendens* and *Secutor insidiator.*

☆ Non-penaeid prawns – *Acetes indicus* and *Nematopalaemon tenuipes.*

☆ Anchovies – *Setipinna taty, Coilia dussumieri, Stolephorus devisi, S. commersonii* and *S. punctifer.*

☆ Other clupeids – *Chirocentrus dorab, C. nudus, Thryssa mystax, T. dussumieri, Reconda russeliana, Pellona ditchela, Ilisha filigera, I. megaloptera, I. melastoma, Tenualosa toil, Escualosa thoracata, Dussumieria elopsoides, D. acuta, Anodontostoma chacunda* and *Megalops cyprinoids.*

Andhra Pradesh, having 974 km of coastal length is situated between 12°41 minute to 22°N latitude and 77°to 84°40 minute E longitude. The continental shelf of 31000 sq. km support a rich fish fauna as below.

☆ Penaeid prawns – *Solenocera crassicornis, S. hextii, Aristaeomorpha woodmasoni, Aristeus semidentatus, Metapenaeopsis* spp., *Metapenaeus affinis, M. brevicornis, M. dobsoni, M. monoceros, Parapenaeopsis stylifera, Penaeus indicus, P. japonicus, P. merguiensis, P. monodon, P. semisulcatus* and *Trachypenaeus curvirostris.*

☆ Carangids – *Rachycentron canadum, Alepes* spp., *Caranx ignobilis, C. sexfasciatus, C. sem, Megalaspis cordyla, Scomberoides commersonianus, S. lysan, S. tol, Selar boops, S. crumenophthalmus, Selaroides leptolepis, Seriolina nigrofasciata, Decapterus russelli, Mene maculate* and *Coryphaena hippurus.*

☆ Perches - *Argyrops spinifer, Priacanthus* spp., *Apogon* spp., *Ambassis* spp., *Lates calcarifer, Sillago sihama, Nemipterus japonicus* and *N. mesoprion.*

☆ Croakers – *Chrysochir aureus, Johnieops sina, Johnius belangerii, J. carutta, J. dussumieri, J. macropterus, Kathala axillaries, Nibea maculate, N. soldado, Otolithes rubber* and *Protonibea diacanthus.*

☆ Elasmobranchs – *Carcharhinus dussumieri, Rhizoprionodon acutus, Sphyrna zygaena, Rhina ancylostoma, Rhynchobatus djiddensis, Dasyatis microps, D. kuhlii, Gymnura* spp., *Himantura bleekeri, Aetobatus* spp., *Aetomylaeus* spp., *Rhinoptera* spp., *Mobula diabolus, Narcine timle* and *Torpedo marmorata.*

☆ Pomfrets – *Pampus argenteus, P. chinensis, Parastromateus niger.*

☆ Ribbonfishes – *Lepturacanthus savala* and *Trichiurus lepturus.*

☆ Sardines – *Sardinella longiceps, S. albella, S. fimbriata* and *S. gibbosa.*

☆ Seerfishes – *Scomberomorus commerson, S. guttatus* and *S. lineolatus.*

☆ Other clupeids – *Anodontostoma chacunda, Dussumieria actuta, Escualosa thoracata, Tenualosa ilisha, Hilsa kelee, Ilisha* spp., *Opisthopterus tardoore, Pellona ditchela, Chirocentrus dorab, C. nudus, Chanos chanos* and *Raconda russeliana.*

☆ Crabs – *Calappa lophos, Scylla serrata, Portunus pelagicus, P. sanguinolentus, Charybdis cruciata* and *Varuna litterata.*

Situated between 8°5 minute to 13°35 minute N latitude and 76°15 minute to 80°20 minute E longitude, Tamil Nadu with 1076 km coast line and 41400 sq km continental shelf is bordered with the Bay of Bengal and Palk Bay in the east, the Gulf of Mannar in the south east, the Indian Ocean in the south and the Arabian Sea in the south west.

The occurrence of fish fauna is the maximum in July-September (30 per cent), followed by October-December (27 per cent). The lean period was from April-June (18 per cent).

The main fisheries available in Tamil Nadu coast were oil sardines (15 per cent), other sardines (11 per cent), silverbellies (11 per cent), penaeid prawns (4 per cent), Indian mackerel (4 per cent), crabs (3 per cent) and *Stolephorus* spp. (3 per cent).

Main species contributing to each category of fishes were as follows;

☆ Oil sardines – *Sardinella longiceps.*

☆ Silverbellies – *Gazza minuta, Leiognathus berbis, L. equulus, L. lineolatus, L. splendens, Secutor insidiator* and *S. ruconius.*

☆ Other sardines – *Sardinella albella, S. fimbriata, S. gibbosa, S. sirm* and *S. leiogaster.*

☆ Perches – *Psammoperca waigiensis, Cephalopholis sonnerati, Epinephelus merra, Terapon* spp., *Priacanthus* spp., *Sillago* spp., *S.sihama, Scolopsis* spp., *Lutjanus* spp. and *Nemipterus bipunctatus*

☆ Other clupeids – *Tenualosa ilisha, Dussumieria acuta, Escualosa thoracata, Hilsa kelee, Pellona ditchela, Coilia dussumieri, Stolephorus butavlensis, S. macrops, Thryssa dussumieri* and *Chirocentrus dorab.*

☆ Indian mackerel – *Rastrelliger kanagurta.*

☆ Penaeid prawns – *Solenocera crassicornis, S. hextii, Aristeus semidentatus, Metapenaeopsis andamanensis, M. stridulans, Metapenaeus affinis, M. brevicornis, M. dobsoni, M. monoceros, Parapenaeopsis stylifera, P. uncta, Penaeus indicus,*

P. japonicus, P. merguiensis, P. monodon, P. semisulcatus and *Trachypenaeus curvirostris.*

☆ *Stolephorus* spp. – *Stolephorus bataviensis, S. commersonii, S. devisi* and *S. indicus.*

☆ Crabs - *Calappa lophos, Scylla serrata, Portunus* spp., *Portunus pelagicus, Charybdis cruciata* and *natator.*

Forty five km of Pondicherry coast lying within 11°56 minute N latitude and 79°53 minute E longitude harbour the following important species.

☆ Oil sardine – *Sardinella longiceps.*

☆ Indian mackerel – *Rastrelliger kanagurta.*

☆ Other sardines – *Sardinella gibbosa* and *S. sirm.*

☆ Carangids – *Rachycentron canadum, Atropus* spp., *Decapterus russelli, Selaroides* spp., and *Nemipterus* spp

☆ Seerfishes – *Scomberomorus commerson, S. guttatus, S. lineolatus.*

☆ Flying fishes – *Exocoetus volitans.*

☆ Penaeid prawns – *Metapenaeopsis stridulans, Metapenaeus affinis, M. dobsoni, Penaeus indicus, P. monodon.*

Situated between 8°18 minute to 12°48 minute N latitude and 74°52 minute to 77°22 minute E longitude, the sea bordering Kerala is rich with marine fauna. Major fish fauna in kerala waters during 2007, were observed is oil sardines (40 per cent), mackerel (11 per cent), penaeid prawns (7 per cent), threadfin breams (5 per cent), scads, *Stolephorus* spp., tunnies and cephalopods (4 per cent each), soles (3 per cent), croakers, ribbonfishes, seerfishes and other carangids (2 per cent each).

Major species occurred under each category were;

☆ Oil sardine – *Sardinella longiceps.*

☆ Mackerel - *Rastrelliger kanagurta.*

☆ Penaeid prawns – *Metapenaeus dobsoni, M. monoceros, M. affinis, Solenocera hextii, S. choprai, Aristeus alcocki, Metapenaeopsis andamanensis, Parapenaeopsis stylifera, Penaeopsis jerry, Penaeus indicus, P. monodon, P. semisulcatus and Trachypenaeus curvirostris.*

☆ Threadfin bream – *Nemipterus japonicus* and *N. mesoprion.*

☆ Scads - *Decapterus russelli,* and *D. macrosoma.*

☆ Stolephorus - *Stolephorus commersonnii, S. punctifier* and *S. devisi.*

☆ Tunnies - *Euthynnus affinis, Auxis rochei, A. thazard, Katsuwonus pelamis, Sarda orientalis, Thunnus albacares,* and *T. tonggol.*

☆ Cephalopods – *Sepia aculeate, S. pharaonis, Sepiella inermis, Loligo duvaucelli* and *Octopus* spp.

☆ Soles - *Cynoglossus macrostomus, C. bilineatus,* and *Synaptura commersoniana.*

☆ Croakers - *Johnieops* spp., *Johnius* spp., *Nibea maculate* and *Otolithes* spp.

☆ Ribbonfishes - *Lepturacanthus savala* and *Trichiurus lepturus*.

☆ Seerfishes - *Scomberomorus commerson* and *S. guttatus*.

☆ Other carangids – *Alectis ciliaris, A. indicus, Alepes djeddaba, A. mati, A. melanopterus, A. para, Atropus atropos, Carangoides malabaricus, Caranx ignobilis, Megalaspis cordyla, Elagatis bipinnulata, Scomberoides commersonianus, S. lysan, Selar crumenophthalmus, Seriolina nigrofasciata* and *Trachinotus blochii.*

The Karnataka coast confined within 11°5 minute to 18°5 minute N latitude and 74°to 78°5 minute E longitude harbour marine resources of oil sardine (33 per cent), Indian mackerel (15 per cent), threadfin breams (7 per cent), stomatopods (96 per cent), ribbonfishes (5 per cent), penaeid prawns (5 per cent) and cephalopods.

Main species contributed to each of the above category were;

☆ Oil sardine - *Sardinella longiceps.*

☆ Indian mackerel – *Rastrelliger kanagurta.*

☆ Threadfin breams – *Nemipeterus bipunctatus, N. mesoprion, N. japonicus.*

☆ Stomatopods – *Oratosquilla neap.*

☆ Other clupeids – *Thryssa* spp., *Stolephorus waitei, Sardinella fimbriata, S. ulbella, S. gibbosa, S. brachysoma, S. melanura, Amblygaster leiogaster, Opisthopterus tardoore, Hilsa kelee* and *Tenualosa* spp.

☆ Ribbonfishes – *Trichiurus lepturus* and *Lepturacanthus savala.*

☆ Penaeid prawns – *Penaeus indicus, Parapenaeopsis stylifera, Metapenaeus dobsoni, M. monoceros, M. affinis,* and *Solenocera choprai.*

☆ Cephalopods – *Sepia* spp., *Loligo duvaucelli* and *Octopus* spp.

Located on the western coast between latitudes 15°43 minutes to 14°53 minute N and longitudes 74°20 minutes to 73°40 minutes E, Goa has a coastal length of 104 km, where the main marine resources were oil sardine (49 per cent), Indian mackerel (20 per cent), carangids (6 per cent), *Auxis* spp., perches, catfishes and other sardines (3 per cent each).

Main species contributed to each category were as follows;

☆ Oil sardine – *Sardinella longiceps.*

☆ Indian mackerel – *Rastrelliger kanagurta.*

☆ Carangids – *Megalaspis cordyla, Decapterus russelli, Scomberoides commersonianus, S. lysan, S. tala* and *S. tol.*

☆ *Auxis* spp. – *Auxis rochei* and *A. thazard.*

☆ Catfishes – *Arius* spp. and *Osteogeneiosus militaris.*

☆ Perches - *Nemipterus japonicus, N. mesoprion, Epinephelus diacanthus.*

☆ Other sardines – *Sardinella albella, S. brachysoma, S. dayi, S. fimbriata, S. gibbosa, S. melanura, Amblygaster clupeoides A. leiogaster.*

Maharashtra with a coastal length of 720 km is located between 15°52 minute to

21°34 minute N latitude and 72°56 minute to 80°30 minute E londitude have the marine resources dominated by penaeid prawns (14 per cent), non-penaeid prawns (12 per cent), oil sardine (9 per cent), Bombayduck (9 per cent), croakers (9 per cent), perches (5 per cent) ribbonfishes (4 per cent) and cephalopods (4 per cent).

Main species contributing to each category are;

- ☆ Penaeid prawns – *Parapenaeopsis stylifera. P. sculptilis, Metapenaeus affinis, M. dobsoni, M. kutchensis, M. brevicornis, M. monoceros, Metapenaeopsis andamanensis, M. stridulans,* and *Penaeus merguiensis.*

- ☆ Non-penaeid prawns – *Acetes indicus* and *Nematopalaemon tenuipes.*

- ☆ Bombayduck – *Harpodon nehereus.*

- ☆ Croakers – *Otolithoides biauritus, Johnius* spp., *Johnieops* spp., *Protonibea diacanthus* and *Otolithes cuvieri.*

- ☆ Perches – *Nemipterus japonicus.*

- ☆ Ribbonfishes – *Trichiurus lepturus.*

- ☆ Cephalopods – *Loligo duvaucelii, Sepia pharaonis,* and *S. aculeate.*

- ☆ Catfishes – *Arius dussumieri.*

- ☆ Sharks - *Carcharhinus* spp., *Scoliodon* spp.

- ☆ Clupeids – *Coilia dussumieri.*

- ☆ Mackerel – *Rastrelliger kanagurta.*

Gujarat, having 1600 km coastline and 2.4 lakh sq km EEZ lying within 20°32 minute to 25 gegree 25 minute N latitude and 68°32 minute to 76°70 minute E longitude have the following major marine resources, like, non-penaeid prawns (13 per cent), croakers (12 per cent), ribbonfishes (10 per cent), penaeid prawns (5 per cent), catfishes (5 per cent), threadfin breams (4 per cent), other perches (4 per cent) and carangids (4 per cent).

The major species contributed to each group are;

- ☆ Non-penaeid prawns – *Acetes* spp., *A. indicus, Nematopalaemon tenuipes* and *Exhippolysmata ensirostris.*

- ☆ Croakers – *Johnius* spp., *Otolithes* spp., *Otolithoides biauritus* and *Protonibea diacanthus.*

- ☆ Ribbonfishes – *Eupleurogrammus* spp., *Lepturacanthus savala* and *Trichiurus* spp.

- ☆ Bombayduck – *Harpodon nehereus.*

- ☆ Cephalopods – *Sepia elliptica, S. pharaonis, Sepiella* spp., *Loligo* spp., *Octopus* spp.

- ☆ Clupeids – *Tenualosa* spp., *T. ilisha, T. toil, Ilisha megaloptera, Sardinella* spp., *Coilia dussumieri, Thryssa mystax* and *Chirocentrus dorab.*

- ☆ Penaeid prawns – *Solenocera crassicornis, Metapenaeus affinis, M. kutchensis, M. monoceros, Parapenaeopsis* spp., *P. stylifera, P. hardwickii* and *P. sculptilis.*

- ☆ Catfishes – *Arius* spp., *Osteogeneiosus* spp. and *O. militaris.*

☆ Threadfin breams – *Nemipterus* spp.

☆ Other perches – *Epinephelus* spp., *Priacanthus* spp., *Lethrinus* spp., *Pomadasys* spp., *P. kaakan* and *Argyrops spinifer*.

☆ Carangids – *Caranx* spp., *Decapterus russelli, Megalaspis cordyla, Scomberoides commersonianus S. lysan* and *Coryphaena hippurus*.

Fish Fauna around Andaman and Nicobar Islands

Andaman and Nicobar group of islands, located between 6 and 14°N latitude and 91 and 94°E longitude in Andaman Sea, have the oceanic research dates back to 1869, when Francis Day, a well known army officer and fishery biologist visited these islands. He recorded the occurrence of 136 species of fish in the Andaman waters. These species were described in his famous volume, " The Fishes of India ", published in 1878. This investigation gained further impetus with the inclusion of a few Surgeon Naturalists in the Marine Survey of India in 1875.

The earliest account of the fishery resources of Andaman Sea came from the results of trawling by Golden Crown during 1908-09. It indicated a " good potential " around the Andaman Islands, but no specific data are traceable. Subsequent to this, there is no record of any regular survey or exploration. But towards the end of forties, a private company, called Andamarine Development Corporation Ltd was set up for exploration of fishery resources of Andaman Sea. Their master fisherman, W.R.Burgess stated that the potentialities are enormous. The company started operation with an Australian built fishing boat, fitted with diesel engine, a flat bottom patrol launch and two dinghies. This did not succeed. Records of the time and reason for the closure of the company are not available. In 1951 another fishing operation started, but without success. In 1952, during the first five year plan period, a fisheries research officer was appointed to start a planned program of development. Later, a nucleus of a fisheries department was established, but the pace of development remained slow.

Apart from the tribals, who are not professionals in the true sense, since they do not fish for profit, there was no endemic community of fishermen in these islands. During the subsequent period, the Government initiated a fishermen settlement scheme, under which fishermen, mainly from the southern states of India were brought to the islands. There has also been some voluntary migration of fishermen from Andhra Pradesh.

Sixtytwo commercially important fish species under 19 groups are stated below with their common names;

Elasmobranchs

☆ Sharks - *Charcharius* spp. (Black shark); *Chiloscyllium indicum* (Cat shark); *Hemigaleus balfouri* (Balfour's shark); *Scoliodon* spp. (Dog shark); *Sphyrna blochii* (Hammer head shark); *S. zygaena* (Hammer head shark).

☆ Skates – *Rhynchobatus dijjiddensis* (Shovel nose); *Pristis cuspidatus* (Pointed sawfish); *P. microdon* (Small toothed sawfish).

☆ Rays – *Amphotistius kuhlii* (Blue spotted sting ray); *Aetobatus narinari* (Spotted eagle ray); *Myliobatis nichofii* (Eagle ray); *Rhinoptera javanica* (Javanese cow ray); *Mobula diabolus* (Devil ray); *Himantura uarnak* (Whip tail sting ray).

Teleosts

Tylosurus spp. (Car fish); *Decapterus russelli* (Scad); *Selar mate* (Scad); *Megalaspis cordyla* (Torpedo trevally); *Chorinemus tala* (Queen fish); *Caranx sexfasciatus* (Six banded trevally); *C. melampygus* (Black tipped trevally); *Elegatis bipinnulatus* (Rainbow runner); *Sardinella melanura* (Black tipped sardine); *S. jussieu* (Tembang); *Harengula ovalis* (Spotted herring); *Pellona ditchela* (Toother shad); *Herklotsichthys punctatus* (Spotted herring); *Chirocentrus dorab* (Wolf herring); *Dussumieria* spp. (Rainbow sardine); *Thrissocles* spp. (Short jaw anchovy); *Anchoviella* spp. (White bait); *Hemirhamphus marginatus* (Half beak); *Leiognathus equulus* (Silverbelly); *Lethrinus* spp. (Bream); *Lactarius lactarius* (Big jawed jumper); *Lates calcarifer* (Cock-up); *Lutjanus* spp. (Snapper); *Liza* spp. (Silverbelly); *Pomadasys* spp. (Grunter); *Polynemus* spp. (Tassel fish); *Johnius* spp. (Jew fish); *Otolithes* spp. (Croaker); *Sciaena* spp. (Croaker); *Epinephelus* spp. (Rock cod); *Sillago sihama* (Silver whiting); *Saurida tumbil* (Lizard fish); *Sphyraena jello* (Barracuda); *Rastrelliger kanagurta* (Mackerel); *R. brachysoma* (Chub mackerel); *Scomberomorus commersoni* (Spanish mackerel); *Indocybium guttatum* (Spotted Spanish mackerel); *Acanthocybium solanderi* (Wahoo); *Auxis thazard* (Frigate mackerel); *Euthynnus affinis* (Little tuna); *Gymnosarda unicolour* (Dog tooth tuna); *Katsuwonus pelamis* (Skipjack); *Neothunnus macropterus* (Yellowfin tuna); *Krishinoella tongol* (Northern bluefin); *Makaira* spp. (Marlin); *Tetrapturus brevirostris* (Short nosed marlin); *Istiophorus gladius* (Sail fish); *Nemipterus japonicus; Upeneoides* spp. (Threadfin bream).

The most important constituents of the fish hauna are in the order of priority, perches, caranx, seer, mullets, mackerel, anchovies, sardines and silverbellies. The low annual landings of sharks, rays and tunnies were due to the fact that almost the entire landings is from the middle and south Andamans due to the concentration of fishing activities in that region. There was comparatively little fishing activity in the Little and Great Nicobar Islands. The fishing activity in the North Andaman was also low being restricted in two centers only.

The best season for fish concentration appears to be the south west monsoon and post monsoon months of July to December. There is a distinct lowering of fish appearance during January to April. The fishery begins to build up from May onwards. This seasonal fluctuation was observed in fisheries of the plankton feeding fishes. Both tunny and seer do not follow this general patterns of fisheries fluctuation. In their cases, the best fishery appears to occur during the months preceding the south west monsoon and the catch of these two are at the lower ebbs during the south west monsoon and subsequent months.

The exploration of Andaman Sea covered mainly eastern seaboard of the Andaman group of islands with a cursory examination of a part of the south western coast of South Andaman, between latitude 10°30 minute N and 13°40 minute N.

Three fishing vessels (one 32.28 m and two 17.5 m) of Exploratory Fisheries Project of Government of India were employed between October 1971 to March, 1973.

A total fishing effort of 2420 hours (including searching time) was put in 32 small squares (100 square nautical miles, each with 10 minutes of latitude and 10 minutes of longitude as its side) of 7 major squares (3600 square nautical miles, each with one degree latitude and one degree longitude at its side).

The total fishing effort was made up of five types of fishing methods, namely, bottom trawling (for bottom fauna), long lining, purse seining, hand lining and trolling. Of the total fishing effort, more than 50 per cent was expended in bottom trawling (1117 hours). Next in order were long lining (609 hours), trolling (243 hours), hand lining (137 hours) and purse seining (315 hours).

Bottom trawling was first tried in the restricted area, on the east coast, for uneven bottom topography. The continental shelf in this region is very narrow, extending only from 3 to 8 km off the east and west coast of the islands with steep slopes, which are not very suitable for bottom trawling. The intervening sea between the islands in the north and south direction, however, the shelf is favourable for bottom trawling. The long stretches of coast line having fringing reef involve great risk for bottom trawling. The echograph survey to locate patches of relatively uniform sea bottom for safe trawling revealed that hardly there are patches, where trawling could be done continuously for more than one and half hours without encountering underwater obstructions.

Andaman Sea are qualitatively very rich in fish fauna. Over 400 species were reported from these areas. But only about 62 species are commercially important, of which most of them are commonly represented in the trawl catches.

The composition of trawl catches and their seasonal distribution is slightly different from each other, in that in the later years, greater sub-divisions in grouping were followed. The composition of elasmobranches varied from 18.1 to 27.4 per cent, catfish from 1.15 to 6.8 per cent, perches from 0.95 to 6.4 per cent. The miscellaneous group constituted nearly 65 to 75 per cent of the total catch in the first three years. During the first three years, the data collected were limited to elasmobranches, catfishes, perches, big and small miscellaneous fish, which did not yield enough informations for drawing meaningful conclusions regarding other species grouped under the latter two categories.

During 1975-76, the data were collected for more groups, like, leiognathids, sciaenids, nemipterids, upeneoides, lizard fish and other small fishes. The later sub-grouping of small miscellaneous as done in 1975-76 showed that leiognathids constitute 28.8 per cent of the miscellaneous group, sciaenids stand next with 12.9 per cent followed by upeneoids (10.0 per cent), nemipterids (2.5 per cent), lizard fish (1 per cent) etc. Other small fishes constitute rest of the miscellaneous class (15.7 per cent). The percentage of quality fish was negligible, the bulk being elasmobranches (18.1 to 27.4 per cent) and small miscellaneous fishes (65.0 to 77.8 per cent).

The relative catch rates of three vessels for different areas indicate certain trends. One peak representing high catches always occur during the months, January to April after the north east monsoon. This period coincides with the pre-south west monsoon months, when the temperature of sea water starts rising. Then there was low catch in the month of May, followed by uncertain catch for next three to four months. The high catch rate recorded during August, 1972 (199.2 kg/hour) and July 1975 (259 kg/hour) may be due to the low effort put there. However, in June and July high catch rates were recorded. During September-October, the catches were generally low. From November onwards, the catch rates tend to rise. But there is apparently no consistent pattern. A catch rate of 100 kg/hour and above was recorded during four months in 1972, two months in the first half of 1973, five months in 1974 and seven months in 1975.

The catch from deeper areas were found to be very poor and erratic. The highest catch rate of 1181 kg/hour was recorded, when a school of red snapper was caught in a single haul. A catch rate of 620 kg/hour was also obtained when operating around 40 meters depth off Little Andaman. The main constituent of the catch was perches. A catch rate of 300 kg/hour was also recorded by the third vessel (17.5 m). The overall catch rate of 32.28 m vessel for the whole period has been only 148.8 kg/hour.

The depthwise distribution of trawl catches from Andaman Sea for the year 1972 is given below;

While the higher catch rates have been observed in 20-40 m (111.0 to 201.5 kg/hour) and in 56-65 m (147.0 to 250.1 kg/hour) depth ranges. During 1975-76 miscellaneous fish catches was highest (90.1 kg/hour) in 20-39 m depth range, leiognathids (33.1-35.6 kg/hour in 40-79 m depth), elasmobranches (25.4-25.7 kg/hour in 40-79 m depth), sciaenids (12.9-18.3 kg/hour), upeneoids (11.8-12.0 kg/hour), catfishes (3.2-3.8 kg/hour), nemipterids (3.0-3.1 kg/hour), perches (0.8-1.5 kg/hour), lizard fish (0.7-1.7 kg/hour in depth range of 40-59 and 60-79 meters. In shallower depth (20-39 m), miscellaneous fish dominated the catch (90.1 kg/hour), with a small percentage of elasmobranches (17.3 kg/hour), catfishes (3.3 kg/hour) and other quality fishes (1.0 kg/hour). Highest catch rate of 111.7 kg/ hour for all fish was obtained from 20-39 meter depth belt, which consisted mainly of miscellaneous fish, elasmobranches, catfishes. No significant variation in the catch rates of elasmobranches, leiognathids, sciaenids, upeneoids, catfishes, nemipterids, perches and lizard fish was observed from 40-59 and 60-79 meter depth belts, where these varieties were predominant.

Comparable results of catch rates obtained by 17.5 m trawlers during 1975-76 in different regions of Indian coast indicate that Andaman Sea is equally rich (100-112 kg/hour from a depth range of 20-79 meter) that of north west coast (126-140 kg/hour), south west coast (77-153 kg/hour), lower east coast (92-119 kg/hour) and upper east coast (61-163 kg/hour) on similar depth belts.

In the long line operated in Andaman Sea, *Charcharius* spp. was the most common variety of sharks caught. Thresher sharks, *Alopias vulpinus*, which are known to prey upon lesser tunas, were also caught frequently, indicating a fairly good concentration of tuna. The scombroid catch was low. Of the scombroids, the marlins and sail fish were the most common. Hooking rates of tuna was very low. Closer to Port Blair harbour, a yellowfin tuna weighing 32 kg was caught, which was the biggest specimen ever caught during the operations. Juveniles of yellowfin and skipjack were caught frequently.

Trolling was found particularly successful for schooling species, like skipjack. A catch of 68 kg was obtained with an effort of 1.33 hours. Hooking of 23 fishes took place in less than 10 minutes emphasizing, the need to chase a school and run through it. The areas east and south of Port Blair were found to be very rich in carangids, with some of them weighing more than 20 kg a piece. Off Little Andaman, a barracuda weighing 22 kg was caught. Varietywise percentage of catch obtained by trolling are, Carangids – 28.0 per cent, Miscellaneous fishes – 18.7 per cent, Tuna – 16.5 per cent, Seerfish – 13.7 per cent, Shark – 10.8 per cent, Barracuda – 7.2 per cent, and perches – 4.0 per cent.

The invisible bank having an area of 250 square miles with a water depth of 10-100 fathoms has been found to be very rich, where catch per hour was between 25 to 40.43 kg in trolling and hand line. The species caught were tuna, barracuda, carangids, sharks, seer, rainbow runner and perches.

In the hand line miscellaneous fish constituted 45.78 per cent of the total catch, fishes belonging to the family Sparidae being the important group. Perches and small sharks constituted 23.02 per cent and 14.69 per cent respectively. Typical pelagic varieties like carangids, tuna and barracuda was also obtained.

Trial fishing conducted by purse seine has not been successful. Only on one occasion, a small quantity of flying fish was caught.

Fish Stock Assessment in Andaman Sea

Based on the estimated organic production of 0.3-0.5 g /sq cm within the 20 m belt and with a potential yield of 0.2-0.3 per cent of the total potential yield in the area covered by latitude 6°to 15°N and longitude 91 to 95°E, was estimated at 50000 tonnes (Kumaran, 1973). Out of which 12000 tonnes will be from 200 m depth zones. Jones and Banerjee (1973) estimated the shelf area to be of the order of 16000 sq km and computed the standing stock of demersal fish at 4000 tonnes. Against this, the computation of the standing stock based on the exploratory study works out to 44576 tonnes.

The swept area of 24 m fish trawl operated from 17.5 m trawler at a trawling speed of 2 knots is taken as 0.03914 sq km. At an average of 110 kg/hour of trawling, the total standing crop works out to 2.786 tonne/sq km. On the basis of this, the total standing crop of the demersal fishery should be of the order of 44576 tonnes, which is almost ten times that of the figure arrived by Jones and Banerjee (1973).

Bottom trawling over the period of survey indicated that though the scope for this method was not of the same level as in the mainland, the catch rate of 109 kg/ hour is comparable to those on the east coast and marginally better than those obtained in the lower east coast.

Small miscellaneous fishes formed the most important group consisting of 65 to 85 per cent of the total catch in different areas of Andaman Sea. Of this, leiognathids, sciaenids and upeneoides were quantitatively more abundant. The per cent of elasmobranches varied from 18 to 27. Next in importance was catfish and perches. The seasonal abundance and variation of the catches show that the period from

October to March is generally more productive than the other periods, but, however, high catch rates have been obtained in July and August also. Off middle Andamans, good catches of perches were obtained with a catch rate of 620 kg/hour.

Sharks and marlin resources are good in Andaman Sea. Areas south of north Andamans and east of Little Andaman including the submerged invisible bank recorded very good hooking rates upto 22 per cent in certain voyages. Monthwise, April, May and July were the best months with 13 per cent hooking rates. Hooking rates of 1.5 per cent for marlin was recorded. Schools of tuna, particularly the skipjack, *Katsuwonus pelamis* were sighted quite frequently. The occurrence of *Alopias vulpinus*, the thresher shark is also indicative of presence of good school of lesser tunas.

Schools of sardines, mackerels, as well as skipjack and other varieties of tuna are very frequently encountered in Andaman Sea.

Specimens of prawns have often appeared in the trawl catches. The availability of prawns in the shallow bays and inlets support the view that there may be sizeable prawn resources along the coast.

Tuna is one of the most important pelagic species in the areas of Andaman Sea. Large-sized purse seine net of about 500 fathoms length can be successfully employed to catch tuna.

Fish Fauna around Lakshadweep Islands

Lakshadweep is the smallest Union Territory of India (8° to 12°3 minute N latitude and 71 to 74°E longitude), lie scattered in the Arabian Sea about 225 to 450 km away from Kerala coast. It consists of 12 atolls, 3 reefs and 6 newly formed or submerged sand banks with 11 out of 36 islands are inhabited, which include Agatti, Andrott, Amini, Bangaram, Bitra, Chetlat, Kadmat, Kavaratti, Kalpeni, Kiltan and Minicoy.

Fishery resources of Lakshadweep are distributed in three distinctly different environments, such as, Territorial waters (20000 sq km), Exclusively Economic Zone (EEZ) (400000 sq km), which borders the territorial waters of each island and Lagoons (4200 sq km).

The fishery resources of this region can be broadly divided into tuna fishery, non-tuna fishery (comprising sharks, seer fishes, rays, perches etc) and marine ornamental fishes. Fishing, particularly tuna fishing (skipjack tuna, caught one by one using pole and line method), is one of the most important occupations of many of the islanders.

Unlike most other fish, which have white flesh, muscle tissue of tuna ranges from pink to dark red. The red colouration derived from myoglobin, an oxygen-binding molecule is expressed in tuna in quantities far higher than most other fish. Unlike other fishes, many tunas can maintain their body temperature above that of surrounding water, making them functionally warm-blooded.

Tunas are grouped taxonomically in the family Scombridae, which includes about fifty species. The most important commercial and recreational fishes are Yellowfin (*Thunnus albacares*), Bigeye (*T. obesus*), Bluefin (*T. thynnus, T. orientalis* and *T. maccoyii*), Albacore (*T. alalunga*) and Skipjack (*Katsuwonus pelamis*) Tuna fish, which is used for domestic consumption as well as for exports, is abundant in the territorial waters of Kakshadweep islands, but hardly 10 per cent of the potential is exploited.

Japan is the biggest tuna consuming nation and is also the leader in tuna farming research. It is the first successor, who bred and raised Bluefin tuna in 1979 and in 2002. They succeeded in completing the reproduction cycle and in 2007, they obtained the third generation. The farm bred is known as Kindai tuna.

Tuna is a transnational migratory species. Tuna in Lakshadweep constituted over 81 per cent of the total fish landings, of which 79 per cent was Skipjack tuna (*Katsuwonus pelamis*). The other major resource, which is grossly under exploited at present is the Yellowfin tuna (*Thunnus albacares*) and its present share is 18 per cent. The remaining tuna landings are constituted by some of the smaller species like Little Tunny (*Euthynnus affinis*), Frigate tuna (*Auxis thazard*).

During the 15 years period (1995-2009), tuna fishery has shown the peak in 1998 and the total catch was estimated at 12308 tonnes. However, compared to the tuna catch of neighbouring island nations, such as, Maldives (148500 tonnes) and Sri Lanka (27000 tonnes), the catch from Lakshadweep is very low.

The tuna fishery, a decade ago, was restricted only to Minicoy Island of the Lakshadweep. But in recent times, the fishery has expanded to other islands as well, bringing a catch of about 7000 to 9000 tonnes per year. The fishing season in Lakshadweep is from October to May. The fishermen, specialized in the pole and line fishing for Skipjack tuna start fishing from early morning. Each pole and line boat will have 10-15 crews depending on the length of the boat. These vessels can operate only in the territorial waters and have limitation with regard to the size of fish hold and fishing duration. This pole and line technique originated from Minicoy and later got popularized in the other islands through a systematic effort. However, at present, operation of pole and line crafts is the most important economic activity in Lakshadweep.

Live bait fishes are used for chumming and attracting tuna shoals and are essential for pole and line fishing. They will first seek for bait fishes in the lagoon using 2-4 mm mesh sized nets. About 21 species of bait fishes are available in the lagoon, which are mainly associated with coral reefs. Bait fishing is carried out from middle of the night to the early hours of dawn, and the most common species in the order of abundance are, *Spratelloides delicatulus, S. japonicus, Apogon sangiensis, A. savayensis* and *Chromis ternatensis*. Good bait fish grounds are found inside the Bangaram lagoon, around Perumal par and Thodu regions of Agatti lagoon. The bait fish will be stored as live in specially designed tanks of the onboard vessel and used to scout for tuna shoals.

Once the tuna shoal is sighted, the boat will approach the shoal and the fishermen will throw live baits to the shoal to keep them together near the boat. Following this, they will swing into action for hooking the fishes, which will continue until the boat is filled or till the shoal gets disaggregated. If the shoal size is good enough, the boat will get filled up in 60 to 90 minutes. If the shoal scouting is unsuccessful, the boat will return to the base after preserving the live baits in floating baskets in the lagoon. It is estimated that, one pole and line boat of Lakshadweep will catch 12-15 tonnes of

Skipjack tuna during the fair season and the average weight of a single tuna is 5 kg. The ideal way to catch large sized deep water Yellowfin tuna is long lining.

One fourth of the landings in Lakshadweep are accounted by fishes other than tuna, which include sharks, perches, carangids, half-beaks and seer fishes.

Fish landings of Lakshadweep from 2002 to 2009 varied between 9149 tonnes (2002) to 11751 tonnes (2006) annually. The average percentage composition of eight year's annual landings (85202 tonnes) by different groups of fish are;

Carangids 1.69 per cent; Perches 1.78 per cent; Rainbow runner 2.09 per cent; Sail fish – 0.99 per cent; Seer fish – 1.16 per cent; Shark – 1.04 per cent, Tuna – 80.52 per cent and others – 10.73 per cent.

Distribution, Occurrence and Biology of some Fishes

Elasmobranchs

Heptranchias platycephalus, the only species of the genus *Heptranchias* recorded from Andaman Sea has elongate body, snout rounded and obtuse. Eyes without nictitating membrane. Spiracles small. Seven pairs of gill openings. A single spineless dorsal fin is present opposite to the anal fin. Lower caudal lobe is well developed. No caudal pit. Teeth dimorphic, jagged and cusped in the upper and comb-like in the lower.

Distributed in the Indian Ocean and Andaman Sea, the genus *Chilloscyllium* has four species, namely, *Chilloscyllium indicum, C. ocellatum, C. plagiosum* and *C. griseum*. They have elongate body, trunk shorter than tail, eyes without nictitating membrane. Nasoral grooves and cirri are present. Spiracles situated below eyes. Five pairs of gill openings. Two spineless dorsal fins, first dorsal behind the pelvics. Anal fin present. One to three dermal ridges on the back. Teeth small and regular, with or without lateral cusps.

Two species namely, *Nebrius concolour* and *N. ferrugineum* belonging to the genus *Nebrius,* inhabitants of Andaman Sea, have elongate body with spindle-shaped trunk region. Snout is short. Eyes small without nictitating membrane. Nasoral grooves and cirri present. Spiracles are minute and situated behind eyes. Five pairs of gill openings. Two spineless dorsal fins, the first dorsal fin is situated opposite the pelvics. Anal fin is present. Teeth multicuspid.

Stegostoma varium, the only species of the genus *Stegostoma* has slender and elongate body. Tail is longer than the trunk. Snout obtuse. Eyes small without nictitating membrane. Nasoral grooves and cirri are present. Labial folds are well developed. Spiracles situated behind eyes. Five pairs of gill openings. Two spineless dorsal fins. Anal fin is present. Caudal fin is very elongate. Teeth trilobed.

Rhincodon typus, is the only species of the genus *Rhincodon* found in Andaman Sea. It is the largest shark in the world growing upto 45 feet in length. Its body is fusiform and massive. Snout broad, flat and short. Eyes very small without nictitating membrane and located just behind the mouth. Nasoral grooves and cirri are absent. Spiracles small. Five pairs of gill openings. Two spineless dorsal fins are present.

Anal fin is present. Sub-caudal lobe is well developed. Several keels are along the sides. Caudal pit is present. Teeth very small, pointed and numerous.

Carcharias tricuspidatus is the only species belonging to the genus *Carcharias* available in Indian waters of Andaman Sea. Its body is fusiform. Trunk about more than twice the tail. Eyes small without nictitating membrane. Nasoral grooves and cirri are absent. Spiracles small behind the eyes. Five pairs of gill openings. Two spineless dorsal fins are present. Anal fin and caudal pit are present. Teeth very large, awl-shaped, smooth except at base, where there exists a basal cusp on either side.

Alopias vulpinus, the only species of the genus *alopias* is fusiform in body shape. Trunk equal to the extraordinarily elongated upper lobe of the caudal fin. Eyes are large without nictitating membrane. Nasoral grooves and cirri are absent. Spiracles minute, behind the eyes. Five pairs of gill openings. Two spineless dorsal fins, the second dorsal very small and equal to the anal fin. Caudal pit is present. Lateral keel on tail is absent. Teeth simple, smooth and sharp-edged. Inhabits in Andaman Sea.

Two species, namely, *Isurus glance* and *I. guntheri* of the genus *Isurus* have fusiform body, pointed snout. Eyes without nictitating membrane. Nasoral grooves and cirri are absent. Spiracles minute, above mouth angle. Five pairs of gill openings. Two spineless dorsal fins, the second dorsal is very small and equal to anal fin. Caudal pit is present. Lateral keel on tail is present. Teeth long awl-like, lanceolate, smooth and without basal cusps.

Scyliorhinus capensis is the only species belonging to sub-genus *Scyliorhinus* found in Andaman Sea. Four species of the other sub-genus *Hemigaleus*, namely, *Scyliorhinus hispidum, S. indicus, S. burgeri* and *S. quagga* are also inhabitant of Andaman Sea. Their body is elongated, head depressed. Trunk slightly shorter than tail. Snout obtuse, short or elongate. Eyes large with nictitating membrane. Nasoral grooves absent or rudimentary. Nasal cirri absent or present. Mouth wide. Labial folds on both jaws or on lower jaw only. Spiracles present. Five pairs of gill openings narrow, not so wide as the orbit. Two spineless dorsal fins, first dorsal fin behind or above the pelvics. Base of the anal fin distinctly longer than the base of the second dorsal. Cuadal pit is absent. Teeth in numerous rows, tri-pentacuspid.

Atelomycterus marmoratum is the only species of the genus *Atelomycterus* available in Indian Ocean and Andaman Sea. Its body is elongate and slender. Trunk shorter than tail. Eyes large, orbit oblong, nictitating membrane present. Nasoral grooves present, cirri absent. Spiracles small, close behind eyes. Labial folds well developed. Five pairs of gill openings. Two spineless dorsal fins, first dorsal fin behind the ventrals. Base of the anal fin equal to the base of second dorsal. Caudal pit absent. Teeth small, tricuspid, median cusp, the longest.

Proscyllium alcocki is the only species of the genus *Proscyllium*, has slender and elongate body. Trunk is shorter than tail. Snout rounded. Eyes large with rudimentary nictitating membrane. Nasoral grooves and cirri are absent. Labial folds rudimentary. Spiracles are small close behind eyes. Five pairs of gill openings are present. Two spineless dorsal fins, first dorsal fin before the pelvics. Anal fin long. Caudal pit absent. Teeth dimorphous on the upper jaw, tetramorphous on the lower jaw.

Physodon mulleri, the only species of the genus *Physodon* has slender and elongate body. Trunk nearly equal to tail. Snout elongate and pointed. Eyes small with nictitating membrane. Labial folds only on the lower jaw. Nasoral grooves and cirri are absent. Spiracles absent. Five pairs of gill openings. Two spineless dorsal fins. Anal fin present. Caudal pit present. Teeth smooth, the central ones are smaller than those at the side, which bear swollen bases with oblique and narrow cusps.

Three species of the genus *Scoliodon,* namely, *Scoliodon sorrakowah, S. palasorrah* and *S. walbeehmi* have elongate and slender body. Trunk nearly equal to tail. Snout elongate and pointed. Eyes moderate with nictitating membrane. Nasoral grooves and cirri are absent. Labial folds on both the jaws. Spiracles absent. Five pairs of gill openings. Two spineless dorsal fins. Anal fin is present. Teeth with smooth edges, all oblique and without swollen bases.

Aprionodon acutidems, the only species of the genus *Aprionodon* has fusiform body. Trunk slightly longer than the tail. Snout pointed. Eyes moderate with nictitating membrane. Nasoral grooves and cirri absent. Short labial fold at the corner of the mouth in the lower jaw. Spiracles absent. Five pairs of gill openings. Two spineless dorsal fins. Anal fin and caudal pit present. Teeth small, narrow, with broad bases, the lower erect, the upper erect or slightly oblique.

Two species, namely, *Hypoprion malcoti* and *H. hemiodon,* belonging to the genus *Hypoprion* have elongated and fusiform body. Trunk slightly longer than tail. Snout acutely pointed or rounded. Eyes moderate with nictitating membrane. Nasoral groove and cirri absent. Short labial fold at the corner of the mouth present or absent. Spiracles absent. Five pairs of gill openings. Two spineless dorsal fins. Anal fin is present. Caudal pit is present. Teeth smooth except at the bases of the upper ones, which are serrated.

Twelve species of the genus *Carcharhinus,* namely, *Carcharhinus temminckii, C. ellioti, C. gangeticus, C. lamia, C. menisorrah, C. limbatus, C. sorrah, C. melanopterus, C. pleurotaenia, C. dussumieri, C. bleekeri* and *C. watu* have elongate and fusiform body. Trunk shorter of longer than tail. Snout pointed or rounded. Eyes moderate with well developed nictitating membrane. Nasal grooves and cirri absent. Labial folds rudimentary or short. Spiracles absent. Five pairs of gill openings. Two spineless dorsal fins. Anal fin present. Teeth serrated at both bases and cusps, teeth in lower jaw non-serrated in some.

Hemigaleus balfouri, the only species of the genus has slender and elongate body. Trunk shorter than tail. Snout pointed. Eyes moderate with nictitating membrane. Nasoral grooves and cirri absent. Labial folds present. Spiracles minute, behind eyes. Five pairs of gill openings. Two spineless dorsal fins. Anal fin and caudal pits are present. Teeth dimorphous, upper inclined, with denticles on basal part of outer edge, lower erect and smooth.

Hemipristis elongates, the only species of the genus *Hemipristis* has slender and elongated body. Trunk is shorter than tail. Snout rounded. Eyes moderate with nictitating membrane. Nasoral grooves and cirri absent. Labial folds present. Spiracles small, close behind eyes. Five pairs of gill openings. Two spineless dorsal fins. Anal fin and caudal pit are present. Teeth dimorphous, upper teeth large, broad, flat and serrated, the lower smooth, slender and curved inwards.

Galeorhinus omanensis, the only species of the genus has long and slender body. Trunk equals to tail. Snout obtuse, depressed. Eyes moderate with nictitating membrane. Nasoral grooves and cirri absent. Labial folds present. Spiracles minute, behind eyes. Five pairs of gill openings. Two spineless dorsal fins. Anal fin present. Caudal pit absent. Teeth monomorphous, oblique, notched and smooth.

Galeocerdo arcticus, the only species of the genus has elongated body. Head depressed. Trunk more or less equal to tail. Snout wide short. Eyes moderate with nictitating membrane. Nasal grooves and cirri are absent. Labial folds present on both jaws. Spiracles small, behind eyes. Five pairs of gill openings. Two spineless dorsal fins. Anal fin present. Teeth large, flat, triangular, notched, oblique and serrated on both edges.

Myrmillo manazo, the only species of the genus has elongated and fusiform body. Trunk more or less equal to tail. Snout produced and pointed. Eyes moderate with nictitating membrane. Nasoral grooves and cirri are absent. Well developed labial fold at each angle of mouth. Spiracles small, behind eyes. Five pairs of gill openings. Two spineless dorsal fins. Anal fin present. Caudal pit absent. Teeth monomorphous, polyserial, pavement like, smooth, obtuse, devoid of distinct cusps.

Triaenodon obesus, the only species of the genus has elongate body. Trunk longer than tail. Snout short rounded. Eyes moderate with nictitating membrane. Nasoral grooves and cirri absent. Labial folds short, not extending along the jaws. Spiracles absent. Five pairs of gill openings. Two spineless dorsal fins. Anal fin and caudal pit present. Subcaudal well developed. Teeth minute, numerous in both jaws, with central and lateral cusps.

Four species of the genus *Sphyrna*, namely, *Sphyrna mokarran, S. blochii, S. tudes* and *S. zygaena* have elongate body. Head T-shaped forming oculonarial expansions on either side. Trunk behind head is compressed. Eyes moderate with nictitating membrane, situated at the oculonarial extremities. Nasoral grooves and cirri absent. Labial folds rudimentary. Spiracles absent. Five pairs of gill openings. Two spineless dorsal fins. Anal fin present. Caudal pit present. Subcaudal lobe produced. Teeth monomorphic, oblique and notched, serrated or non-serrated.

Centroscyllium ornatum, is the only species has elongate and fusiform body. Trunk longer than tail. Snout depressed. Eyes large with nictitating membrane, orbit elongated. Nasoral grooves and cirri absent. Labial folds well developed in both jaws. Spiracles are large behind eyes, higher, superior. Five pairs of gill openings. Two dorsal fins, each with a well developed spine anteriorly. Anal fin absent. Caudal pit absent. Caudal fin truncated or painted. Teeth small, raptorial, multicuspid.

Six specis of the genus *Rhinobatos*, namely, *Rhinobatos thouniniana, R. granulatus, R. armatus, R. obtusus, R. annandalei* and *R. lionotus* have depressed and elongated body. Disc triangular, slightly rounded and under behind. Tail depressed, nearly equal to the trunk. Snout triangularly pointed. Nostrils oblique and wide. Spiracles wide just behind the eyes, with fold on hind edge. Five pairs of gill openings on the ventral side. Two spineless dorsal fins behind the pelvics and closer to the caudal than snout end, pelvics closer to the pectorals than to the dorsals. The rayed portion of the pectoral fins not continued to the snout. Anal fin absent. Teeth obtuse with indistinct, transverse ridges.

Rhina uncylostoma, the only species of the genus has depressed and elongated body. Disk sub-triangular, obtusely rounded in front. Tail depressed, nearly equal to trunk. Snout broad, obtusely rounded. Nostrils slightly oblique and wide. Spiracles large without posterior folds and about an eye diameter and a half behind the eyes. Five pairs of gill openings on the ventral side. Two spineless dorsal fins, first dorsal fin opposite pelvics and nearer to snout end than to caudal end. The rayed portion of pectoral extends only up to spiracles. Anal fin absent. Teeth obtusely rounded, each with several longitudinal ridges.

Rhynchobatus djiddensis, is the only species of the genus has depressed and elongated body. Disk triangular, longer than wide. Tail depressed, nearly equal to trunk. Snout triangularly pointed. Nostrils oblique. Spiracles large, close behind eyes and with two small folds on hind edge. Five pairs of gill openings on the ventral side. Two spineless dorsal fins, first dorsal opposite pelvics and nearer to snout end than to caudal end. The rayed portion of pectorals extends only upto spiracles. Anal fin absent. Teeth obtuse, pavement-like, dental surfaces undulated.

Four species of the genus *Pristis*, namely, *Pristis microdon, P. pectinatus, P. cuspidatus* and *P. zijsron* have elongated and moderately depressed body. Rostrum very much produced and saw-like. Nostrils oblique. Eyes without nictitating membrane. Spiracles large, behind the eyes. Five pairs of gill openings on the ventral side. Two spineless dorsal fins. Pectorals moderate, front edge quite free neither joining with head nor reaching snout. Anal fin absent. Rostral teeth strong and large, set in sockets on either edges of the blade-like snout. Oral teeth small, pavement-like.

Zanobatus schoenleinii, the only species of the genus is found in Indian Ocean and Andaman Sea. Its disk is wider than long, partly rounded. Tail depressed, slender, nearly half total length. Rostral cartilage small. Snout short, obtuse. Nasoral grooves rudimentary. Nostrils transverse, internarial space about two-third mouth width. Spiracles close behind eyes. Five pairs of gill openings on ventral side. Two spineless dorsal fins. The rayed portion of the pectorals continued to the snout to form sub-circular disk. Anal fin absent. Teeth very small.

Five species of the genus *Raja*, namely, *Raja mamillidens, R. reversa, R. johanhis-davisi, R. powelli* and *R. andamanica* have sub-circular to quadrangular disk. Tail not whip-like, without spines and without fold along either side. Snout produced and pointed. Eyes prominent. Nasoral grooves present. Five pairs of gill openings on the ventral side. Two spineless dorsal fins. The rayed portion of the pectorals reaches beyond the eyes, but not upto the snout. Anal fin absent. Teeth small, tessellate, flat to sharply pointd.

Two species of the genus *Taeniura*, namely, *Taeniura meyeni* and *T. lymma* have rounded disk with head not distinct from it. Tail compressed, longer than body, with midcaudal serrated spines and without any lateral folds. Mouth with buccal processes. These are small flaps of skin across the floor of the mouth of sting rays. No rostral cartilage. Nostrils slightly oblique. Spiracles wide, behind the eyes. Five pairs of gill openings on the ventral side. Dorsal fins absent. Rayed portion of the pectoral fins united anteriorly. Anal fin absent. Subcaudal rayless, below terminal end of the tail. Teeth small, tessellate, grooved transversely.

Fourteen species belonging to the genera *Dasyatis* and four sub-genera, *Himantura, Amphotistius Dasyatis* and *Pastinachus*, namely *Dasyatis pastinaca, Pastinachus sephen, P. bennetti, Amphotistius zugei, A. jenkinsii, A. imbricate, A. marginatus, A. kuhlii, Himantura bleekeri, H. favus, H. microps, H. alcochi, H. uarnak* and *H. gerrardi* have oval to rhomboidal disk. Tail elongate, whip-like, with serrated caudal spines, with or without dermal fin folds, not terminal in position, but behind spines, without lateral folds on caudal base. No rostral cartilage. Nasoral grooves present. Nostrils slightly oblique. Spiracles large, behind the eyes. Five pairs of gill openings on the ventral side. Rayed dorsal fins absent. Rayed portion of the pectoral fins united anteriorly. Anal fin absent. Teeth flattened or with central point or transverse ridge.

Urogymnus Africana, found in the Indian Ocean, has sub-circular disk, which is profusely tuberculated. Tail feeble, about as long as disk length without caudal spine. Nasoral grooves rudimentary, cirri present. Rostral cartilage absent. Spiracles large, close behind eyes. Five pairs of gill openings on the ventral side. Rayed dorsal fins absent. Rayed portion of the pectoral fins united anteriorly. Anal fin absent. Teeth tessellated, flattened, rhomboid.

Four species, namely, *Aetoplatea tentaculata, A. zonnurus, Gymnura peocilura* and *G. micrura* of genus *Gymnura* have much wider disk than long. Tail short, slender, with serrated spine. Rostral cartilage absent. Nasoral grooves present. Spiracles large, close behind eyes. Five pairs of gills openings on the ventral side. Rayed dorsal fins absent. Rayed portion of pectoral fins united anteriorly. Anal fin absent. Teeth minute, numerous in broad bands, each tooth with one to three cusps.

Four species of the genus *Aetomylaeus*, namely, *Aetomylaeus maculatus, A. nichofii, A. nichofii cornifera* and *A. milvus* have lozenge-shaped disk about twice as broad as long. Tail whip-like, much longer than disk and without caudal spine. Head moderately conspicuous, rostal fins forming a unilobed snout. Eyes lateral. Nasoral grooves present. Spiracles large, behind eyes. Five pairs of gill openings on the ventral side. First dorsal fin small, situated at basal part of tail. The second dorsal and anal fins absent. Rayed portion of the pectorals falciform and extending only upto posterior region of the orbit. Teeth in three rows of which the lateral narrower than the central ones.

Aetobatus narinari, is the only species of the genus has lozenge-shaped disk, about twice as broad as long. Tail whip-like, larger than the length of the disk and with a serrated caudal spine. Head conspicuous, rostal fins forming a unilobed pointed snout. Eyes lateral. Nasoral grooves present. Spiracles large, about twice eye diameter and laterally situated about an eye diameter and a half behind eyes. Five pairs of gill openings on the ventral side. First dorsal fin small, situated at basal part of tail. Second dorsal and anal fins are absent. Rayed portion of the pectorals falciform extending upto the anterior margin of the spiracles. Teeth in a single row.

Four species, namely, *Rhinoptera javanica, R. sewelli, R. jayakari* and *R. adspersa* of the genus *Rhinoptera* have lozenge-shaped disk, about twice as broad as long. Tail whip-like, longer than disk, with basal serrated spine. Head some what conspicuous, rostral fins forming a bilobed snout. Eyes prominent, lateral. Nasoral grooves present. Spiracles large, behind eyes, open laterally. Five pairs of gill openings on ventral side. First dorsal fin above basal part of the tail. Second dorsal and anal fins absent. Rayed portion of the pectorals falciform and not joined with the rostral fins in the front. Teeth wide, angular, flat in pavement, median row widest.

Mobula diabolus, is the only species of the genus has lozenge-shaped disk, about twice as broad as long. Tail short whip-like in the young, sometimes one and half times the length of the disk, and in the adults, a little more than half the disk length, with or without serrated spine. Head conspicuous, broad and flat with two curled cephalic horns. Mouth inferior, well behind the head. Eyes large, lateral. Nasoral grooves present. Spiracles moderate, behind eyes. Five pairs of gill openings on the ventral side. First dorsal fin small, triangular above and between pelvics. Second dorsal and anal fins absent. Rayed portion of the pectorals falciform extending upto the post-orbital region. Teeth small, numerous in both jaws or absent in the upper jaw.

Manta birostris, is the only species of the genus has lozenge-shaped disk, about twice as broad as long. Tail whip-like, about as long as disk length, without serrated caudal spine. Head greatly depressed, broad and flat with two cephalic horns, rarely curled. Mouth large, terminal in front of head. Eyes prominent, lateral. Nasoral grooves present Spiracles moderate behind eyes. Five pairs of gill openings on ventral side. First dorsal fin small, above and between pelvics. Second dorsal and anal fins absent. Rayed portion of pectorals falciform, extending upto the post orbital region. Teeth small, numerous, in pavement, usually on lower jaw and sometimes on both jaws.

Three species of the genus *Narcine*, namely, *Narcine timlei, N. indica*, and *N. brunnea* have sub-circular disk with head not distinct from it. Tail with lateral folds, slightly shorter than the length of the disk and without serrated caudal spine. Snout broadly rounded, twice the inter orbital distance. Rostral cartilage present. Nasoral grooves present. Spiracles large, situated close behind the small eyes. Five pairs of gill openings on the ventral side between the electric organs. Two spineless dorsal fins on tail. Anal absent. Pelvics well developed. The rayed portion of the pectorals continued to the orbital region. Teeth in narrow bands.

Two species, namely, *Torpedo mormoratus* and *T. sinus-persici* of the genus have widely circular disk with head not distinct from it. Tail very short with lateral folds and without serrated caudal spine. Snout broadly rounded and equal to inter-orbital distance. Rostral cartilage present, but reduced. Nasoral grooves present. Eyes well developed. Spiracles moderate, close behind eyes. Five pairs of gill openings on the ventral side between electric organs. Two spineless dorsal fins on tail. Pelvics well developed. The rayed portion of the pectorals extends to the orbital region. Anal fin absent. Teeth small, in pavement, irregularly rhomboidal with the crown obliquely pointed.

Chimaera mons strosa has elongate body and shark-like in form tapering posteriorly to a point at tail. Head large compressed and without beak. Eyes large or moderate, lateral. Mouth inferior. Nasoral grooves present. Spiracles absent. One gill opening on either side of pharynx, containing four gill slits and four gills covered over by a skinny operculum. Two dorsal fins, the first dorsal with a strong spine anteriorly, the second dorsal long and low. Pectorals large, free and low. Pelvics abdominal, many rayed. Anal fin small, distinct or not distinct from sub-caudal. Mature males with trifid or rarely bifid claspers. Skin naked, devoid of placoid scales. Teeth united to form bony plates, laminae or tritors, 4 tritors in the upper jaw, 2 tritors in lower jaw.

Harriotta indica, has elongated body tapering to a long tail with a filamentous tip. Head with long rostral proboscis. Snout depressed. Eyes moderate lateral. Mouth inferior. Spiracles absent. One gill opening on either side of pharynx, containing four gill slits and four gills covered over by a skinny operculum. Two dorsal fins with a spine anteriorly, second dorsal low. Pectorals large free. Anal small and distinct from subcaudal. Supra-caudal moderately high, upper edge without spines. Mature males with a single clasper. Teeth with tritors.

Rays and Skates along Tuticorin Coast

Tuticorin coast is known for the rich elasmobranch resources, consisting of sharka, rays and skates. Unusual heavy landings of rays and skates (23.3 tonnes) consisting of 9 species was noticed on 15.9.2009. They were caught from a depth of 52 m off Tuticorin.

Species	Per cent in Catch	Length (cm)	Sex Ratio (M:F)
Fam–Dasyatidae (Sting ray)			
Himintura bleekeri	45	59-106	73:27
Himintura uarnak	10	67-109	64:36
Himintura marginatus	3	42-69.6	69:31
Dasyatis centroura	4	67.9-97	68:32
Pastina sephen	9	94-125	91:9
Fam–Myliobatidae (Eagle ray)			
Aetobatus narinari	10	113-172	78:22
Fam–Mobulidae (Devil ray)			
Mobula mobular	11	162-220	81:19
Fam–Rhinopetridae (Cow nose ray)			
Rhinoptera javanica	7	141-166	83:17
Fam–Gymnuridae (Butterfly ray)			
Gymnura poecilura	1	36-47	69:31
Skates			
Fam–Rhinidae (Wedge fishes)			
Rhina ancylostoma	27.1	117-136	90:10
Fam–Rhinobatidae (Guitar fish)			
Rhinobatos granulatus	43.2	52-79	86:14
R. obtusus	19.2	49-72	79:21
R. annandalei	10.5	53-82	84:16

Sharks in Maharashtra Waters

Rhincodon typus, commonly known as whale shark is a pelagic species. Though whale shark is the largest fish, it mainly feeds on plankton. There is no regular fishery for this species in India, except in Gujarat. They are occasionally trapped in different gears, such as, trawl net and gill net.

A female shark measuring 3.1 m and weighing 0.4 tonnes was caught by a trawler from a depth of 45-65 m on 23.3.2009. From 8.1.1980 to 23.3.2009, nine whale sharks of 3.1 to 20.8 m length weighing 0.4 to 11 tonnes were caught from Maharashtra waters.

A male whale shark, measuring 8.2 m in length, weighing 3 tonnes was caught in a pair trawl net operated 30 km away in Gulf of Mannar at a depth of 55 meter.

Another male whale shark, measuring 10.58 m and weighing 1.95 tonne was caught by a trawler from Maharashtra coast at a depth of 45 to 68 m, landed at Versova, Mumbai. The whale shark is an endangered species.

A female *Galeocerado cuvier*, commonly called tiger shark was caught by a trawler on 17.01.09 from a depth of 45-65 m in Maharashtra waters. Usually they occur upto depths of 140 m. The tiger shark, which measured 4.2 m in total length is probably the largest record from Maharashtra waters. It weighed 1.1 tonnes. Between 20.4.85 to 23.3.09 seven tiger sharks of 2 to 4.2 m in length weighing 0.1 to 1.2 tonnes were caught from a depth of 25 to 300 m.

Hound Shark from Calicut Coast

Occurrence of a hound shark, *Mustelus mosis* was noticed in May, 2010 from the landings of multi-day trawlers operating from Beypore. *M. mosis* belongs to the family Triakidae, inhabiting the continental shelves of western Indian Ocean at depths ranging between 200 and 250 meters. They are small to medium sized sharks and grow upto 150 cm. Hound sharks are distinguished by possessing two large spineless dorsal fins, an anal fin and oval eyes with nictitating eyelids. They are found throughout the world in warm and temperate waters, where they feed on fishes and invertebrates on the sea bed and mid-water.

Quagga Shark from Cochin

One of the world's rarest shark, *Halaelurus quagga* of the family Scyliorhinidae, was collected from Cochin and Quilon landing Center of Kerala. This is the second specimen from India and available third report in the world.

Bramble Shark in the Veraval Coast

Bramble shark, *Echinorhinus brucus*, is exclusively marine. It is also known as sluggish bottom shark, sometimes occurring in shallow waters, especially on the continental and insular shelves and upper slopes upto a depth of 900 meters.

They have dorso-ventrally compressed head and the body is long cylindrical, covered with scattered, large protruding thorn-like denticles, two small spineless dorsal fins placed far back on the body, just before the tail and five pairs of gill slits.

It has no anal fin and has thick caudal peduncle (tail stalk). The teeth are star-cusped and similar in both the jaws. The skin of its back and sides is sparsely strewn in large scales with either one or two sharp points. The colour of the species is dark grey with metallic reflections withour darker blotches. They can reach a maximum length of 3.94 m (13 feet) and weigh 227 kg. There are records on the occurrence of this species in the Western Atlantic, Mediterranean, Pacific Ocean, Australia and NewZealand, besides in Arabian Sea, off Veraval coast near Pakistan border. The species belong to the sub-class Elasmobranchii under the class Chondrichthys.

Reproduction and Breeding of Elasmobranchs

Elasmobranchs have a highly evolved reproductive system. It provides for internal fertilization, that is, followed by the production of the young, thereby eliminating the larval stages. They exhibit three types of reproduction, oviparous, viviparous and ovo-viviparous. Batoid sharks include saw fishes, skates, electric rays, and sting rays.

Rhynchobatus dijiddensis was previously believed to occur throughout a wide part of Indo-Pacific region. But recent evidence have shown that it has a species complex of four different species. The complex includes *Rhynchobatus australiae*, the white spotted guitar fish, *Rhynchobatus springeri*, Broad nose wedge fish, and *Rhynchobalus iuevis*, smooth nose wedge fish.

With these as separate species, the giant guitar fish has come to be restricted to Red Sea and tropical western Indian Ocean upto Eastern Cape of South Africa.

The fish reaches upto a length of 3.1 m, weighing as much as, 227 kg. The colour underneath is white and overall dark grayish or olive above. Large individuals lack the distinct white spots. The species is ovo-viviparous and a female can give birth upto ten young ones at a time. The fish is hardy and remains alive out of water for quite sometime. This species commonly occur in inshore waters and in shallow estuaries and mainly feeds on crabs, lobsters, bivalves, small fishes and squids.

Two fishes were caught accidentally, while trawling at a depth of 30-40 m at 60-70 km towards north-east coast of Mumbai from coastal waters.

In Guitar fish, a pair of ovaries and uteri are present and are fully functional. The embryos feed initially on yolk, then receive additional nourishment from the mother by indirect absorption of uterine fluid, enriched with mucus, fat or protein through specialized structures.

The mature females apart from fully grown embryos and large rounded eggs full of viscous yellow yolk, amidst numerous follicular cells has the mature ova. The female of 225 cm total length, had seven fully developed embryos in its ovary. The total length of the embryos ranged between 279 to 290 mm. Yolk sac of the embryo was large, measuring 82 to 88 mm in diameter and the cord length ranged from 60 to 70 mm. Five mature ova between 65 to 72 mm were also observed. Out of 7 embryos, 5 were males and the other two were females.

Bamboo shark, *Chyllocilium griseum*, dweller of reef area of size range from 538 to 590 mm laid 38 egg capsules (Mermaid's purse) at a temperature from 25.5 to

28.2°Celsius, water salinity 29 to 35 ppt and pH 7.7 to 8.5. Hatching started from 118th day. The yolk attached to the juveniles lasted for 24 days, although the young may start feeding from 18th day onwards.

Perches

Sea Bass, *Lates calcarifer*

Sea bass are found in coastal waters, estuaries and lagoon, usually at a depth of 40 m and mainly feed on crustaceans and fishes. They are distributed around Sri Lanka, extending to Arabian Sea, Eastern Indian Ocean and the Western Central Pacific.

The species is observed in the fish catch throughout the year in Mumbai waters, but in small numbers with large sized fishes caught during April-June. About 80 to 100 kg of sea bass per gill netter landed on 25.4.2008 from a fishing ground off Worli at a depth range of 15-20 m. Other species landed included *Protonibea diacanthus, Otolithoides biauratus* and *Eleutheronema tetradactylum*. The total length of *L. calcarifer* ranged between 80-121 cm with a mode in the length group 90-99 cm. The species is generally caught in good numbers during full moon and new moon days, indicating that for maximum exploitation of the species, lunar cycle may be followed.

A similar high catch was recorded from Orissa (2002) in the month of February. A total of 192 sea bass with a total weight of 4816 kg was caught by two operations of shore seine. The fishery of sea bass is seasonal and the maximum length of the species is 200 cm. A large sized sea bass measuring 106 cm and weighing 7.5 kg was caught from Karwar coast by shore seine in 1998. In the present catch, the maximum size recorded was 121 cm.

In view of its easy adaptability to low saline waters, the fish is cultured in recent years in Thailand, Singapore and Philippines. Cage culture of sea bass was initiated in Indian waters.

The seabass occur in Andaman Sea. Culture in tide-fed brackish water ponds in Andaman and Nicobar Islands with 148 to 320 mm sized fingerlings was done, the overall monthly growth was 14.5 mm (25 g) and 18 mm (44 g). The size at harvest ranged from 274 to420 mm with an average of 360 mm and weight ranging 220 to 880 g with an average of 527 g.

The perches in Andaman Sea constitute about 15 to 18 per cent of the average annual landings representing the important genera like, *Lethrinus, Lates, Lutjanus, Pomadasys, Epinephelus, Pristipomoides* and *Aprion*. Thirty three species have been reported in the catch and the main fishing season is from August to November. Perches for the most part, are fished by hooks and line. A small percentage is also caught in gill nets. Off Little Andaman, very good catches of perches were obtained on the limited hauls by the trawl nets.

Otolithoides biauritus and *Protonibea diacanthus*

Two hundred seventyfive (3300 kg) of *Otolithoides biauritus* were landed by a multi-day trawler operating 35-40 km off shore Salya in the Bay of Kutch at a depth of 25-30 m. The length of the species was between 76 cm and 136.8 cm and a major share

Figure 25.1: *Psudanthias* sp.

Figure 25.2: *Chelidoperca investigatoris* (Alcock, 1895)

Figure 25.3: *Lutjanus argentimaculatus*

Figure 25.4: Scorpion fish, *Scorpaena scropha*

**Figure 25.5: Lateral View of Giant Grouper *Epinephelus lanceolatus*
Maintained in the Marine Research Aquarium**

of the catch (70 per cent) belonged to length around 120 cm. The average weight of the fish landed was 12 kg.

The day before (26.3.2008), twenty eight (560 kg) of *Protonibea diacanthus* was landed by a multi-day trawler operating at the same fishing ground and depth. Their length ranged from 91.2 cm to 152 cm and the majority of the catch belonged to higher length classes. On an average each fish weighed 20 kg.

Honeycomb Grouper

Epinephelus merra (Honeycomb grouper) attain sexual maturity at 30 cm length. Fishes above 30 cm form pairs and natural spawning was obtained (7 times) during

August-September. The periodicity of spawning ranged from 3 to 12 days, but the interval of majority spawning ranged between 3 to 4 days. The approximate number of eggs in the different spawning ranged from 11220 to 63020. The eggs hatched on the same day of spawning. The average length of newly hatched larvae is 1.5 mm. Availability of sufficient density of copepods is the critical factor for the survivality of the larvae. The larvae start metamorphosing from 40th day onwards and the metamorphosis is completed by 60th day. The young ones ranged a total length of 20 to 64 mm, and the majority is in the length range of 30 to 49 mm.

Grouper

The trawl catch, operating at a depth range of 14 to 80 m, Off Neendakara, Kerala, consists of adults and juveniles of penaeid and non-penaeid prawns, ribbonfishes, sardines, scianids, decapterus, mackerel, flat fishes, whitebait, lizard fishes, threadfin breams and a wide variety of miscellaneous fishes and benthic biota.

During August to December, 2005, a large number of juvenile groupers, *Epinephelus diacanthus* was caught by trawlers (average catch/hour was 1.5 kg), estimated catch 316 tonnes (8301763 numbers) during this period. Total length of groupers ranged from 85 to 180 mm, weighing 10-78 g. The juveniles feed on crabs, stolephorus and squilla. The coastward migration of the juvenile groupers was for feeding, since Neendakara being the feeding ground in coastal shallow waters.

During October-November, 2004, large number of juvenile (40-70 mm) grouper, *Epinephelus diacanthus* were caught by dol nets at New Ferry Warf, Mumbai. Although landing of *E. diacanthus* is a regular phenomenon in Mumbai, the occurrence of *E. fasciatus* is reported to be rare. Their length ranged from 30-60 mm and they have migrated to shallow waters for feeding.

Epinephelus lanceolatus is the largest reef dwelling fish in the world. Being such a large predator, it is rare even in areas unexploited by fishing. A juvenile (236 mm total length) was caught by the gill net at Korapuzha estuary, Elathur, near Calicut in June 2010. The species is included in the IUCN red list of the threatened species under the vulnerable category.

Groupers of the genus *Epinephelus* are tropical and sub-tropical reef associated fishes found in the Andaman Sea. The grouper, *Epinephelus marginatus*, inhabitant of coral reef ecosystem had been caught in bigger size in estuarine system, revealed its long term migration, which could be due to change in their environment, such as, climate, current etc.

Fortyfive cm long *Epinephelus marginatus* has elongate body, sub-equal and slightly large nostrils, naked maxilla, four rows of sub-equal teeth in mid-lateral part of lower jaw. The dorsal fin has 11 spines and 15 rays, anal fin has 3 spines and 9 rays, pectoral fin has 17 rays and caudal fin has 18 rays and the same is round in shape. The entire body from head to tail is found with pale greenish yellow, silver grey blotches and irregular white patches. The marginal layer of the fins is dark greyish in colour, followed by yellowish gold. It is the most common species in South African waters, where it is well known as the "yellow belly". The species is very rare in Indian coastal waters and does not have much of commercial value compared to other species.

Snapper

Ten pinjalo snapper, *Pinjalo pinjalo* were caught off south west of Mumbai coast from a depth of 40-60 m of which a female specimen of *P. pinjalo*, measuring 678 mm weighing 3.941 kg. The ovary of the female weighing 91 g was partially spent. The ova diameter ranged from 0.31 to 0.42 mm.

A high catch (4.5 tonnes) red snapper, *Lutjanus argentimaculatus* was obtained in purse seine net on 23.10.09 from a fishing ground south of Mumbai upto Ratnagiri in the depth range of 40-50 m. The fishes were large weighing 1.5 to 2.5 kg each. Other species of snapper like, *L. johni* was also caught along with red snapper. Appearance of red snapper in Mumbai waters are common during October-November. Owing to the fast growth of the species, delicate fish flavour, it is a culture species in Indonesia, Thailand etc.

Serranid Fish

The family Serranidae comprising of 62 genera with 449 species into three sub-families, namely, Anthiinae, Epinephelinae and Serraninae. Among the genera, *Epinephelus* has the largest number of species and are most commercially important.

The boulenger's anthion, *Sacura boulengeri*, a very rare anthias, was previously known only from six specimens, five collected from Gulf of Oman in 1963 and one from Sindh (Pakistan) in 2004. Between 2005-06, *S. boulengeri* was reported from several landings in India, namely, Mumbai, mangalore and Neendakara (Kerala). They have been caught from mangrove area by gill net. All 7 specimens were male as identified by their characteristic bright colour of golden mauve and lavender. The body of the fish is ovate, laterally compressed with a lunate caudal fin. The third spine of the dorsal fin is very prolonged, as also are third and the fourth dorsal soft rays.

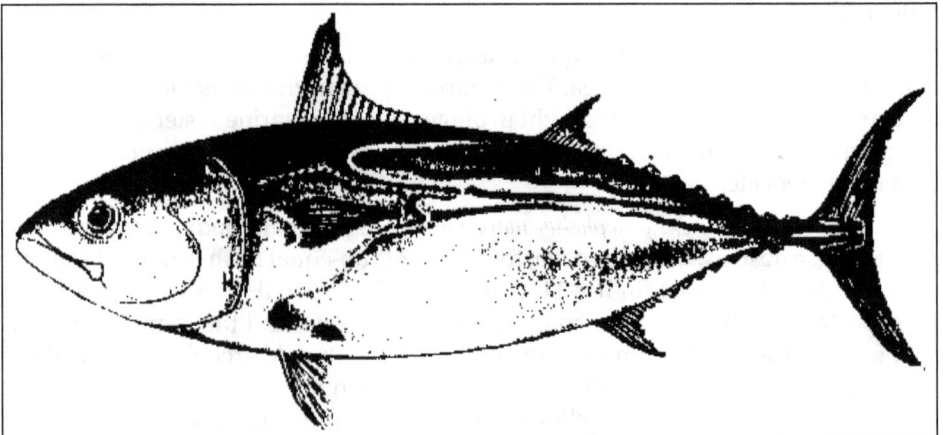

Figure 25.6: *Euthynnus lineatus*

Figure 25.7: *Sardinella longiceps*

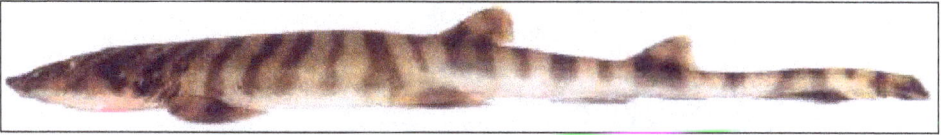

Figure 25.8: Quagga Shark, *Halaelurus quagga*

Figure 25.9: *Pseudanthias* **sp.**

An anthine fish belonging to genus *Pseudanthias* (Serranidae) was collected during early 2008 from Cochin and Quilon, Kerala, which was confirmed as a new species.

Ghol

Protonibea diacanthus, locally known as "ghol" were landed by multi-day trawler operating 50-60 km from Okha at depths of 60-70 m for 5 to 6 days. On 14.10.08, 210 numbers (4165 kg) and on 15.10.08, 165 numbers (2500 kg) were landed with length ranged from 75 to 150 cm. Majority (80 per cent) of the catch had a length of 120 cm. The average weight was 20 kg.

Threadfin Bream

Five species of threadfin bream, a major demersal resource off Visakhapatnam region, namely, *Nemipterus mesoprion, N. japonicus, N. delagoae, N. luteus* and *N. tolu*

Figure 25.10: Tuna, *Euthynnus affinis*

generally contribute to the fishery of the region. A sixth species *N. zysron* has been recorded in the catch for the first time at Visakhapatnam fishing horbour on 15.7.2008. The species has slightly elongated body (compared to other species of the genus) and the presence of yellow stripes in front of eye through nostrils and from upper lip to beneath the eye. It has a single dorsal fin with 10 spines and 10 rays, anal fin with 3 spines and 7 rays. Upper lobe of the caudal fin produced into a long yellow trailing filament. Body colour is reddish in the upper part, silvery below, sides below lateral line with distinct yellow stripes along the middle of each scale row. Head pinkish. Dorsal fin pale yellow with bright yellow margin. Pelvic fins hyaline with a yellow axillary area and axillary scales. Caudal fin pinkish, upper and lower lobes pale yellowish, filament yellow. Though the species is reported to have widespread distribution in the Indo-Pacific from north-western Australia, the Indo-Malay

Archipelago and Andaman Sea, it is being reported for the first time along Andhra Pradesh coast. The total length of the species ranged from 18-21.5 cm and is known to attain a maximum length of 25 cm.

Five specimens of *Nemipterus bipunctatus* was landed by a trawler in Mangalore on 23.11.2010.

Cat Fishes

Heavy landings (around 28 tonnes) of cat fish, *Arius dussumieri* was observed from elevn "dol net" units on 23.3.2009, with an average of 2545 kg per dol net at Rajapara landing center in Saurashtra region. Weight of each fish varied between 3.6 to 5.2 kg, with an average weight of 4.65 kg. The dominant size group with the mode at 70-74 cm and the range between 56-92 cm was observed. Females were found to be dominant in the population and the 42 per cent of the female were in the advanced ripe condition, 17 per cent were mature and 8 per cent are in the maturing stage.

Majority of the fishes fed on *Acetes*. Second important food item was *Coilia dussumieri* (found in 50 per cent of the fish). The other important food items were *Chirocentrus dorab*, seer fish, carangids, ribbonfish and non-penaeid prawns.

Spawning migrations of *Arius thalassinus* and *Arius tenuispinis* towards shallow waters of less than 10 m depth have been reported during the south-west monsoon from the west coast. It appears that *A. dussumueri* also would have migrated towards the shallow waters of the Gujarat coast during February- March for breeding.

An unusual bumper catch (1.6 tonnes) of juvenile cat fish, *Arius caelatus* was observed in Pondicherry from 12 km off the coast and 14 km north of Pondicherry port, at a depth of 34 m. The water was clear and the drift/current was moving towards northern direction. The length ranged between 18.3 to 33.2 cm and weight ranged 100 to 260 gram.

On 25.8.2008, bumper landing of catfishes were observed at the Malpe Harbour by purse seiners. Heavy landings of catfishes were observed in 1980's, after which catch showed a sharp decline raising concern about extinction of the species, *Arius thalassinus* along the coast. Three purse seiners on 25.8.2008 caught about 59 tonnes of catfishes in a single operation from a depth of about 54 m. The size range of fishes were 9.7 to 12.2 cm and the majority of fishes were found to be matured.

Clupeids

Oil Sardine

Oil sardines, *Sardinella longiceps* forms 10-18 per cent of total fish landings in India, mainly caught along the south west coast. Except for Kerala and Karnataka, oil sardine fishery is not a major resource in other states. Oil sardine is known to occur in the Indian waters in large schools in the inshore waters.

Sporadic instances of heavy landings of sardine was recorded earlier along both east and west coasts, such as, Pondicherry, Chennai, Cuddalore, Pazhapam, Rameswaram, Pamban, Srikakulam, Tuticorin, Uchila and Ullal including Saurashtra. The fishing season for oil sardine is generally during June to December, when about 90 per cent of the annual catch is obtained, though it occurs throughout the year. Even though, the period of spawning of the species extends from May to November, its peak is from June to August. Oil sardines have a very high fecundity ranging from 37000 to 80000 with the egg diameter ranging between 1.20 to 1.23 mm. Distribution of oil sardine is restricted to certain localities, having rich production of phytoplankton, which form the main food of the species.

In the past several decades, the oil sardine fishery has shown fluctuations, spatially, seasonally and annually. Out of the many reasons for fluctuations, one of the reasons might be changes in diatom production. An increase in the strength of the monsoon over its critical limit would be favourable for an increase in the catch and below the critical value, the catches were found to decline. In general, the south west monsoon and the resultant biological, oceanographic and meteorological conditions seem to be responsible for the catch fluctuations to a large extent. Resources potential of oil sardine of the west coast is high despite its inherent fluctuations.

The shoals of oil sardine can either be migrating from south west coast of India or from the offshore. The wind-driven surface currents of the west coast, sea water temperature and salinity appear to influence the oil sardine migrations.

Oil sardines never formed a sizeable fishery along the Maharashtra coast earlier, and of late, the Indian oil sardine, *Sardinella longiceps* has started appearing in large quantities in the dol net catches. Unusual and unprecedented landings of oil sardines by dol netters were observed at Arnala during January and February, 2007. The length-frequency studies of the sardine catch revealed that the length of sardines ranged between 77 and 178 mm with a mode at 160-169 mm. Most of them were lean with the head comparatively looking longer than the body and showed a starved appearance. The catch of oil sardines by trawlers also increased substantially.

On 14.5.08, a record catch of *Sardinella sirm* was made from a depth of 32-44 m, north west of Neendakara. The catch was 800-1000 kg per fiber boat. Next day the fiber boats landed 800-1600 kg from a depth range of 38-56 m off north west and west of Neendakara. The fishery was not observed for subsequent two weeks and again reappeared on 2.6.2008 at a depth range of 32-38 m in the same area. Following day (3.6.2008) from the same area 800-1100 kg per fiber boat was landed. The fishery of *S. sirm* dwindled after 3.6.2008 with the arrival of *Sardinella longiceps* and other lesser sardines.

Sardines contribute to a significant portion (about 15 per cent) of the annual fishery from Andaman Sea. The genera contributing to this group are *Sardinella, Harengula, Pellona, Dussumieria* and *Herklotisichthys*. Of these, *H. punctatus* is the most common and most important species constituting as much as 75 to 80 per cent of the sardine group in the fishery. Schools of sardines occur in the inshore waters throughout the year, but the peak season is from July to December. The commercial fishing of sardine is restricted mostly to the eastern side of the South Andaman. A very little fishing efforts for sardines has been put in Nicobar and west of Andaman due to coral and barrier reef, which are seemed to be the abode of tertiary resources of *Sardinella* of considerable quantities (Cushing, 1971).

Hilsa: The Indian Shad

Three species under the genus *Tenualosa*, namely, *Tenualosa ilisha, T. toli* and *T. kelle* have been recognized from coastal waters of which *T. ilisha* forms commercial viable fishery, whereas, *T. toli* and *T. kelle* are scarcely available in Indian waters.

The average growth of young Hilsa after hatching was 28-32mm, 48-63 mm, 64-73 mm, 74-83 mm, 84-93 mm, and 94-102 mm at the end of 1, 2, 3, 4, 5, and 6 months respectively. The maximum size of Hilsa recorded so far is 600 mm, where females are larger than males.

The age at maturity of the species was 2+ or 3+ years. Higher values of gonado-somatic index were observed during September-March, with peak in October.

The unfertilized mature eggs of hilsa are almost spherical in shape and demersal in nature. The diameter of hilsa eggs range from 0.70-0.89 mm. The fecundity of fish of 253-451 mm size was 250000 to 1600000.

Hilsa normally inhabits the foreshore area of the sea and lower region of estuaries. The fish prefers to inhabit this region due to the presence of sub-surface oxygen, relatively low salinity, strong tidal action, high turbidity, heavy siltation and rich growth of plankton.

Most of the stocks of hilsa are anadromous, breeding much beyond tidal limits. Hilsa is highly salinity tolerant. Hilsa moves on the surface in the foreshore region of the sea, whereas in the river, they move in deeper zones near the bottom. The species move in shoals and the peak upstream migration commence with south west monsoon (July and August and continue upto October or November). Hilsa ascends into Hoogly river for spawning and their progeny, generally after attainment of size range from 80-110 mm start downstream migration towards the coastal region, which commences from February onwards and continues upto June.

The spawning of this species is seasonal. The spawning season of hilsa has been during the period between August and October. Hilsa of Hoogly estuary spawn during August-March with peak in October-November and February-March.

Milk Fish

Chanos chanos, milk fish, is a fish of open sea, a swift and powerful swimmer has moderately compressed oblong body with terminal mouth, small scales, 75 to 80 along the lateral line. Scales form a sheet along the base of dorsal and anal fin. Auxillary scales above and below pectorals and ventrals. Body is silvery green above, bright silvery along the sides, whitish on the abdomen. Dorsal and caudal fin, tip of anal and inside of pectoral fins are pigmented. Maximum length observed is 150 cm.

Chanos chanos is widely distributed in the tropical and sub-tropical areas of Indian Ocean, including Andaman Sea. It is also found in tropical and sub-tropical areas of Pacific Oceans and Red Sea eastward to the Paumotu Islands,from Southern Japan to Southern Australia, along the Pacific coast of North America from San Francisco to the waters of Mexico.

Being an euryhaline species, disease resistant, a herbivore (feeding mostly on algae) and exhibiting good growth rates, it is considered as one of the fishes, best

Figure 25.11: Adult Mullet Fish

**Figure 25.12: Dissected Out Matured Mullet Fish
Showing Ripened Ovary in the Abdomen**

suited for culture in brackish water ponds. It spawns, annually or bi-annually, off shore, in depths between 20 to25 m and larvae seek clear coastal waters for survival and growth. One year old fish of about 20 cm long and 200 g in weight, move out to sea to mature. They live upto six years and adults may reach weights upto 20 kg. The adults seldom enter shallow water, except during spawning season, at which time some of them are caught in nets and fish traps. The milk fish is capable of making

Figure 25.13: Iced Mullet Fish Ready to Transport

Figure 25.14: Harvested Mullet Fish

tremendous leaps out of the water, and it is said that a one meter fish can easily jump a distance of 10 meters, while clearing obstructions of 2 to 4 meters above the surface of water.

Fry and fingerlings of milkfish often enter freshwater rivers and a few individuals may find their way into lakes. These fish may grow to a large size in fresh water, but never reach sexual maturity there and must return to the ocean to spawn. Fish pond owners have failed to produce sexually matured milk fish in ponds.

Females are said to contain 1.5 to 7 million eggs. They spawn in sea near the coast. Eggs are pelagic, 12 mm in diameter and embryos hatch out within 24 hours.

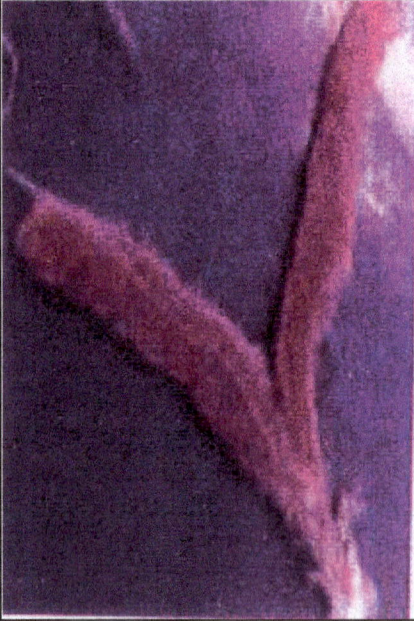

Figure 25.15: Developing Ovary   **Figure 25.16: Ripe Ovary**

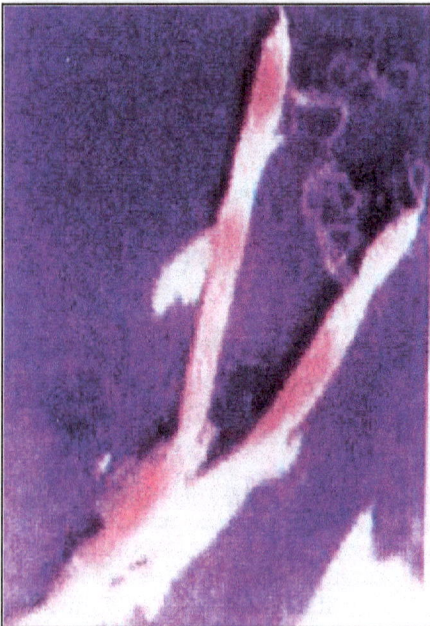

Figure 25.17: Spend and on Recovery   **Figure 25.18: Testis**

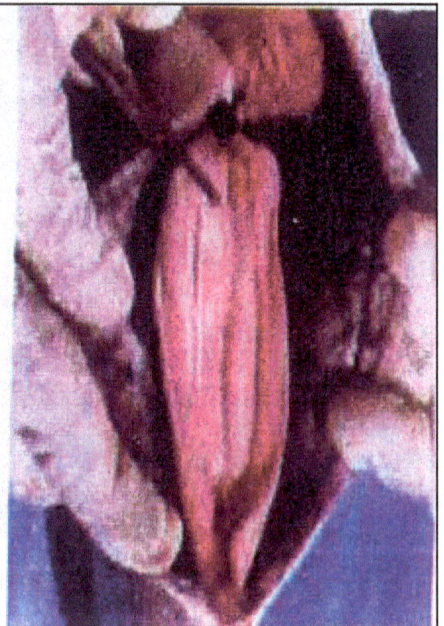

Larvae, 12 to 15 mm in length occur periodically in large quantities along sandy coasts and estuaries. They enter estuaries and tidal streams during the advancing flood tide. It is said that they are most abundant three days before and three days after full moon, although they are common at every high tide and small numbers may be obtained at all times of the day or night.

In India, the period of occurrence of milkfish fries are from March to August and from October to December.

Milkfish fry begin to feed when the yolk sac is absorbed on attaining 12 to 15 mm in length and their first food in ponds consists of blue green algae and associated organisms, called lab-lab in Philippines. When the fry grow between 50 to 75 mm long, the food preference begins to shift to filamentous green algae, known as lemut.

Milkfish fries are available in the coastal areas of Andaman Sea. As such, the culture of milkfish was undertaken in brackish water tide fed ponds to explore the feseability of introducing and establishing milkfish farming in them. At harvest after 13 months, the milkfish is reported to grow 234 to 435 mm in length, weighing 85 to 500 gram, with an average weight of 204 gram. The overall growth rate was 22 mm and 16 gram per month. The survival rate was 30 per cent, with a production rate of about 600 kg/ha/13 months of farming.

Mullets

Mullet is a commercially important fish that occurs in seas, estuaries, backwaters, lagoons and even in the reservoirs of salt works, that are totally disconnected from sea. The mullet fishes form the main stay of the marine littoral fisheries, providing livelihood to traditional fishermen. These fishes are capable of tolerating high salinity of 91.34 ppt. they are euryhaline, eurythermic and capable of living under extreme environmental conditions. Besides being non-predacious and predominantly detritus feeders, they are efficient converters of natural foods into fish flesh and therefore, are farmed in tropical and sub-tropical waters of the world.

Among eight species of mullets, the grey mullet, *Mugil cephalus* is the most common one in Andaman Sea. It has a wide distribution, good farming traits, highlighted by its growth, consistent with its known maximum size, the fish commands a demand because of its delectable taste.

Seven maturity stages in female and six maturity stages in male are recognized. The ovary is a paired one. Each of this pair is a sac like elongated structure. The anterior portions of each of these are free, but are united at the posterior end. The colour of ripe ovary is deep yellow, but reddish during developing stage and spent ones are flaccid in reddish colour and in a regressive and recovering stage. The length and weight of ripe ovaries are in the range of 120-170 mm and 160-630 g respectively.

The length at first maturity was recorded at 376 and 403 mm in male and female respectively. The spawning season of the species is prolonged, fromSeptember to February with a peak during October to December.

In *Mugil cephalus*, the fecundity ranges from 16 lakh to 72 lakh eggs in fishes of total length size range of 410 to 622 mm. The absolute fecundity of the fish depends on its length and weight. The relative fecundity range from 17.6 to 24.17 lakhs and the average was 21.34 lakh per kg body weight.

The mullets contribute to almost as much as seer fish (7.01-7.08 per cent) in the total landings of Andaman Islands. The main genera are *Mugil* and *Liza*. Though they occur almost throughout the year, the important fishing season appears to be from July to December, synchronizing with the low salinity and high plankton production period. They are mostly caught in gill nets and boat seines.

During June, 2005, bag nets operated at a distance of 3-5 km from Chennai Fisheries Harbour at a depth of 5-10 m, there was heavy catch of mullet, *Mugil cephalus*. Higher catch rate was noticed in the months of September-October, January-February and May-June. The catch comprised of immature fish with length range 105-139 mm and weight range of 12-28 gram.

Subsequently, fish collected after 12 days indicated that the fish were in pre-adult stage (180-244 mm) and 70-130 gram weight range. Females were dominant in the catches.

Scombroids

Tunas

Tunas are grouped taxonomically in the family Scombridae, which includes about fifty species,the most important commercial and recreational fishes are, Yellowfin (*Thunnus albacares*), Bigeye (*T. obesus*), Bluefin (*T. thynnus, T. orientalis, T. maccoyii*), Albacore (*T. alalunga*) and Skipjack (*Katsuwonus pelamis*).

Unlike most other fishes, which have white flesh, muscle tissue of tuna ranges from pink to dark red. The red colouration derived from myoglobin, an oxygen-binding molecule, is expressed in tuna in quantities far higher than than most other fish. Unlike other fishes, many tunas can maintain their body temperature above that of surrounding water, making them functionally warm-blooded.

Little tuna species constitute almost the entire catch of Andaman Sea. Long tail tuna account for 70 per cent and are caught all the year round mainly with gill nets, trolling line and seine nets by artisanal fisheries. Meteorological conditions are favourable from October to May during the northeast monsoon. Dissolved oxygen levels are low throughout the year, which explains the scarcity of yellowfin and skipjack.

The distribution pattern of long tail tuna, *Thunnus tonggol* is limited to the north and east of Indian Ocean including Andaman Sea. Eventhough long tail tuna is a coastal species, it is not found in turbid water or water with low salinity. This species is caught by artisanal fisheries equipped with trolling lines, gill nets or long lines along the coast.

Artisanal fisheries land longtail tuna, measuring 40-70 cm from Andaman Sea. The maximum length recorded is 130 cm in India.

Sivasubramaniam (1981) specifies that the size at first sexual maturity is attained by individuals measuring 45-50 cm. Little is known about long tail spawning periods. Ripe ovaries were found in a 7.3 kg female measuring 81 cm caught in September off the coast of India. Females measuring 44-49 cm (total length) produce an average of 1.4 million eggs per spawning.

Longtail is one of the most voracious tuna species. Klinmuang (1978) concludes that long tail tuna age-length relationship is 30 cm at 1 year, 47 cm in 2 years, with a monthly growth rate of 1.5 cm during their second year.

Yellowfin tuna, *Thunnus albacares* is widely distributed in Indian Ocean, but less abundant in Andaman Sea. Young individuals are rather concentrated in the upper waters near the equator (10°N to 10°S), where they are caught by commercial purse seining fleet and by artisanal fisheries, using pole-and-line and trolling techniques. The adults disperse more, covering approximately the first 150 m. They are known to dive to greater depths, and are generally caught by long lining and purse seining.

Yellowfin reproductive cycles has been studied basing on the long line catches.Fifty per cent of the fish observed to reach first sexual maturity, when measuring 120-140 cm, but some rare individuals display the same maturity, when measuring only 80 cm. Females in advanced ovarian development (capable of reproduction) produced gonado-somatic indexes (GSI) levels between 1.5 and 2.5. Spawning intensifies between January and March in the Central Indian Ocean (from 5 to 15°S and from 70 and 85°E). Later between April and June, only the yellowfin tuna caught in the vicinity of Sri Lanka are spawning.

The feeding habit of this species around Maldives and the Chagos Islands are almost exclusively comprise of flying fish, small young tunas and squids. Off Somalian coasts, squid, scombroids predominate in the stomach contents. The proportion of shell fish found in yellowfin tuna stomachs increases as they swim closer to the coasts. The areas with highest repletion indexes are those where thermal gradients are high.

Yellowfin tuna grows 54 cm at 1 year, 92 cm at 2 years, 120 cm in 3 years, 140 cm at 4 years and 154 cm at 5 years. Yellowfin juveniles measuring 45-70 cm fork length have growth rate of 3 cm per month. The growth rate of Indian Ocean yellowfin tuna is 3cm per month at 57-76 cm fork length and reaches 4.3 cm per month at 88-101 cm fork length.

The sex ratio of yellowfin was found equal (1:1). In yellowfin measuring more than 140 cm, they registered a 52 per cent male predominance.

There are two stocks of yellowfin in the Indian Ocean, located both sides of 100°E. During their life-span, yellowfin is exploited by various surface fisheries, both commercial and artisanal. Long liners catch mainly large yellowfin of average size approximately 120 cm. Live bait pole-and-line fisheries exploit smaller fish averaging 60 cm. The large yellowfin caught generally come from free swimming schools, whereas, the smaller ones are often caught near drifting wreckage.

Yellowfin, characterized by its large size, deepest body near the middle of the first dorsal, has 26 to 34 gill rakers located on first arch. Some large specimens have very long second dorsal and anal fins, which can become well over 20 per cent of the fork length. Pectoral fins are moderately long, usually reaching beyond the second dorsal fin origin, but not beyond end of its base, usually 22 to 31 per cent of fork length. No striations on ventral surface of liver. Swimbladder is present. Vertebrae 18 precaudal plus 21 caudal. The back of the fish is metallic dark blue in colour, changing through yellow to silver on belly. Belly is frequently crossed by about 20 broken nearly vertical lines. Dorsal and anal fins and dorsal and anal finlets are bright yellow. The finlets are with a narrow black border. The species is epipelagic, oceanic, inhabit above or below the thermocline. The thermal boundaries of occurrence are roughly 18 and 31°Celsius.

Schooling of yellowfin occurs more commonly in near surface waters, primarily by size, either in monospecific or mulit-species groups. In some areas, larger fish greater than 85 cm fork length, frequently school with porpoises. Association with floating debris and other objects is also observed.

Skipjack tuna (*Katsuwonus pelamis*) have a fusiform, elongated and rounded body. Their teeth are small and conical in a single series. The fish have numerous gill rakers, 53 to 63 on first gill arch. Two dorsal fins separated by a small inter space, not larger than eye. The first dorsal with 14 to 16 spines and the second dorsal followed by 7 to 9 finlets. Pectoral fins are short with 26 to 27 rays. Inter pelvic process is small and bifid. Anal fin is followed by 7 or 8 finlets. Body of the fish is scaleless except for the corselet and lateral line. A strong keel on each side of caudal fin base is present between two smaller keels. The species do not have swimbladder, but have 41 vertebrae. The back of the fish is dark purplish blue, lower sides and belly silvery with 4 to 6 very conspicuous longitudinal dark bands, which in live specimens may appear as discontinuous lines of dark blotches.

An epipelagic, oceanic species with adults distributed within the 15°Celsius isotherm, overall temperature range of occurrence is 14.7 to 30°Celsius, while larvae are mostly restricted in waters with surface temperatures of atleast 25°Celsius. Aggregations of this species tend to be associated with convergences, boundaries between cold and warm water masses, upwelling and other hydrographical discontinuities. Depth distribution ranges from the surface to about 260 m during the day, but limited to near surface waters at night.

Skipjack tuna, generally weighing less than 5 kg are found throughout all intertropical waters. This species are found throughout the Indian Ocean as far as 40 to 45°S in the west as well as off South Australia. This is a vast area with an extremely heterogenous environment. Skipjack tuna larvae can be found in most parts of the Indian Ocean, from 36°S on the western side to 30°S in the east. They are not found further north than 11 to 15°N. Large larvae concentrations have been observed south off Sri Lanka, close to the Lakshadeep Islands.

Skipjack juveniles are seldom seen in the open sea and consequently are rarely caught there. The area of distribution of skipjack juveniles (3-35 cm in length) has been determined 15°N and 35°S.

Apart from a few cases of hermaphrodism, found among skipjack tuna, off the Lakshadweep Islands, the species is heterosexual. The length at first maturity has been observed between 40-50 cm total length off Minicoy Island.

Central Indian Ocean skipjack tuna spawn from January to April and June to early September, with maximum activity in January and June. Peak gonado-somatic index (GSI) are registered from October to April in the eastern Indian Ocean, between Indonesia and Australia. Peak spawning thus corresponds to the southern summer.

There is no defined peak of spawning periods, except possibly for December. Sexual rest seems to last from May to September, but according to Marcille and Suzuki (1974), skipjack tuna appear to spawn all year round. Observation on larvae abundance in northeast Madagascar coast suggests a continuous spawning season throughout the year and peak from November to May.

Variations in fecundity from 151900 to 1977900 eggs for fish ranging between 41.8 cm and 70.3 cm (total length) have been observed off Lakshadweep Islands.

The principal prey of skipjack tuna are shellfish, fish and mollusks. The growth rate and age of skipjack tuna (examining the growth rings on the first spine of first dorsal fin) have been found as 3 years for fish 40-45 cm and 4 years for fish 40-60 cm.

In the Indian Ocean, the sizes of skipjack tuna varies according to the fishing techniques used. Fish caught by long lining in deep waters are generally larger than those caught by trolling, pole-and-line fishing or seining in shallow waters. However, large fish have been caught occasionally in surface waters off Lakshadweep Islands and Sri Lanka.

The gregarious nature of tuna causes them to swim in schools, which can be observed from the surface. This behaviour is considered generally as being a protection measure and can take different forms according to area or even according to the hour of the day. Precise knowledge of the schooling pattern is needed to determine the fishing gear to be used.

Surface schools, can be spotted by observations by trained personnel, swim freely or aggregate with flotsam. Free swimming schools, swim close to the surface or sub-surface and are accompanied by birds. To qualify a fish school according to the behaviour of the individuals comprise of;

Breezer – the presence of tuna indicated by a rippling on the surface, caused by fish swimming sub-surface in the same direction. This situation signals very often the presence of large schools.

Finner – the fish do not surface entirely, only dorsal fins can be seen surfacing from time to time.

Jumper – single fish jump out of the water and dive head first (fish behaving in this way are generally thought to have momentarily lost contact with the rest of the school).

Smoker – the fish collectively jump out of the water, generating a choppy sea. This behaviour is characteristic of mixed schools, made up of small tunas.

Boiler or foamer – This term is used when the preceding situation becomes highly accentuated and is usually caused by large fish preying on anchovies or euphausiids. Even from a distance the fish can be seen to jump in a disorderly fashion causing the water to foam.

Schools can also aggregate around floating objects (flotsam, wreckage) or anchored objects (FADs) and mammals (whales, dolphins) or sharks. These schools may be or may not be accompanied by birds and may be composed of one or more species.

Deep sea schools may be free swimming or may aggregate with floating objects and can be detected acoustically.

1. Only detected below flotsam, compact in form, comprised of adult skipjack mixed with small yellowfin and bigeye

2. Both free swimming and aggregated schools, comprise of two parts, one above the other. The upper one is compact and lower one is longer. Sometimes two sections are separated, made up of tuna mixed in size and species. The smaller individuals form the upper stratum in a compact formation and the larger ones, the second deeper ones.

3. Typical free-swimming formation comprised of large fish may be occupying a thick layer of water and registering an oblong diamond shape on the echo-sounder.

In the western Indian Ocean, the sizes of school observed vary considerably from one season to the other. During the northeast monsoon (December –March) and during inter-monsoon periods (April-May and October-November), the yields obtained per positive sets are similar and ranging from 21-25 tonnes/set. During the south west monsoon (June-September) levels rose to 35 tonnes/set.

Thunnus obesus, the bigeye tuna is a large species, deepest body near the middle of the first dorsal fin base. 23 to 31 gill rakers are present on the first arch. Pectoral fins are moderately long, 22 to 30 per cent of the fork length in large individuals over 110 cm fork length, but very long in smaller individuals, though in fish shorter than 40 cm, they may be very short. In fish, longer than 30 cm, ventral surface of liver is striated. Swim bladder and a total of 39 vertebrae are present. The lower side and belly of the fish is whitish in colour. A lateral iridescent blue band runs along sides in live specimen. The first dorsal fin is deep yellow. The second dorsal and anal fins are light yellow. Finlets are bright yellow edged with black.

Bigeye tuna is epipelagic and mesopelagic in ocean waters, occurring from the surface to about 250 m depth. Temperature and thermocline depth seem to be the

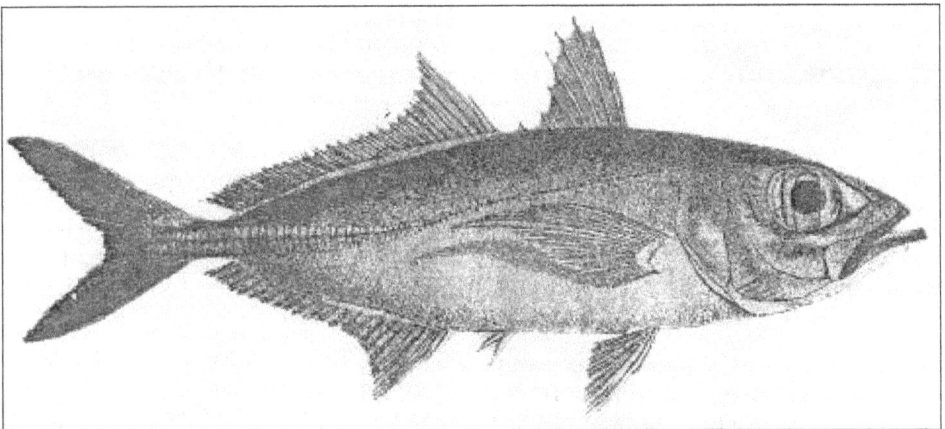

Figure 25.19: The Horse Mackerel (*Caranx crumenophthalmus*)

main environmental factors governing the vertical and horizontal distribution of bigeye tuna. Water temperatures in which the species has been found range from 13 to 29°Celsius, but the optimum range lies between 17 and 22°Celsius. This coincides with the temperature range of the permanent thermocline. In fact, in the tropical western and central pacific, major concentrations of *T. obesus* are associated with the thermocline rather than with the surface phytoplankton maximum. For this reason, variation in occurrence of the species is closely related to seasonal and climatic changes in surface temperature and thermocline.

Juveniles and small adults of big eye tuna school at the surface in mono-species groups or together with yellowfin tuna and/or skipjack schools may be associated with floating objects.

The annual average landing of coastal tuna in India is around 51000 tonnes (Little tuna-49 per cent, frigate tuna and bullet tuna-23 per cent, longtail tuna-13 per cent, skipjack-6 per cent). The estimated resource potential of oceanic tuna is 2.13 lakh tones (yellowfin-54 per cent, skipjack-40 per cent and bigeye-6 per cent). Already industrial fishing of yellowfin and bigeye started since last three years (2007-2009) from Visakhapatnam coast. Boats equipped with long line of 45 km with 600 hooks started operation at a depth of 45-60 m along Indian coast.

Quality of tuna varies from region to region. It is affected by biological and non-biological factors like, feeding, age, malnutrition, disease, fishing and killing methods, handling and processing, chilling and storage methods, sorting and grading techniques. Tuna is physiologically a warm blooded fish, less susceptible to temperature changes, when it is alive. When it is dead, however, it causes difficulties in chilling. Coupled with histamine problems, dark meat content, stress induced release of high amount of lactic acid in flesh, exposure to high ambient temperature (burnt flesh) turns the red tuna meat into white pale and flabby meat with poor semi-cooked appearance.

Seer Fish

Day (1889) recorded five species of *Scomberomorus*, occurring in Indian Seas. Seers constitute about 5 to 8 per cent of the total annual catch from Andaman Sea. They are caught mostly by gill nets. In trolling lines however, 13.7 per cent of the catch was represented by the seer fish. Though the group is constituated of two genera and three species, namely, *Scomberomorus guttatus, S. commersonii* and *Acantho cybium solandri*, the commercial fishery is mostly made up of former two species. The latter is occasionally caught near the reef and do not constitute a significant fishery. The best season for the fishery is from March to August synchronizing with that of the southern portion of the west coast of Malayasia.

Scomberomorus guttatus is the commonest seer in Tamil Nadu waters.

Biology

Several transparent spherical eggs of an average diameter of 1.17 mm occurred in the surface tow-net collections between September and January. Unsegmented yolk and single oil globule, situated ventrally has been observed in the newly hatched larva. The oil globule has an average diameter of 0.34 mm, was covered on the outer

Figure 25.20: Seer Fish, *Scomberomorus linceolatus*

half by yellow stellate cells and on the inner half, facing the embryo with black pigment spots cover almost the entire embryo. Eyes and auditory vesicles were well marked. The head and tail of the embryo were not free from yolk mass.

After 3 to 4 hours, the embryo started wriggling inside the egg, the tail becoming free from yolk mass. The larvae on hatching (average length 2.8 mm) float with the yolk above and oil globule uppermost. There were 14 preanal myotomes. Black branching chromatophores were seen on the anterior end of the head, behind the eyes, and a few were scattered almost dorsally along the myotomes. The gut was not visible posteriorly and was bent.

In the juveniles of 24-28.5 mm, the general shape of the body was different. These juveniles resembled the adult in the unpaired fins and the course of the newly developed lateral line, and could be recognized as belonging to the genus *Scomberomorus*.

The young post-larvae of 3.8 mm were found to have fed only on the nauplii of copepods (*Oithona* sp. and *Euterpina* sp.). The juveniles of 24 to 28.5 mm length were found to have fed on larger animal planktons, such as, copepods, megalopa, zoea larvae and ostracods.

The change over from planktonic diet to one composed of mid-water or nektonic organisms is undoubtedly associated with the fish (378-485 mm) changing its habitat from the inshore to the offshore waters. They were found to have fed on small crustacean, such as, megalopa larvae, *Penaeus* sp. and *Metapenaeus* larvae. From February to July fishes in offshore waters were found to have changed over to nektonic diet feeding mainly on young teleosteans (*Leiognathus* spp., *Lates calcarifer*, *Caranx* spp., *Lactarius lactarius* and larval eel).

The fish do not move in shoals even when they migrate for spawning or feeding purposes.

The habit of *S. guttatus* of being in the surface water near the shore from July to February, when they breed, and descending to midwater during summer months of March to June has been observed.

Flat Fishes

Eleven species of flat fishes (Cynoglossids) belonging to two genera, *Paraplaguisia* and *Cynoglossus* were collected from commercial fishing trawlers in coastal waters of Indian Ocean.

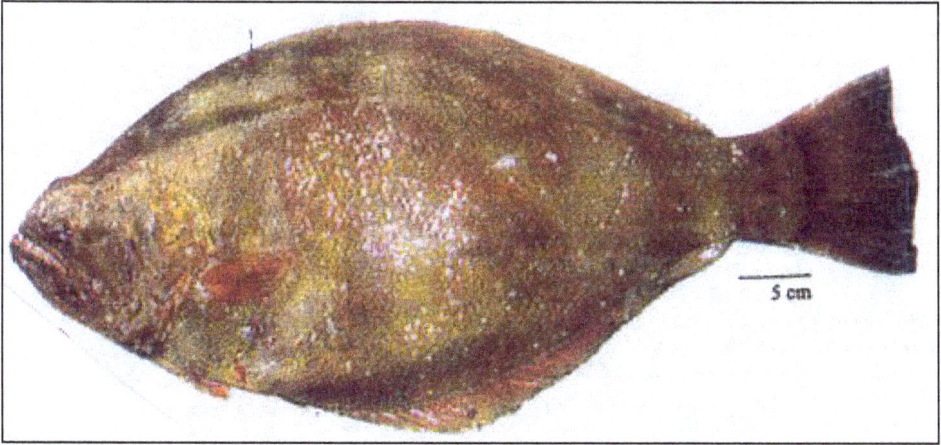

Figure 25.21: Flat Fish, *Psettodes erumei*

Psettodes erumei occurs in some quantities in Bombay-Saurashatra and south eastern coast. *P. erumei* is a carnivorous fish subsisting mainly on fishes (81.73 per cent) crustaceans (11.93 per cent) and mollusks (6.34 per cent). Their spawning season is observed to be a prolonged one, extending over a period of 7 months from November to May as against the short and restricted spawning of this species in west coast of India. The animal being a bottom dweller is able to feed on both pelagic and demersal forms.

Growing to a maximum size of 60 cm in total length, and inhabiting the sand and mud bottom of coastal waters upto a depth of 100 m, *P. erumei* is commercially the most valuable flat fish in India. This species is distributed in the tropical belt between 26°N to 21°S and 43°E to 156°E in the Indo-west Pacific from Red Sea and East Africa to Japan and Australia.

The species in Andaman Sea is mid-level carnivores (trophic level 4.0-4.5), showing medium or high level of resilience. Elimination of predatory fish communities has been reported from oceans around the globe due to rampant fishing activities, with potentially serious consequences for ecosystems.

Pauly (1994) recorded heavy depletion of flat fishes, including *P. erumei* in the gulf of Thailand, since the introduction of trawling, and hypothesized that the tropical flat fish maintain low biomass when their environment is undisturbed, but may increase in their recruitment if external disturbance, such as, fishing removes their competitors. The increase in fishing effort, though resulted in larger biomass and diversity of flat fish landings, may have resulted in the local extinction of top level carnivores, such as, *P. erumei* from the coastal waters.

Long lived demersal species tend to decline faster than the short lived pelagic species. Pauly (1994) hypothesized that flat fishes in general are over-adapted to and too smsll for the niche they occupy. Since *P. erumei* is a strict carnivore and a fish occupying a higher trophic level than sharks and rays in Indian coastal waters, any possible local extinction may create serious repercussions on the delicate food web of

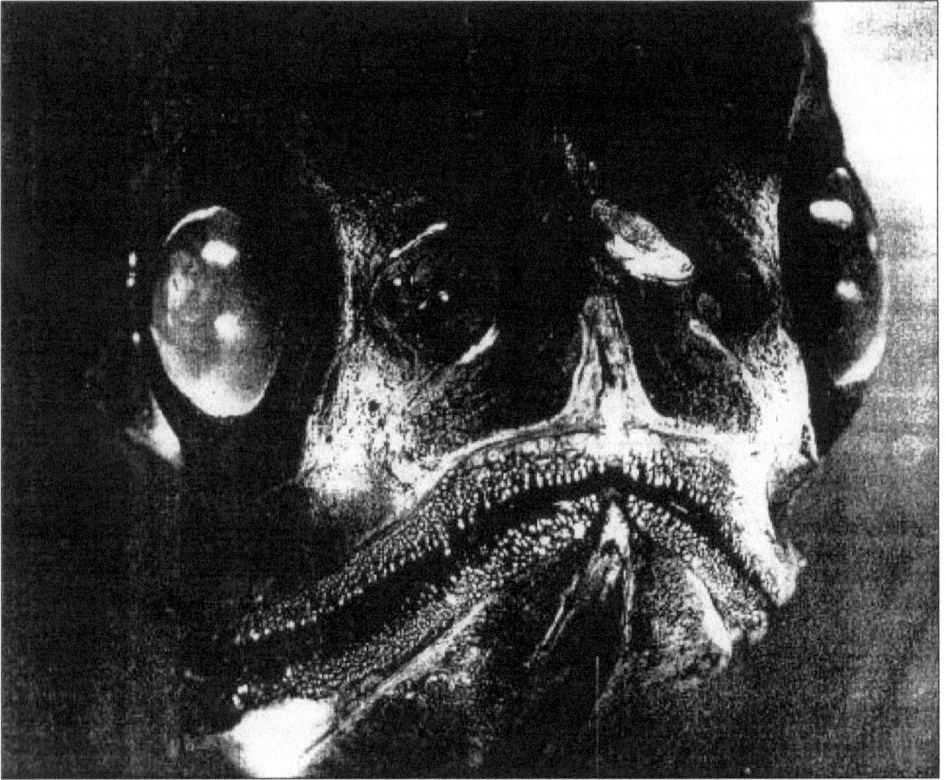

Figure 25.22: Deep Sea Lantern Fish

the Indian coastal waters, which already indicate a gradual transition in landings from long-lived, late maturing high trophic level piscivorous demersal fish towards short-lived early maturing low trophic level planktivorous pelagic fish. The non-availability of *P. erumei* is often masked by the presence of other species lower in food chain and other mid-level carnivores, including other species of flat fishes.

Since there is no targeted fishing for flat fishes along the Indian coasts, they are landed mainly as by-catch of shrimp trawlers. Since all fishery management studies concentrate on target species, the disappearance of even commercially valuable species may go unnoticed.

In August 2010, a large sized *P. erumei* measuring 60 cm was observed at Cochin Fisheries Harbour at Kerala. During recent years, the fishery of Indian Halibut (*P. erumei*) was very sporadic and catches were represented by relatively smaller fishes below 30-40 cm along the south west coast of India. The reported maximum size of the species is 64 cm total length.

Pelican Flounder

Pelican flounder, *Chascanopsetta lugubris* are distributed in the Eastern Atlantic, Gulf of Guinea round the Cape to Natal, South Africa, Western Atlantic, Florida, USA and northern Gulf of Mexico to Brazil, Indo-Pacific, off eastern coast of Africa and off India and Sri Lanka to Japan. The species belong to the family Bothidae, was collected from Puthiappa Fisheries Harbour on 16.8.2010. The fishes measuring 218-222 mm in length and weighing 48-51 gram were caught by a trawler operated off Calicut at a depth of 160 m. The body of the species is elongated, laterally compressed and eyes are on the left side. It is having a uniform dark grey colour, maxilla long extending backward well beyond the posterior edge of the eye. Large mouth with small teeth and gill rakers absent, Dorsal fin rays 115-118, origin of fin well in front of eyes, anal fin rays 71-82, pelvic fin bases unequal in length, that on eyed side much longer. Scales small, cycloid on both sides and 189 in lateral line. It is having greyish colour in eye side, fins dusky, peritoneum black, visible through thin abdominal walls. Blind side uniformly light. It is usually found in the deeper continental shelf to depths of almost 1000 meters.

Sole

Sole fishery is important only along Malabar coast, although these fish occur in some quantities along the adjoining coast of South Kanara as well. Many species of sole including *Cynoglossus semifasciatus* are also represented on the east coast of India. *C. semifasciatus* is the only flat fish that occurs in large shoals.The species is known to occur in shoals from August of one year to February of the following year.

C. semifasciatus is a small fish, growing commonly to about 15 cm in length, the maximum size so far recorded being 17.5 cm. The body is flattened and leaf-like, one side lying on the substratum and being white in colour. Both the eyes are found on the side away from the substratum and this side is pigmented with irregular brown half-bands across the body. The mouth is narrow and symmetrical. The dorsal and anal fins are long extending along the margins of the body. There are two lateral lines on the ocular side and none on the blind side, which is the right side of the animal Scales are ctenoid on both sides.

The bulk of the commercial catches of soles consisted of individuals of one year age group, the older individuals being negligible inproportion. The products of spawning of a particular fishery season grow up to commercial size and directly enter the fishery in the next season. The best season for the fishery is immediately after the south west monsoon, September being always the month of peak commercial catches.

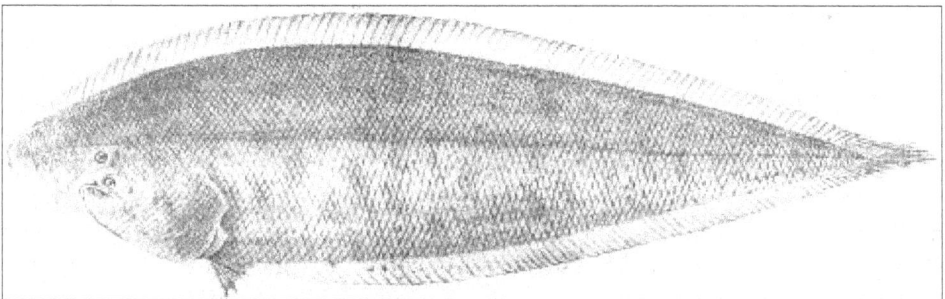

Figure 25.23: The Sole (*Plagusia* sp.)

The species is a bottom feeder and a carnivore. It favours a diet of polychaetes, amphipods and small lamellibranches. *Prionospio pinnata*, a polychaete is its chief food.

The sexes are recognizable in the Malabar sole of 6 to 7 cm total length. There is no external marks to distinguish between two sexes, but in case of females with gonads in advanced stages of maturity, the ovaries can be seen through the body wall, when the fish is held against light.

The main spawning grounds of the species lie somewhere in deeper waters and the migration out of and into inshore grounds is taking place in batches. Spawning migration commence sometime in October with the weakening of the commercial fishery

The smallest specimen with ovary developed to fifth stage was noticed, when the fish grows to 10 cm in total length. Normally, the ovary is not ripe below a size of about 12 cm in total length. The ripe ovarian egg is spherical. Several oil globules are present in each egg. Ripe and full ovaries of specimens measuring 15.6 and 15.9 cm were found to have 42200 and 65900 ripe and ripening ova. Spawning commences in the species in October and continues upto May with varying intensities in between. Eggs have been found in offshore plankton samples in November and the post larvae were collected in April.

The 2.8 mm larva has a symmetrical and laterally compressed body. The head is large and there is a clear operculum. The mouth is anterior and slit-like. The eyes are large and prominent. There are groups of dark pigment spots characteristically distributed over the body. The positions of brain, auditory vesicle, nerve cord and vertebral column can be easily visible.

In the post-larva, measuring 4.5 mm in total length, the body become more elongated and lanceolate. The mouth is prominent and anterior, with very minute teeth on the jaws. There are distinct fin rays in the dorsal and anal fins, but all he rays are not distinctly formed in the caudal fin. The pelvic fins are clearly seen, but their fin rays are faint. The pectoral fin continues to be without any fin rays. Air bladder is distinctly visible through the skin and there is a dense group of pigment spots in the region of air bladder.

The first sign of approaching metamorphosis is the further development of the rostral projection into a prominent parrot beak like structure, rostral-beak.The post-larva is found on the bottom, still in the vertical position with the head touching the substratum and the rest of the body disposed at an angle to the substratum. It look as if the larva is rubbing its head on the bottom and jerking the tentacle sometimes. Ocassionally it is lying on its right side and is showing occasional jerky movements. It do not assume the vertical position again after this. The lower jaw is moving constantly and rhythmically and the tentacle is lashing slightly now and then. The larva show occasional movements of the entire body also, but slowly and still lying on the right side only. The first signs of asymmetry is noticed when the right eye slightly shifted its position. The subsequent stages in shifting of the eye are reached rapidly, and in about two hours time, the eye is completely on the left side. The right eye gradually moves towards the dorsal edge of the head, where the rostral beak comes more and more in apposition with the front of the head and finally when the eye has reached the left side, the entire beak presses against the head and finally fuses with it just leaving the eye clear of the left side. Within 12 hours, the fusion of the beak is complete and the appearance of the head is very much as in adult. The eye does not shift further towards the original left eye, once it has completely shifted on to the left side of the body. The shifting of the eye is invariably taking place during nights.

The main phases of metamorphosis is over with the shifting of the eye and the correlated adjustments in the structure of the skull. The abdominal bulge disappear following the metamorphosis. Clear scale centers are noticed in the skin on the ninth day of metamorphosis. The light brown pigment spots of the body arrange themselves into irregular transverse bands across the body. Both the lateral lines are well formed in larvae of 23 mm total length.

Carangids

Carangids form a part of pelagic fisheries, consisting of horse mackerels, round scads, queen fishes, trevallies and jacks. While some of them attain large sizes, others grow to smaller sizes. The smaller ones are abundant in the form of big shoals.

Next to perches, Carangids (horse mackerel, trevallies and scads) form the important group in artisanal fisheries of Andaman and Nicobar Islands. This group is represented by six genera, namely, *Decapterus, Selar, Megalapsis, Chorinemus, Caranx* and *Elegatis*. July to November is the best fishing period for them. Most of the catches are obtained with gill nets and boat seines. An appreciable quantity is also taken through hook and lines. Carangids represent 28 per cent of the catch obtained by trolling. The areas east and south of port Blair are very rich in Carangids, some of them weighing more than 20 kg each.

Carangids are gonochoristic and in the case of most of them, there is no apparent external difference between the sexes. Females are oviparous. Some species spawn pelagically, whereas others spawn close to the shore. Carangids are generally described as fast swimming carnivores and pursuit predators. Carangids exhibit the carangiform mode of swimming. The anterior one-half to two-thirds of the body bends only slightly while swimming and most of the thrust is generated in the rear third. The characteristic narrow caudal peduncle and the forked caudal fin in high aspect ratios serve to increase the efficiency of the tail. The scutes, when present probably help in reinforcing the narrow peduncle.

The carangids are widely distributed in the Atlantic, Pacific and Indian Oceans in the temperate, sub-tropical and tropical regions of the northern and southern hemispheres. In India, carangids are available almost throughout the year and they constitute a major fishery. In the recent years, the production of carangids is increasing and it forms about 4.5 per cent of the total marine fish production and 8-9 per cent of the pelagic resources.

There were maximum landings of Scad in 1992 (117810 tonnes) and then the catches dropped to regain again in 1995 to 103063 tonnes. In the subsequent years, the catches declined and in 2009 they were 50759 tonnes. On an annual average,

D. russelli contributed to 31.55 per cent to the total carangid landings, 2.94 per cent to the pelagic landings and 1.615 to the total marine fish landings.

Carangids have been receiving attention in various parts of the world for the past 30 years or so as potential aquaculture candidate species.

The Indian Scad, *Decapterus russelli* is a commercially important fish in Kerala, TamilNadu, Andhra Pradesh, Karnataka and Gujarat. It is the dominant species in trawl landings at Verabal, Mangalore, Kochi and Visakhapatnam, as also in purse seine and ring seine net at Kochi. *Megalaspis cordyla* is the dominant species in gill net landings at Veraval, Kochi and Visakhapatnam and in purse seine and gill net landings at Mangalore. At Tuticorin, *Caranx ignobilis* is the dominant species caught in paruvalai and hooks and line. *C. carangus* is dominant in podivalai and *Decapterus* spp. and *S. leptolepis* are the dominant species in the trawl landings.

On January, 2008 two giant sized female black banded trevally, *Seriolina nigrofasciata* were landed at Verabal, caught by hook and line from a depth of 80-90 m. The fishing area was off Jaffrabad at 100 to 110 km in the south east of Verabal. The fishes have a total length of 110.8 and 104 cm, weighing 9.4 and 7.4 kg respectively. The maximum length and weight of the species, caught earlier in the Persian Gulf and the Oman Sea was 70 cm and 5.2 kg respectively.

Mackerel

The genus *Rastrelliger* have four species dwelling in Indian Oceans. They are *Rastrelliger faughni*, recognized by short gill rakers, 21-25 present on lower part of first gill arch, *Rastrelliger brachysoma*, identified by long gill rakers with more than 30 on lower part of first gill arch, head length distinctly shorter than body depth. Body depth 28.5 to about 34 per cent in fork length, depth ratio about 2.7-3.2 without dark stripes or regular rows of dusky spots along upper half of the body. In *Rastrelliger kanagurta*, however, head length is distinctly greater than body depth. Dark stripes or rows of dusky spots are present along upper half of the body. Body depth is 23-27 per cent in fork length and depth ratio about 3.4-4.0. Popularly known as short mackerel, *Rastrelliger neglectus* have been met in Mayanmar waters Smaller specimens of *R. neglectus* come close to the shore in May and June. *R. neglectus* is a coastal or inshore form and the species from Mayanmar waters has a bigger length than the fishes from Thailand. The predominance of females were more pronounced among *R. neglectus* as they constituted 62.4 per cent, whereas males only 37.6 per cent.

Indian mackerel, *Rastrelliger kanagurta*, widespread in the Indo-West Pacific from South Africa, Seychelles and Red Sea, east through Indonesia and off northern Australia to Melanesia, Micronesia, Samoa, China and the Ryukyu Islands. The species has entered the eastern Mediterranean Sea through Suez canal.

The fish has elongated slightly compressed body with pointed snout. Its front and hind margins of eye is covered with an adipose eyelid. Its teeth in upper and lower jaws are small and conical and teeth is absent from vomer and palatine bones, on roof of mouth. Body of the fish is moderately deep and its depth at margin of gill cover is 4.3 to 5.2 times of length. Its head is longer than depth of the body. Maxilla is partly concealed, covered by the lacrimal bone, but extending to about hind margin of eye. Its gill rakers are very long, visible when mouth is opened, 30 to 46 on lower limb of first arch. Two widely separated dorsal fins, interspace atleast equal to length of first dorsal fin base, have 8 to 11 spines in the first dorsal The second dorsal and anal fins have 12 rays. Anal spine is rudimentary. Five dorsal and five anal finlets are present. Pectoral fins are short with 19 or 20 rays. Scales behind head and around pectoral fins are larger and more conspicuous than those covering rest of the body, but do not have well developed corselet. Two small keels on each side of the caudal peduncle are present, but no central keel between them. Swimbladder is present. Back of the body is blue green in colour with narrow dark longitudinal bands on the upper part of the body and a black spot on body near lower margin of pectoral fin. Dorsal fin is yellowish with black tips, caudal and pectoral fins are also yellowish, but other fins are dusky.

Indian mackerel, an epipelagic neritic species occur in areas, where surface temperature are atleast 17°Celsius. Schooling of the species occurs by size. The spawning season around India seems to extend from March through September. Spawning take place in several batches.

Schools of mackerel are very frequently encountered around Andaman and Nicobar Islands and are represented by two species, namely, *Rastrelliger kanagurta* and *R. brachyspma* of which the former is a commercially important one. Mackerel constitute about 8 per cent of the total annual landings. They are caught by gill nets and boat seines. The fishery shows a bimodal fluctuations, one peak being March to June and another from September to December.

Juveniles of the species feed on phytoplankton, diatoms, and small zooplankton, such as, cladocerans, ostracods, larval polychaetes etc. With growth, they change their dietary habits, a process that is reflected in the relative shortening of their intestine. So adult Indian mackerel prey primarily on macro-plankton, such as larval shrimps and fish. Their longevity is believed to be at least four years.

The species attain first maturity at about 23 cm length. They grow to a maximum length of 35 cm, but common size is 25 cm.

The Indian mackerel, *Rastrelliger kanagurta* is a very important pelagic resource along the west coast of India, showing wide fluctuations from year to year.

The mackerel has a prolonged spawning season, and the species in advanced stage of maturity occurred almost throughout the year except during October to February. Their peak spawning periods are around July and January. Mackerel measuring less than 200 mm length are immature and maturing. Fishes measiring more than 210 mm comprised of all stages of maturity. The normal modal size of one-year-old mackerel appears to be 120-150 mm and the fish measure 210-230 mm at the end of second year. The length of mackerel at the end of first year, second year, third year and fourth year of their life is found to be 110-150, 210-240, 250-270 and 280-290 mm respectively.

Juveniles of Indian mackerel, *R. kanagurta*, of 86-142 mm size weighing 5 to 22 g were recorded from Bombay waters (Lat. 18-19°N and Long 72-73°E) appears to be isolated. The food of these young mackerel consisted mainly of planktonic organisms, like *Coscinodiscus* and copepods.

Bombay Duck

Bombay duck, represented by a single species, *Harpodon nehereus*, is a semi-transparent, like gelatine with minute black or brown dots along the head, back and sides. The abdomen is silvery white. The fins are transparent, but in some specimens there are black dots.

The length of fish head is one-sixth of the total length. Bombay ducks are elongated and rather compressed fishes in which the head is thick and short and are provided with very short rounded snouts. The scales are small and transparent and can only be distinguished with great difficulty in fresh fish. The cleft of mouth is very wide. There are teeth in a band in both the jaws. The structural features, namely, the gelatinous consistency, large mouth and re-curved teeth on the lower jaw, indicate that the fish is a deep sea form, but it is not known definitely, whether it is a demersal or a pelagic fish

The fish attains 40 cm in length. This is a shoaling fish swimming near the surface and is very common at Bombay, but decreases in numbers down the Malabar coast.It is not common in the Coromandel coast, but north of Tamil Nadu along the Andhra coast it reappears in large numbers and finally is very abundant in the rivers and estuaries of Bengal. This migration has been shown to have nothing to do with reproduction of the species, but in some way connected with the marked changes in the physical conditions of the shore waters.(variation of salinity along the coast, the 70°F isotherm barrier of July and distribution of various food components).

Females of *H. nehereus* predominate the catches throughout the year except in the months of July and August, when male predominates. In view of the continuous breeding habits of the species, the recruitment to the stocks is also continuous and is not restricted to certain periods.

The fish appears to attain an average length of 127 mm at the end of the first year and 210 mm at the end of second year. The frequent appearance of the "0" year class shows that the recruitment is continuous and is more pronounced during the post-monsoon period, that is, September to December.

Ribbon Fishes

Ribbon fishes occur in considerable magnitude along the east coast of India, particularly along Tamil Nadu and Andhra coasts. A limited occurrence exists in the Hoogly-matlah estuaries. The most dominant species caught is *Trichiurus lepturus*, which grows upto 1.5 m. Average commercial size is 75 cm. the other species, *Lepturacanthus savala, Epleurogrammus intermedius* and *E. muticus* are smaller in size. Large shoals of *T. lepturus* are caught from the inshore waters of peninsular India during August and October. Ribbon fishes breed more than once in a year. *L. savala* is the predominant species in Hoogly-Matlah estuaries. Commercial size of the species is 15-50 cm. *E. intermedius* is abundant in Palk Bay and *E. muticus* along Orissa coast. Commercial size of *E. intermedius* is 14-35 cm and the life span about 4 years. Commercial size of *E. muticus* is 35-50 cm. Ribbon fishes are carnivorous, the food consisting of crustaceans and fishes. Spawning of ribbon fishes appear to be in offshore waters.

Ribbon fishes of the genus *Trichiurus* occupy an important place among Indian marine fishes with respect to the magnitude of the fishery they support. They can be easily identified by its elongated ribbon shaped and laterally compressed body tapering towards the posterior end. There is no caudal fin. The dorsal fin extends from the hind edge of the preopercle along the whole of the back and is many rayed. Ventrals are in the form of scales and anal spines are minute and sometimes concealed under the skin. The entire body is scaleless and is coated with a shining silvery substance. The cleft of the mouth is deep with teeth in jaws and palatines, those on the premaxillaries being arched and strong. The lateral teeth are lancet-shaped. Lateral line is just below the middle of the body and is very clear.

E. intermedius around 42 cm have immature gonads, where testis is slender and transparent in males, while in females ovaries are also slender, transparent and eggs are not visible to the naked eye. Around 46 cm length, the gonads become more mature, when testis is slightly enlarged and opaque in case of males, while in females the ovary is slightly enlarged, eggs having granular appearance. In matured fishes of 48 cm length, the testis become greatly enlarged with pale white colours. Ovaries in females also enlarge greatly, with yellowish eggs visible to the naked eye. Average size of the largest eggs at this stage become 1.1 mm in diameter. The minimum size at which the species mature is 47-48 cm in length.

The species is a voracious carnivore and a selective feeder. Immature fishes of 17-45 cm length show a tendency to feed on prawns and small fishes. Adult forms (46-62 cm long) feed on fewer prawns and smaller fishes. The mature and spent fishes have been found to be voracious feeders, even resorting to cannibalism, whereas individuals prior to spawning, show a definite slackness in feeding.

A 43-45 cm long matured fish produce about 6000 to 6500 eggs; while 48 to 50 cm long fishes have 14900 to 15500 matured eggs. The occurrence of gravid specimens measuring 48 cm and above during the months of April and May and their reappearance in large numbers in spent condition in July, lead to the conclusion that the spawning of the species should have taken place towards the end of June and that the spawning is restricted to a definite and short period.

In the Indian waters, the three species shoal separately at different places. On the east coast, *E. intermedius* predominates in the catches at Tamil Nadu and north Andhra, north of Visakhapatnam. *L. savala* has been reported to be more common on the coast of central Andhra, Visakhapattinam to Masulipattam, whereas in the south, near Gulf of Mannar and Palk Bay, in addition to the two above mentioned species, which occur in very small numbers, the third species, *T. muticus* is also quite common. In the southern part of west coast consisting of Travancore-Cochin and Malabar and in the northern part comprising, the Konkan, Bombay and Ratnagiri coasts, *E. intermedius* and *L. savala* respectively are the common species caught.

Pomfrets

Pomfret contribute one of the main groups of table fish in India. They are found everywhere in the tropical seas of the Indo-Pacific region. Though pomfret occur all along the coast of India, the main area of abundance are Gujarat and Bombay coasts on the west and Orissa and lower West Bengal on the east coast. Of these, the major landings are from Gujarat fishing centers. The following three species contribute the fishery in the Indian waters.

Pampus argentius (Silver pomfret), *Pampus chinensis* (Chinese pomfret), *Parastromateus niger* (Black pomfret).

Maharashtra landed about 36 per cent, closely followed by Gujarat (26 per cent) and Andhra Pradesh (15 per cent). Kerala, Tamil Nadu, West Bengal, Orissa and Karnataka also landed small quantities of pomfrets, their percentage to total all India catch being about 8, 7, 6 and 3 in the order of abundance.

Maximum landings of pomfrets are recorded during October to December in Maharashtra and the minimum during July to September. *Pampus argenteus* and *Parastromateus niger* are the dominant species comprising the pomfret fishery.

In Gujarat, the pomfret fishing season is in second and fourth quarters (April to June and October to December). In some years, the maximum catch was during the fourth quarter, while in other years, the highest landings were obtained in the second quarter. The lean fishing season for pomfret fishery is in the first and third quarters (January to March and July to September).

The dominating species of pomfrets along Gujarat coast are, *Pampus argenteus, Pampus chinensis* and *Parastromateus niger*. Large quantities of small sized *P. argenteus* are caught in dol net in April-May and September-October, mostly in the size range of 20-99 mm length (84-86.3 per cent). The fish in the size range of 100-199 mm and above 200 mm were 13.1-15.5 per cent and 0.5-0.6 per cent respectively. Though the young ones are present throughout the year in the narrow coastal belt of 45 km, at a depth of 15-35 meters, large number of these, however, enter the fishery during January-March and September-October. The juveniles grow 50 mm in 4 months and 60 mm in five months.

The silver pomfret, *Pampus argenteus* is the principal constituent of the catches from the Gulf of Mannar, whereas, black pomfret, *Parastromateus niger* is occasionally caught. The main area producing pomfrets are shallow waters upto a depth of 50 m. Their monthwise catch rate (kg/hour) was 19.30 in October, 12.60 in November, 112.90 in December, 28.66 in February and 42.90 in March. The maximum yield recorded was 113 kg per hour in December followed by 43 kg per hour in March. The aggregate catch rate of 24 kg per hour indicates resource of significant magnitude of pomfrets. Both silver and black pomfret caught were in their prime size 21-30 cm (94 per cent and 85 per cent) of the catch respectively.

Cobia

Cobia (*Rachycentron canadum*), known as lemon fish or ling, belonging to family Rachycentridae, occurs in tropical warm waters world wide, except for the eastern Pacific region. It is pelagic, but also found over shallow coral reefs and off rocky shores and occasionally found in estuaries.

Though the species has a wide distribution, it never formed a commercial fishery. But of late, the landings of this species have increased considerably from Mumbai waters, though stray landings were occasionally observed prior to 2007. In 2007, the estimated landings were 110.6 tonnes. The species was observed throughout the year with major peak during September-October and a minor peak in February. The total length of fishes ranged from 123 to 205 cm and weight ranged between 29 to 95 kg.

Cobia has an elongated, sub-cylindrical body with a broad depressed head. The mouth is large, terminal having a projected lower jaw, with villiform teeth in the jaws, on the roof of the mouth and the tongue. The first dorsal fin consists 7-9 strong spines. The second dorsal fin starts at the center of the body and almost reaches the tail. The anal fin is similar to the second dorsal fin, but shorter. The caudal fin is almost round in the juveniles, but becomes lunate, when the fish grow older. The pectoral fins are pointed. The scales are small and embedded in thick skin. The lateral line is slightly wavy anteriorly, but straight at the posterior. While the adult cobia is chocolate brown colour, they are dark brown on the back and paler down on the sides. The fins are brown and the belly is white silvery in colour. A black lateral band, as wide as the eye runs from the snout to the base of the caudal fin, bordered by two paler bands.

Cobia is a migratory pelagic species. It inhabit at wide temperatures ranging 16.8-32°Celsius. The salinity range in which the cobia is found is 22.5 to 44.5 ppt, but adaptation to lower salinities seem possible.

The species is found in the tropical, sub-tropical and temperate warm waters of the West and East Atlantic, the Gulf of Mexico, the Indian Ocean, the Red Sea, the Persian Gulf and the West of Pacific. It can be found from the north of Japan to the south of Australia, from South Africa to Morocco and from north of Argentina upto the south of Canada.

Eggs and larvae are usually found off-shore, but juveniles are more common in shallow coastal waters near beaches, islands and bays and river mouths. Adults are most commonly found in coastal and continental shelf waters, sometimes in estuaries. Adults are pelagic, but often stay close to the structures (buoys, drifting objects, pilings).

Cobia is gonochoristic. External differences between sexes are limited. When mature cobia females have a higher condition factor than the males, having a full soft belly, compared to the more slender hard body of the males. In the Chesapeake Bay, cobia male mature from the second year onwards at a minimum size of 51.8 cm and a minimum weight of 1.14 kg. Females mature from the third year onwards at a minimum size and weight of 69.6 cm and 3.27 kg respectively. In the warmer region, maturation does occur earlier.

Cobia is a multiple batch spawner and the spawning season is extended over several months. Depending on the location, spawning takes place between May and August in spring and summer (Chesapeake Bay, Gulf of Mexico), but possibly year round in Indian Ocean.

During the spawning season, the adult migrate away from the shore and form spawning aggregations. Eggs and sperms are simultaneously released and fertilization is external. Spawning cobia and fertilized eggs have been found 40-80 km off the coasts in the Gulf of Mexico, as well as in North Carolina waters.

The fertilized eggs are pelagic, with a large oil globule, yellow in colour and mottled with melanin pigment. They take about 1.5 days to hatch.

The larvae are 3-3.5 mm long and colourless immediately after hatching, carrying a yolk sac. Five days after hatching, the larvae absorb the yolk sac and reach a length of 4-5 mm and start exogenous feeding. They have limited swimming abilities at this stage.

During the first 30 days, the larvae undergo metamorphosis and they start looking like adults at a later stage. After metamorphosis, they look brown in colour with a black and a white stripe running from snout to tail.

After about two months of larval life too; the shape of the juvenile remains the same, but the colour bands along the body become more pronounced, one of these is a black central line along the lateral surface, bordered by two gold or white bands. When the fish reach adulthood, they live solitary, or in groups of 2-8 fish.

Cobia is a carnivorous bottom dwelling predator. Its larvae feed on zooplankton, especially copepods. In nature, juvenile and adult cobia feed on wide range of fish species, crustaceans and benthic invertebrates. The preferred prey items include crabs and shrimps. Although known as pelagic fish, a large part of its food items is found close to the bottom. Food is often swallowed whole. In farming, cannibalism is a common phenomenon among juveniles.

One 8.5 kg female carried 1.4 million eggs (160000 eggs/kg body weight). Their immature eggs are clear nucleated cells, 0.1-0.3 mm in diameter. Matured eggs are of 1.09-1.31 mm diameter with a 0.29-0.44 mm diameter oil globule.

Cobia are caught in hand operated hook and lines, using baits of tuna, cephalopods etc, targeted to catch sharks. The species of sharks caught along with cobia are *Carcharhinus limbatus, C. sorrah* and *Galeocerdo cuvieri*. A similar high catch of cobia in purse seine was recorded in 2004, but smaller in size (100-119 cm).

Sucker Fish

Sucker fishes, generally attach themselves to many different marine vertebrates including sharks, rays, tarpons, barracudas, sail fishes, groupers, marlins, sword fishes, jacks, basses, ocean sun fish, sea turtles, whales and dolphins. The family Echeneidae to which the sucker fish (*Remora brachyptera*) belong is divided into two sub-families, four genera and eight species, seven of which occur in the Western Indian Ocean.

While the common shark sucker (*Echeneis naucrates*) is found along Mumbai waters, a single specimen of an unusual species, stouter than others, with long sucking plate was caught in September, 2009 by a trawl net operated at a depth of about 30-40 m, off north west coast at a distance of 70-80 km from the coast. The sucker fish, identified as *Remora brachyptera* looks very much similar to cat fish. The species is also called as "Spear fish remora" is generally found in the body of billfishes and sword fishes and rarely on other fishes

The species is distributed in the Western Atlantic, Nova Scotia, Canada to Brazil, Eastern Central Atlantic to Madeira Island.

The body of this fish is cylindrical and elongate, possessing an oblong cephalic disc with 14 to 17 transverse laminae and the cranium is depressed. Posterior end of the sucking disc is before the posterior end of pectoral fins. The jaws are broad with the lower one projecting beyond the upper. The dorsal fin has 27 to 44 rays and the first gill arch has around 21 gill rakers. He second dorsal fin and anal fin are long and the anal fin has 22 to 28 soft rays. The pectoral fins are rounded and the caudal fin is forked. The fish is light reddish brown above and darker below. The dorsal and anal fins are with faint white margin with the upper and lower posterior tips of caudal fin slightly pale. The maximum length of the species recorded was 55 cm, weighing 2.5 kg, largest ever recorded.

Moontail Bullseye

Moontail bullseye, *Priacanthus hamrur* has been forming a part of trawl fishery from a depth beyond 100 m off Mangalore coast. Not preferred as edible species in Karnataka, the bullseye landing was forming 1.48 per cent and 1.31 per cent of the trawl landings in 2005 and 2006 respectively. But due to heavy demand from processing industries for preparation of surumi, the species was made targeted fishery from 2007 (7946 tonnes) forming 10.8 per cent of total trawl landings. The heaviest landing of the species was recorded during October-December (5750 tonnes).

Earlier threadfin bream (*Nemipterus mesoprion*) of required size was considered as priority species for surumi preparation. In 2006, threadfin bream catch especially that of *N. mesoprion* juvenile composition was 10 per cent by number, which increased to 51 per cent in 2007, leading to heavy exploitation of juveniles and the mean size of the species came down from 148 mm in 2006 to 136 mm in 2007.

The size range of *P. hamrur* in 2007 was found to be between 105 and 410 mm and the majority of the catch was in the size range of 140-170 mm, which were juveniles.

Tetradons

Lagocephalus inermis, belonging to the family Tetraodontidae was considered as menance by the fishermen during previous years (2006), as it caused damage to the other fishes and net. In 2007, the fish gained attention as a new fishery resource along Mangalore coast of Karnataka. The landings of 74 tonnes in 2006 increased to 488 tonnes in 2007 by multi-day trawlers going for fishing for 8-10 days from a depth range of 20-100 m the fish is caught during night trawling and is said to be moving in shoals. The peak fishery was in December, 2007 with a landing of 222 tonnes.

The length of the fish ranged from 27 to 420 mm with modes between 340 and 360 mm and weight ranged from 260 to 760 gram, with an average weight of 510 gram. All the fishes were in matured condition and the largest size recorded for the fish was 900 mm (Standard length).

Puffer fishes are known for its poison, tetradotoxin (TTX), which is nurotoxic and considered to be the most toxic poison found in nature. It is generally known that liver and ovary concentrate the toxin in most cases, but lesser amount could also be found in skin, muscle and blood. Although the fish is poisonous, Japanese prepare a dish from the puffer fish, which is called "Fagu". Despite the risk involved, Fagu remains a Japanese delicacy and is the most popular dish in Japanese cusine.

Over the years, during the post-monsoon period, moderate quantity of puffer fish, *Tetrodon inermis* used to appear along Kerala coast. It appeared in big shoals in and around the sea off Kollam during 2005. The shoal was entrapped in the trawlers operating along the shore at a depth of 60-90 m, targeting *Loligo duvaucelli*.

Spotfin Porcupine Fish

Diodon hystrix, a rare spotfin porcupine fish was caught in a ring net operating at 40-60 m depth during July from Malabar region. The species, under the family, Diodontidae is widely distributed in tropical and temperate waters. There are more than 19 species in 8 genera under this family. The specimen measured 78 cm in length and weight 3.5 kg. It has a greenish brown back and a white undr side. It has small dark spots which cover its body and fins. Its distinctive spines are usually held flat to its body. But when disturbed, it is capable of inflating its body to stick out the spines for protection. The fish can grow to a maximum length of 3 feet, although they are usually found to be 1-2 feet. They are mostly found near cave openings in recesses in coral reefs to depths between 6 and 150 feet. They are solitary and nocturnal and feed on hard shelled invertebrates like, sea urchins, gastropods and hermit crabs. They are poisonous to eat.

Silver Pompano

Silver pompano, *Trachynotus blochii* has got fast growth rate, good meat quality. The species is able to acclimatize and grow well even at a lower salinity of about 10 ppt.

Wild pompano of 250 to 500 gram attain maturity and can be induced to breed indoor with photo period control facility (14L : 10D). The female was 39.8 cm in length and 2.3 kg in weight. The length of the males ranged from 30.7 to 35.7 cm and weight from 1.75 to 2.1 kg.

They feed on squids, fish eggs. At fully matured condition, the intra-ovarian eggs attain more than 500 micron diameter. The fecundity of the female was 1.3 lakhs. Fifty per cent fertilization was noticed, when induced with HCG of 350 IU per kg body weight. The eggs hatched after 18 hours of incubation at a temperature range of 30-31°Celsius. The newly hatched larva measured 2.0 mm in total length. The critical stage mortality of the larvae occur during 3-5 days past hatch (dph), the subsequent mortalities were negligible.

On 35 day past hatching (dph) the fingerlings were of 33-40 mm.

Lizard Fish

Lizard fish landings of Visakhapatnam coast is usually dominated by *Saurida undosquamis*, *S. tumbil*. But the occurrence of other lizard fishes, like, *S. micropectoralis*, *S. longimanus* and *Trachinocephalus myops* are recorded.

Occurrence of *T. myops* along the east coast is seasonal. During the year 2006-07, the annual average landings of *T. myops* was 2.4 per cent of the total lizard fish

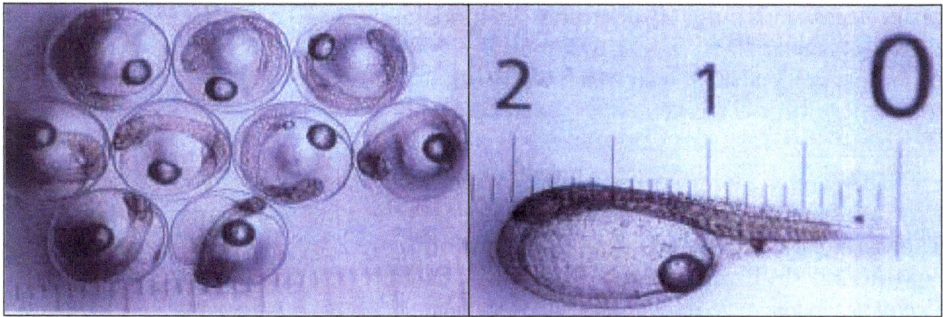

Embryonic Development

Newly Hatched Larva

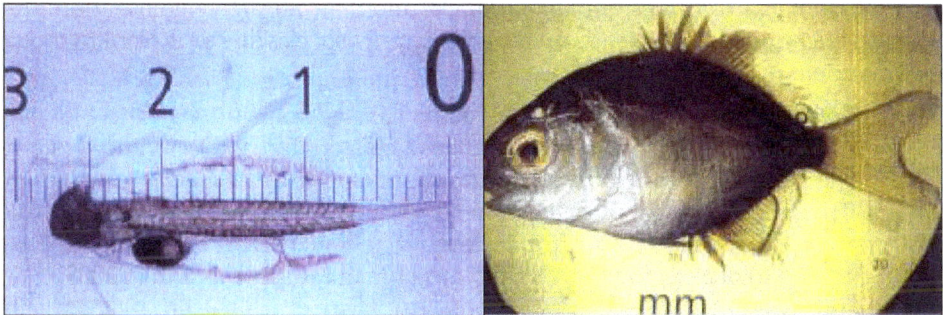

Larva on 2 dph

Fingerling on 28 dph

Figure 25.24: Larval Development of Silver Pompano

Figure 25.25: Silver Pompano, *Trachynotus blochii*

landings at Visakhapatnam. Females ranged in size from 136 to 228 mm and males from 142 to 145 mm in total length.

On 25.2.2008, a heavy landing of lizard fis (1000 kg) along with silverbellies by trawlers were recorded from 10-20 m depth off Neendakara. *Saurida undosquamis* among lizard fish and *Leiognathus elongates* were dominant species. The length range of *S. undosquamis* was 110-170 mm with modal length 150 mm. The length range of *L. elongates* was 20-70 mm with a mode of 50 mm. Most of the lizard fish were in immature stage and their stomach were full.

Slope Fish

The family Symphysanodontidae comprises of small to medium sized bony fishes, commonly known as, banquelovelies, slope fishes and insular shelf beauties. They are caught in trawls operating in depths from 50 to 500 m, mainly on the continental shelf and slopes. The family is comprised of ten species, namely, *Symphysanodon anderioni, S. berryi, S. katayamai, S. parini, S. mona, S. maunaloae, S. octoactinus, S. rhax, S. typus* and one yet to be named, species known only from the stomach contents of *Latimaria chalumnae*. Despite their attractive colouration, the insular shelf beauty is not taken as preferred aquarium species due to its planktivorous feeding habit.

S. typus is known to occur in the Pacific Ocean, from Indonesia to Hawaii. Eshmeyer (2009) gives its distributional ranges as eastern Indian Ocean and western Pacific. The present record is the first confirmed occurrence of the species from Indian waters. The specimen was collected from trawl net at Neendakara, Kerala, on the west coast of India, on 19.11.2005 from a depth zone of 60 m. Three years later, large quantities (28 kg) of *S. typus* was caught off Cochin in the same month in 2008.

Species of *Symphysanodon* are mostly distinguished on the basis of number of scales on the lateral line, number of gill rakers and pelvic and caudal fin morphology. Characters typical of *S. typus* are 49-55 lateral line scales, 10-12 gill rakers, including rudiments on upper limb, 25-28 on lower limb, 36-40 total on first gill arch. Length of anal fin base is 15-21 per cent standard length. Pelvic fins 22-26 per cent standard length. Caudal fin lobes slightly produced. Pectoral fin rays usually 16, can be 15-18.

Body bright fuscia or orange red with a broad mid-lateral yellow band. Dorsal fin yellowish, upper lobe of caudal fin orangish red, while lower lobe conspicuously yellow.

Cornet Fish

Cornet fish (*Fistularia patimba*) belonging to the family Fistulariidae was found to occur off Verabal coast, Gujarat. A large specimen was caught in trawl net from a depth of 60 m. The body of the fish is elongate having depressed mouth at the end of a long tubular snout, which is hexagonal in cross section and teeth in jaws with small ridges diverging anteriorly and the inter-orbital space flat. Dorsal and anal fins are short, based and opposite with 15 rays and 15 dorsal segmented rays, pectoral fins with 15 rays, pelvic fins small and abdominal with 6 rays.

Lateral line is arched running anteriorly almost along the middle of back and then bending down to side and continuing posteriorly to an elongate filament produced by the middle of two caudal fin rays. Posterior lateral line is ossified without spines. The colour of the fish is brown, becoming lighter to silvery below. Dorsal and anal fins are with orange cast becoming transparent at base and caudal filament white. The landed fish had a total length of 1480 mm weighing 2529 gram.

Sun Fish

The slender sun fish, *Ranzania laevis* of family Molidae is the only member of the genus *Ranzania*, found globally in tropical and temperate seas. They are oceanic and found in depths ranging from 1 to 140 m. The slender sun fish is the most elongate species and the maximum reported size is 100 cm total length. *R. laevis* is a distinctively shaped fish, they lack true tail and instead have clavus.

One specimen of *R. laevis* measuring 550 mm total length was landed at Chinnapalam, Pamban, south east coast of India on 11.11.2011. The fish was an incidental catch of gill net. In fresh condition, the fish is silver coloured with dark grey above, brownish green on side with dark stripes.

A sun fish, measuring 620 mm in total length weighing 7 kg was caught in a shore seine on 25.1.06 from Gulf of Mannar near Rameshram. The landing of this species from this region was reported earlier in 1998. These are the most colourful and rarest of the ocean sunfishes and they idle at the surface of the sea mostly. They have a smooth and thinner skin and a vertically oriented mouth with lips produced forward like a funnel. They lack a caudal fin and the rudder like structure at the caudal region, called clavus. The dorsal and anal fins are large and look like paddles. They are flapped synchronously from side to side and can propel the fish at a surprisingly good speed.

The commonly occurring species of sun fish from Indian waters including Andaman Sea are, *Mola mola, Ranzania laevis, R. truncate, R. typus, Masturus lanceolatus, M. oxyuropterus.* These species are distributed world wide in tropical and temperate seas and is generally known as ocean sunfish.

The body of *R. laevis* is scaleless, elongated, laterally compressed and covered with extremely thick elastic skin. Body is dark grey above, silvery-grey brown on sides and pale or dusky below in colour. There is no lateral line on the body. The fish has a tiny slit-like mouth, placed vertically. Each side of head is with two tiny nostrils. Dorsal and anal fins are with short base and lacking in spines. These are placed posteriorly nearer to caudal. In swimming, these fins are flapped synchronously from side to side and they can propel the fish at surprisingly good speed, thus providing the power for locomotion. Caudal fin is present in post-larvae, but it is reabsorbed through metamorphosis at a later stage of life and replaced by a rudder like structure, called "clavus". Pectorals are small, rounded at base and directed upward. Pelvic fins are absent. Gill openings are reduced to small holes at the base of the pectoral fins. The species is completely oceanic, epipelagic, generally solitary and feeds on planktonic crustaceans.

Threadfin Group

Eleutheronema tetradactylum, under the family Polynemidae for the first time reported along Chennai coast. Usually *Polynemus indicus* group of threadfin only is landed at Chennai coast. 100 kg of *E. tetradactylum* was landed by mechanized trawlers at Chennai harbour. The size range was 250-289 mm and the weight range was 140-200 gram. The species has been previously reported from Cochin.

Leather Jacket

Unusual heavy landings of unicorn leather jacket, *Aluterus monoceros* was observed between September to December, 2010 in Chennai, particularly in the month of November, 2010, when 15 tonnes of fish was landed by trawlers. The size range was 380-469 mm and the weight range 550-950 gram.

Goat Fish

388 tonnes of goat fish (*Upeneus taeniopterus*) were caught from depths of 60-80 m, which contributed 66 per cent of the goat fish catch in June, 2005 along Chennai coast. The total length of the species ranged from 90-162 mm with modal lengths 110-119 mm and 120-129 mm.

Similar landings were recorded (164.2 tonnes) in June, 2004. The size of the fish in the catch ranged from 120-129 mm.

Dolphin Fish

On 29.10.05, an unprecedented 12 tonnes of dolphin fish, *Coryphaena hippurus* was caught by gill nets, operating as surface drift nets upto 40 m depth of Marud, about 100 km, south of Mumbai. The fish is sexually dimorphic with mature males, having a pronounced bony crest in the frontal portion of the head. The size range of males were 736-1435 mm (modal length 935 mm) and of females 676-1037 mm (modal length 835 mm). Males formed 65 per cent of the catch, while females 35 per cent. There were no juveniles in the catch.

Sand Tile Fish

Reef associated sand tile fish, *Malacanthus latoviltatus* belonging to the family Malacanthidae, a commercially important rare ornamental fish (female), measuring 33.8 cm in total length, weighing 292.5 gram was caught in shrimp trawl off Pamban in Gulf of Mannar at a depth ranging from 10-50 m. The species was earlier recorded from Lakshadweep and Andaman and Nicobar Islands.

Moluccan Sweeper

On 6.7. 2008, out of 620 kg landed 41 kg of *Pempheris moluca*, a rare vatiety was landed from Kerala waters constituting 7 per cent of the total landings. The species was caught from a depth of 5 m. Average length of fishes was 10 cm and average weight 20 gram. *P. moluca* comes under the family Pempheridae (sweepers). It generally lives away the coral reefs in rocky areas and feeds mainly on crustaceans.

Oil Fish

Thirteen specimen of oil fish, *Ruvettus pretiosus* were recorded in the trawl landings along with heavy catch (200 tonnes) of *Coryphaena* sp. at the Cochin on 29.8.2008. The largest specimen was 360 mm with a weight of 260 gram. Earlier, in November, 1968, *R. pretiosus* measuring 310 mm and weighing 220 gram was collected from the continental slope, off the Quilon Bank. The flesh of *R. pretiosus* contains more oil.

The diagnostic features of the species delineate the skin as very rough, scales small and cycloid interspersed with rows of spinous, bony tubercles. Ventral keel rigid and scally on the belly between pelvic and anal fins. Body elongate and uniformly dark brown, tips of the pectoral and pelvic fins black, lateral line single and obscure. Fang like teeth in front of upper jaw, vertebrae 32.

The oil fish grows to a large size and the maximum weight recorded 63.5 kg and 200 cm total length. *R. pretiosus* is oceanodromous and bathypelagic, found in the depth range of 100-800 m, but usually occurs in 200-500 m along the upper continental slope of Indian Seas, and is known to migrate to surface at night. It is widely distributed in the tropical and temperate seas of the world. The fish is rated under the category of very high vulnerability.

The species is stated to form minor commercial fishery along the coast of Tuticorin in the Gulf of Mannar contributing nearly 2 per cent of the Gempylidae family during November-December and are caught by the trawls and large-meshed gill net. The other two species of importance reported along the Tuticorin coast are *Lepidocybium flavolerunneum* and *Neoepinnula orientalis*, though *Thyrsitoides Marley* is also available sporadically in small numbers along with tuna landings. Previously, *R. pretiosus* has appeared in the catch only during the post-monsoon season, but recently the time of occurrence has been advanced to the monsoon season signifying its availability in deeper waters relatively more in numbers.

Anchovies

The anchovies constitute an important fishery representing 6 per cent of the total landings in the inshore areas of Andaman Islands, of which *Thrissina baclena* is the most important. The other genera, which form the complex group of Anchovies are, *Anchoviella*. The best fishing period is June to November and they are caught by shore seines and boat seines.

Silverbellies

The group represented by two genera, namely *Leiognathus* and *Gazza* account for about 5 to 6 per cent of the total annual catch. In trawl catch, they represent about 28 to 29 percent. Boat seines and shore seines account for the major portion of the catch and the main fishing season is from June to December. More than 90 per cent of the fishery is composed of *Leiognathus*.

Moray Eels

Anguilliform fishes, collectively known as eels are characterized an elongate body, a small cross-sectional area and an absence of pelvic fins. These characteristics enables eel to live in confined spaces, promoting reclusive and enigmatic lifestyles. Within the Anguilliformes, moray eels (Muraenidae) are a monophyletic group of predators that occupy coral crevices of shallow reefs. Despite high species richness (roughly 200 species constitute approximately one-quarter of eel diversity world wide) and their role as top predators of many coral reef ecosystems, the mechanism of feeding in morays are poorly understood, particularly with respect to the large prey

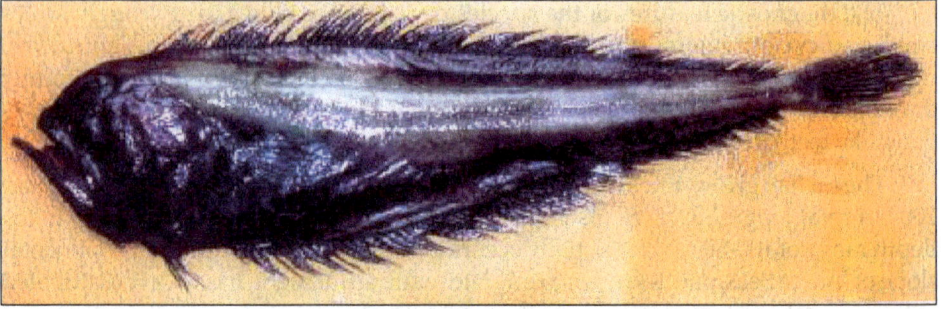

Figure 25.26: *Chascanopestta lugubris*

Figure 25.27: *Nemipterus jison*

they are known to eat. Given the striking morphological and ecological features of morays, understanding the basis of their feeding performance may provide insight into their successful radiation on coral reefs.

Moray eels display much less-effective suction-feeding abilities. Moray eels (*Muraena retifera*) overcomes reduced suction capacity by launching raptorial pharyngeal jaws out of its throat and into its oral cavity, where the jaws grasp the struggling prey and transport it back to the throat and into the oesophagus. The extreme mobility of the moray pharyngeal jaws is made possible by elongation of the muscles that control the jaws, coupled with reduction of adjacent gill-arch structures. The discovery that pharyngeal jaws can reach up from behind the skull to grasp prey in the oral jaws reveals a major innovation that may have contributed to the success of moray eels as apex predators hunting within the complex matrix of coral reefs.

New Records

Specimen of red alfonsino, *Centroberyx rubricaudus* of 21.5 cm total length was caught by a fishing vessel operated in Andaman Sea. This is the first report of the species from Indian waters. This fish has been reported earlier once from southern

Figure 25.28: Flying Gurnard (*Dactyloptena peterseni*)

Figure 25.29: *Roa jayakari* (Norman, 1939)

coast of Taiwan. Species in the genus *Centroberyx*, have their habitat on the shelf and upper slope to a depth of 700 m. Of the seven species recorded, three species (*C. lineatus, C. affinis* and *C. gerrardi*) are distributed in the Indo-Pacific and three from

Indian Ocean (*C. spinosus* on the South African shelf; *C. druzhinini* on the Saya de Malha Bank and *C. australis* on southern Australia) and one *C. rubricaudus* in North-West Pacific, endemic to Taiwan. The report of *C. rubricaudus* is the first record from the east coast of India and the first from Andaman Sea indicates a marked extension of the distribution from its type locally.

Ten species of fish belonging to various families have been collected and identified from Karnataka coast during the year 2011. With this newer findings, the fin fish diversity of Karnataka stands at 422 species.

1. *Horabagrus brachysoma* (Gunther's cat fish0
2. *Aluterus scriptus* (Leather jacket file fish)
3. *Neoharriotta pinnata* (Sickle fin Chinese)
4. *Canthidermis maculate* (Ocean trigger fish)
5. *Calionymus margaretae* (Margaret's dragonet)
6. *Ostichthys acanthorhinus* (Spinesnout squirrel fish)
7. *Scorpaenodes parvipinnes* (Low finscorpion fish)
8. *Aesopia cornuta* (Unicorn sole)
9. *Apogon queketti* (Spotfin cardinal)
10. *Terapon theraps* (Large scaled terapon)

Less Important Fishes

According to FAO discard database, during 1992-2001, yearly average discards were estimated as 7.3 million tonnes of which Indian Ocean accounted for 9 per cent.

Mesopelagic fishes are common by-catch in many of the world fisheries, targeting deep sea shrimp species. Shrimp trawl fisheries account for more than one third of the estimated total global discards from fisheries. In most cases, the weight of the by-catch exceeds that of the shrimp catch and is comprised of hundred species of fish and invertebrates that dwell in the same ground with shrimps.

The mesopelagic zone has been defined in different ways, based on depth, temperature and light regimes. Depth seems to be the best criterion and mesopelagic fish can be defined as fishes that live in the mesopelagic zone, that is, between 200 and 1000 m depth. Many fish families fall within this definition, but generally the Myctophidae, Neoscopilidae, Bathylandae, Chiasmodontidae, Trichiuridae, Nomeidae and others seem to be fairly important in some areas.

Myctophids are the most species-rich family (32 genera with 240 species) of mesopelagic communities with an estimated biomass of about 70-200 million tonnes found in all oceans from near surface to deep waters. They are thought to migrate to the productive epipelagic zone, which contributes to their abundance in the open sea. About 55 species of mictophids are known from the Arabian Sea including its southern part of the Indian Ocean with an estimated biomass of 100 million tones of *Benthosema pterotum*.

In the Quilon Bank (8-11°N and 74-76°E), Kerala coast, myctophid and other mesopelagic fishes have been found to occur with shrimps, which mostly inhabited the uneven bottom surface. The multi-day trawlers (9-15 days) operating at a depth range of 350 to 450 m specifically targeted deep sea shrimps like, *Aristeus alcockii*, *Heterocarpus woodmasoni, Heterocarpus gibbosus, Plesionika spinipes* and *Metapenaeopsis andamanensis*.

Trawling operations are mainly carried out during early morning and late evening from 4 to 6 hours at a trawling speed of 2 knots. In each operation, by-catch contributed 20 to 40 per cent and some times exceeded more than 80 per cent.

The major components in the by-catch belonged to the families, Rhinochimaeridae, Echinorhinidae, Centrophoridae, Squalidae, Stomiidae, Sternoptychidae, Gonostomatidae, Ateleoopodidae, chlorophthalmidae, Ipnopidae, Evermannellidae, Neoscopelidae and Myctophidae.

Among lantern fishes, benthopelagic myctophids dominated. Among the myctophids, *Diaphus watasei* was the most dominant species. *Diaphus garmani* was recorded for the first time in Indian waters. The size range of *D. watasei* was from 7 to 13 cm with prominent mode at 10 cm. Stomach of most of the species were found empty.

Rare Fish

A few brightly coloured rare specimen were observed along with the landings of threadfin breams and lizard fish on 13.8.2010 at Puthiappa landing center, Calicut.

They were *Chelidoperca investigatoris, Roa jayakari* and *Pseudanthias* sp. During the period of capture, sea condition was rough with surface to bottom churning and inshore waters remained murky. Water temperature was between 24-25°Celsius.

Blue Ringed Angel Fish

A blue ringed angel fish, *Pomacanthus annularis* was caught in a trawl net from a depth of about 40-50 m off Versova on 15.11.2009. The fish measured 300 mm in total length and was in a very healthy condition. The fish feed on marine algae and shrimp.

Marine Ornamental Fish

The tomato anemone fish, *Amphiprion frenatus*, belonging to the family, Pomacentridae, inhabitant of Andaman Sea, is popular in the marine aquarium industry. The adult of the species is bright orange-red colour, with one white vertical stripe just behind the eyes, joined over the head, whereas the juveniles are drak red with three vertical white bands and black pectoral fins. They are very hardy and easy to feed as they eccept a variety of food. Usually they grow upto 14 cm in length. Howerer.the females grow significantly larger than males. In natural habitat, they live in association with sea anemone species, namely, *Heteractis magnifica, Entacmaea quadriolour, Heteractis aurora* and *Macrodactyla dorensis*. Being a protandric hermaphrodite, males are much smaller than females.

The species respond to captive breeding at an environmental parameters of 27-29°Celsius temperature, 32-34 ppt salinity, 4.6-6.2 mg/l dissolved oxygen and 8.1-8.8

pH. After rearing for a period of 3-5 months, one pair found to grow ahead of others and the pairs thus formed are separated for spawning in breeding tank, providing tiles or earthen pots for egg deposition.

Few days prior to spawning, the male selected a suitable site near the sea anemone for laying the eggs, and cleared off the algae and debris with its mouth. On the day of spawning, both the male and female spent considerable time for cleaning the site, indicating that breeding may occur within few hours. The spawning was noticed between 06 00 hours and 15 30 hours during day time, and the spawning lasted one to one and half hour. In each spawning 200 to 600 capsule shaped eggs are laid at an interval of 15 to 30 days, depending on the size of the fish. The egg size ranged between 1.2 to 3 mm in length, with a width of 0.8 to 1 mm. The eggs adhered to the sides of the earthen pot with stalk.

During the incubation period, both the male and female carefully looked after the eggs during day-time, which involve two basic activities, namely fanning by fluttering the pectoral fins and mouthing to remove the dead or weakened eggs and dirt. No nocturnal care was noticed.

The newly spawned eggs are reddish brown or red in colour for the initial two days and as the embryo developed, it turned to black from third to fifth day and later turn to silvery on 6th and 7th day of incubation. At this stage, the glowing eyes of the developing larvae inside the egg capsule are clearly seen. Male guard the hatching embryo for a longer time. When incubated at a water temperature of 27-29°Celsius, the hatchlings emerged on completion of 7th day of incubation. The peak hatching take place at sun set hours at a water temperature of 27-29°Celsius. The newly hatched larvae measured 1.5-3.5 mm in length and each has a transparent body, large eyes, visible mouth and a small yolk sac.

Out of sixty species collected from the Andaman Sea (East coast of Little Andaman) five species, namely, *Chaetodon vagabundus* (butterfly fish), *Monodactylus argenteus* (Silver bat fish), *Caesio caerulaurea* (Blue and gold Fusilier), *Acanthurus triostegus* (Convict surgeon fish) and *A. lineatus* (Blue striped surgeon fish) were known as ornamental fishes.

The ornamental fishes occurring around the Little Andaman Island have the potential to promote commercial fisheries development in this area. Occurrence of the ornamental fishes in Little Andaman waters is due to the diversity of marine habitats, such as, mangroves, creeks, muddy shores,coral reefs etc.

Butterfly Fish

The body of the fish is compressed and some what rounded and very deep with the depth being 1.5 to 1.6 times of its standard length. Ctenoid scales cover the body and there are 29 to35 scales along the middle line. The majority of the trunk scales are relatively large and rhomboid. The fins include a single continuous dorsal fin with 12 to 13 spines and 23 to 26 rays, an anal fin with 3 spines and 19 to 22 rays and a rounded caudal fin. The spinus dorsal base is longer than the soft dorsal base. The mouth is terminal and protruding and bears 5 bristle-like teeth in both jaws. There are 4 to 6 gill rakers on the upper limb and 11 to 15 on the lower limb of the first gill arch respectively.

The overall body colour of the fish is pearly white, becoming yellowish posteriorly. The body in front and at the top has about 8 streaks or lines directed obliquely upwards and backwards. This set of streaks is met an angle by another set of 12 line running downwards and backwards. A broad black bar running along the bases of the dorsal, caudal, and anal fins and the other which has a narrow white edge on the anal fin. The caudal fin is yellow.

The species occurs throughout tropical Indo-West Pacific region. The fish inhabits shallow coral rocky reefs down to 30 m depth and is usually encountered in pairs. The diet of the species includes small crabs, worms and other small invertebrates. The vagabond butterfly fish attains a total length of 23 cm. Five species of butterfly fish was recorded around Great nicobar Islands.

Silver Moony

Silver moony, *Monodactylus argenteus*, belonging to the family Monodactylidae is wide spread in distribution and occur in the tropical Indo-Pacific region,Omani waters and in the entire coast line except in the Arabian Gulf. They have been found in the Andaman Sea also.

The body of this diamond-shaped fish is strongly composed and very deep, its depth being 1.3 to 1.5 times of its standard length. The mouth which is small and terminal bears very fine villiform teeth in each jaw. The dorsal and anal fins are symmetrical with dorsal fin comprising 7 or 8 spines and 26 to 30 soft rays and the anal fin with three spines is small and inconspicuous. The pectoral fins are short and rounded and have 16 rays each. The large caudal fin is truncate to slightly emarginated. The body is covered with small deciduous ctenoid scales, which are arranged in 52 to 60 series along the lateral line. Scales also extend out to the fins. There are 15 to 28 gill rakers on the first gill arch with 6 to 8 rakers occurring the upper limb and 19 to 22 on the lower limb of the gill arch.

The silver moony is silvery to silvery-grey in colour, with its fins having yellowish tips. The dorsal and anal fin lobes are dusky to black. Young specimens have two dusky bars across the head which fade with age.

The silver moony is gregarious and is commonly found in small but dense shoals in estuaries, wharfs and shallow reefs. It can tolerate wide fluctuations in salinity and may occur in fresh water also. It mainly feeds in the water column and its diet consists of planktonic organisms and of detritus. It is believed to attain sexual maturity at about 15 to 18 cm in length and to lay a few number of large eggs. It uses estuaries as a nursery area. The silver moony is a small fish, growing upto 25 cm in total length.

Gold Fusilier

The body of blue and gold fusilier, *Caesio caerulaurea* is elongate, slender and slightly compressed. Dorsal and ventral profile equally convex. Anterior region of the head and lower region of preoperculum and operculum are scaleless. Interorbital space, slightly concave, width about 1.3 times of eye diameter. Eye moderately large, adipose eye lid not well developed, covering only orbital margin of eye. Eye diameter is greater than snout length. Snout slightly blunt. Small terminal mouth within thin lips. A series of minute conical teeth are present in both jaws. No teeth on palatine and vomer. Maxilla reaching below anterior margin of pupil. Head and body covered with small ctenoid scales. On head, supratemporal bands of scales, interrupted at dorsal mid-line by a scale zone 'V" shaped scaleless zone anteriorly at mid-line intruding between supratemporal band of scales. Lateral line runs horizontally. Dorsal and anal fins almost completely scaled. Pelvic fin with auxillary scales. Dorsal fin with minute first spine.Caudal fin is deeply forked.

Distributed in Andaman Sea, Indo-Pacific region including Malaysia. The upper half of head and body of the fish is bright metallic blue and lower half pinkish in colour. Golden horizontal band runs from head to caudal fin base just above lateral line. A black blotch at upper base of pectoral fin and broad black band on each lobe of caudal fin are visible.

The species is very common in coral reef areas and usually seen in large schools. Prefers swimming at the surface of water or under floating objects and boats. Feeds mainly on zooplankton, crustaceans and small fishes. Common size is about 20 cm and maximum size 35 cm. Usually caught in purse seines and traps.

Convict Surgeon

Acanthurus lineatus, commonly known as convict surgeon fish, under the family, Acanthuridae, is found in the Indo-Pacific, the South China Sea and Western Indian Ocean southward to Mozambique and in Andaman Sea. Its deep and compressed body is 2.2 times of standard length. Mouth small with spatulate teeth, close set, with denticulate edges. Caudal peduncle with lancet-like spine folded into deep grooves on each side. Caudal spine long and venomous. Stomach thin-walled. Caudal fin deeply lunate, with filamentous upper and lower rays.

Body and head with alternating blue and yellow black-edged stripes except for lower one-fourth, which is bluish white. Dorsal fin with pale blue and yellow stripe. Anal fin grey, yellow at the base and light blue at edge. Caudal blackish with a grey crescent centro-posteriorly, front with bluish-white edge and black margins. Pectoral fins pale, pelvic fins orange-yellow, whitish along lateral margin and black submarginally.

The species inhabits in inshore waters, feeding on benthic algae on coral reefs or rock areas. Known to be aggressive territorial fish. Common size at capture are 25 cm, although maximum length 30 cm. Caught by cast nets, gill nets, spears and trap.

Blue Striped Surgeon Fish

The body of *Acanthurus triostegus*, commonly known as Blue striped surgeon fish is high and compressed. Its body depth is 1.8 times of standard length. Dorsal profile of head, before eye is concave, above eye convex. Mouth with small, teeth spatulate, close-set, with denticulated edges. Dorsal fin is continuous, intermembrane without notch. Caudal peduncle with a lancet-like spine which folds into a deep groove. Caudal fin truncate to slightly emarginated.

The species is found in Andaman Sea, East Africa extending to the east coast of Mexico, Japan, Indo-Pacific and the South China sea

The fish is white-greenish dorsally and white ventrally. Five vertical bars, first from the nape passes through eye downward. Second from first spine of dorsal fin to base of pectoral fin. Third, fourth and fifth bars below dorsal fin. There is also one dark vertical bar on caudal peduncle.

The fish inhabits in inshore waters and feeds on algae. Dwells either singly or in large groups. Spawning occurs at dusk, from 12 days before to 2 days after full moon. Eggs are pelagic and spherical, with a single oil droplet. Average diameter of the eggs is about 0.68 mm. Eggs hatch after 20 hours. Juveniles live in very shallow waters of tide pools or reef flats and grow at about 12 mm per month. Common length caught is 17 cm and maximum length is 24 cm.

Chapter 26

Benthic Macrofauna

Macrobenthic fauna of the Andaman Sea in the depth zone of 30 to 2750 m, cpmprised of 20 faunal taxa. Of these, polychaetes are the most dominant group and contributed 83 per cent, comprising of species belonging to 17 families mentioned below. The distribution of macrofauna was substrate specific with maximum (410/cubic m) population in clayey deposits and minimum (266/square m) in sandy coralline areas.

An extensive area (between lat. 8-15°N and long. 91-94°E) in the depth range of 30 to 2750 m covering 41 stations were sampled in February, 1979.

Benthic macrofauna, in varying proportion was represented by 20 taxa. Crustaceans, as represented by eight faunal groups were qualitatively important. Polychaetes were the most dominant (83 per cent) followed by amphipods (3.5 per cent), cirripeds (3.0 per cent) and stomatopods (1.5 per cent). Remaining taxa were present in inconspicuous percentages of 0.9 (mysids) to 0.1 (actiniaria, corallimorpha, pycnogonids, gastropods etc).

Table 26.1: Distribution of Macrobenthos in Different Parts of the Andaman Sea Showed a Well Demarcated Depth Zonation, Composition and Prevalence (in paranthesis)

Sl.No.	Macrobenthic Taxa	Depth Zone (m)			
		0-200 (per cent)	201-1000 (per cent)	1001-2000 (per cent)	More than 2000 (per cent)
1.	Porifera	2.00 (10.00)	-	-	-
2.	Hydrozoa	0.34 (10.00)	-	2.28 (10.00)	-
3.	Actiniaria	-	0.64 (10.00)	-	-
4.	Corallimorpha	0.34 (10.00)	-	-	-

Contd...

Table 26.1–*Contd...*

Sl.No.	Macrobenthic Taxa	Depth Zone (m)			
		0-200 (per cent)	201-1000 (per cent)	1001-2000 (per cent)	More than 2000 (per cent)
5.	Polychaeta	78.04 (100.00)	82.21 (100.00)	84.17 (100.00)	100 (100.00)
6.	Echiuroidae	-	3.78 (10.00)	-	
7.	Sipunculidae	0.34 (10.00)	-	-	-
8.	Mysidacea	2.38 (20.00)	-	-	-
9.	Euphasiidae	0.68 (20.00)	0.71 (10.00)	-	-
10.	Amphipoda	5.41 (60.00)	7.63 (30.00)	-	-
11.	Stomatopoda	0.68 (20.00)	3.81 (10.00)	2.24 (10.00)	-
12.	Paenidae	4.08 (30.00)	-	-	-
13.	Cirripedia	-	-	9.09 (10.00)	-
14.	Anomura	2.38 (20.00)	-	-	-
15.	Pycnogonidea	0.34 (10.00)	-	-	-
16.	Gastropoda	-	0.64 (10.00)	-	-
17.	Bivalvia	-	-	2.28 (10.00)	-
18.	Scaphopoda	1.97 (10.00)	0.58 (10.00)	-	-
19.	Ophiuroidea	0.68 (10.00)	-	-	-
20.	Pisces	0.34 (10.00)	-	-	

Fifteen taxa, in varying proportions (0.34 to 78.0 per cent) were observed upto 200 m depth; whereas in deep waters (more than 2000 m), only one taxa was observed. In the intermediate depth zones of 201-1000 m and 1001-2000 m, 8 and 5 taxa respectively were observed. Except polychaetes, which were present in increasing percentage of 78 to 100 from shallow to deeper waters, only few of the other taxa were found in more than one depth zone. Most of the faunal groups were restricted to a

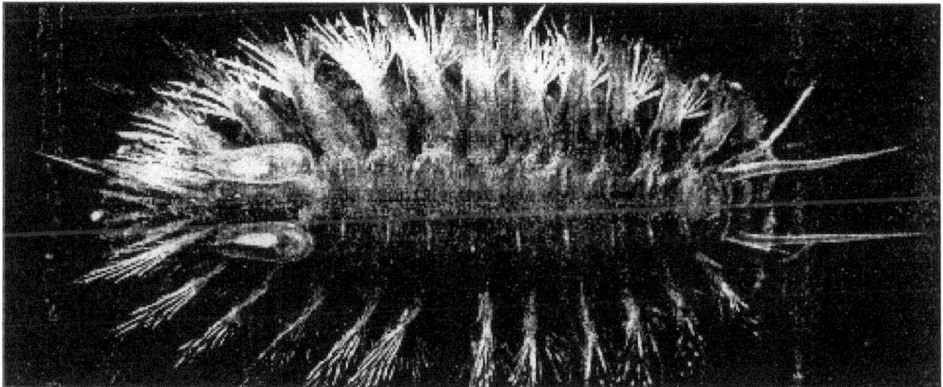

Figure 26.1: Deep Sea Polychaete

single depth zone. Similarly the occurrence as assessed from the percentage of prevalence also indicated that polychaetes were present at all stations in all the depth zones, and in all types of bottom deposits, with high prevalence. Other taxa, like stomatopods,though present in mostof the depth zones, had decreasing percentage of prevalence towards increasing depths.

Abundance of macrofauna, varied from 80 to 882 animals per square meter, with a fairly high average of 326/ sq m. In the bottom deposits of clay and its admixture with sand or silt, the fauna was richer (420/sq m) than in the sandy deposits and in coralline areas (266/sq m). Animal distribution, in relation to depth indicated that mean population count was 293/sq m upto 200 m depth zone; decreased to 220/sq m in 201-1000 m followed by sudden rise upto 430/sq m in 1001-2000 m and the highest value of 764/sq m in the deeper waters above 2000m. Thus the abundance of benthic macrofauna seems to be regulated more by the nature of bottom deposits, rather than by the depth of water column. As the depth above 1000 m were characterized by clayey deposits in contrast to sandy deposits in waters below 1000 m and consequently the observed disparity in population density.

Standing stock of the benthic macrofauna varied within a moderate range of 0.4 to 20.1 gram/sq m with a mean value of 3.4 gram/sq m at two shallow areas 10°38 minute N and 92°31 minute E (33 m) and 10°39 minute N and 92°31 minute E (40 m), high biomass values upto 72.8 gram/sq m were observed, mainly due to the presence of large epifaunal animals, like, prawns and fishes or onfaunal animals like sponges.

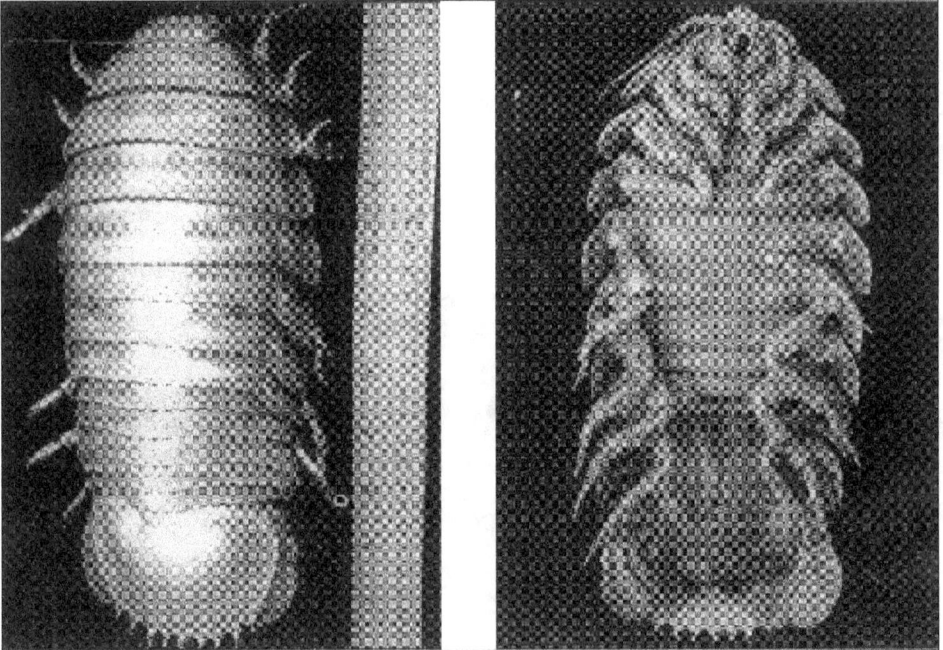

Figure 26.2: Deep Sea Isopod, *Bathynomus giganteus*

Biomass distribution in relation to depth indicated that high mean values (4.4 g/sq m) were found in shallow waters (less than 200 m). With increasing depths, the mean biomass showed a sharp decline by about 100 per cent in the next depth zone of 201-1000 m. there was a 33 per cent increase in the 1001-2000 m depth zone followed by further 200 per cent decrease in the deeper waters (more than 2000 m). Biomass production (mean value) was comparatively higher (4.1 g/sq m) in sandy and coralline deposits than in the clayey substrate (2.4 g/sq m).

With a view to assess the magnitude of macro-benthic biomass production in the Andaman Sea, an attempt has been made to compare the results of the 1979 investigation with earlier reports from geographically corresponding areas in the Indian Ocean.

Earlier, Neiman has reported the macrobenthos of shelf region from Mayanmar coast, which forms the northern and eastern part of the Andaman Sea. Similarly, Ansari *et al* (1973) have studied the macro and the meio-benthos upto 1750 m depth, off the southeast and central east coast of India (Bay of Bengal), which forms the western and to certain extent the southern limits of the Andaman Sea.

There is no comparable data for depths above 200 m from the eastern Andaman Sea. The observations on the macro-benthic biomass upto 200 m depth, indicates a fairly high mean value of 6.5 g/sq m, as compared to the reported value of 1.5 g/sq m from the Mayanmar coast. Even the standing stock in deeper waters (more than 200 m), as observed in 1979 investigation, is reasonably high (3.3 g/sq m).

Comparison of data of the western Andaman Sea with the corresponding areas in Bay of Bengal reveals that the macrobenthic biomass, either from the shallow (200 m) or from deep (2000 m) waters are about 300 per cent less than these reported from the Bay of Bengal. Even a comparison between the macro-benthic biomass upto 2000 m depth of the corresponding areas (between lat. 7 and 15°N) in the Arabian Sea, Bay of Bengal and the Andaman Sea respectively, clearly indicates that mean biomass is highest (12 g/sq m) in the Arabian Sea, lowest (3.4 g/sq m) in the Andaman Sea and fairly high (9.5 g/sq m) in the Bay of Bengal.

The most striking feature about the variations in environmental factors is the sharp decrease in bottom temperature and consistently low oxygen content in the bottom waters of Andaman Sea are the cause of low macrobenthic biomass production in the Andaman sea.

The sharp fall of about 20°Celsius in the water temperature from 30 to 300 m depth and further rapid decrease upto 2000 m has an important role in the distribution and production of macrobenthos. The depth range of 100-300 m represents the upper boundary of the deep sea macrobenthos.

Dissolved oxygen content also display an interesting pattern of distribution. The value which was 4.82 ml/l at 30 m sharply decreases to 2.52 ml/l at 80 m and went on rapid decline till near depletion (0.43 ml/l) is attained at 300 m depth. A slow recovery upto 1.75 ml/l at 2000 m follows. As compared to the distribution of dissolved oxygen in the Arabian Sea and Bay of Bengal, the Andaman Sea is an oxygen deficient environment, which causes a high physiological stress resulting in considerable impoverishment of the bottom life.

Sharp stratification of water is a characteristic feature of Andaman Sea. The very stable stratification, except for the shelf region, impedes influx of nutrients to the surface and oxygen to the sub-surface layers. The limited mixing between sub-surface and bottom waters has resulted in making the Andaman Sea an area of low biological productivity. The primary productivity of the overlying waters, which is an important source of nourishment to benthic organisms, varies from 120-615 mg C/sq m/day with an average value of 273 mgC/sq m/day and is considerably lower than in most of the other regions of the Indian Ocean. Accordingly, the organic carbon in the sediment is low (0.87 per cent) and similarly the meiobenthos. Thus, the Andaman Sea, because of peculiar hydrographic conditions and of its adverse effects on the processes of biological productivity, becomes a region of low benthic production.

The available polychaetes in the Andaman Sea are;

☆ Family – Aphrodittidae: *Aphroditta talpa* (Quatrefages), *Aphrogenia alba* (Kinberg), *Admetella longipedata* (Mcintosh), *Allmaniella ptycholepis* (Grube), *Panthalis oerstedi* (Kinberg), *Polyodontes maxillosus* (Ranzani), *Psammolyce zeylanica* (Willey), *Sthenolepis japonica* (Mc Intosh)

☆ Family – Palmyridae: *Bhawania goodie* (Webster)

☆ Family – Amphinomidae: *Amphinome rostrata* (Pallas), *Chloeia amphora* (Horst)

☆ Family – Phyllodocidae: *Eulalia albo-picta* (Marenzeller), *Phyllodoce fristedi* (Bergstrom)

☆ Family – Hesionidae: *Hesione intertexta* (Grube)

☆ Family – Pilarcidae: *Synelmis albini* (Langerhans)

☆ Family – Syllidae: *Syllis (Syllis) gracilis* (Grube), *S. (Typosyllis) armillaris* (Muller)

☆ Family – Neredae: *Ceratonereis anchylochaeta* (Horst), *C. mirabilis* (Kinberg), *Nereis (Nereis) coutierei* (Gravier)

☆ Family – Eunicidae: *Eunice afra afra* (Peters), *E. afra paupera* (Grube)

☆ Family – Maldanidae: *Asychis gotoi* (Izuka), *Maldane sarsi* (Malmgren)

☆ Family – Sabellaridae:*Idanthyrsus pennatus* (Peters), *Lygdamis indicus* (Kinberg)

☆ Family – Pectinaridae: *Pectinaria antipoda* (Schmarda), *P. (Amphictene) crassa* (Grube)

☆ Family – Ampharetidae: *Amphicteis gunneri* (Sars)

☆ Family – Terebellidae: *Terebellides stroemi* (Sars), *Streblosoma persica* (Fauvel), *Thelepus cincinnatus* (Fabricius), *Eupolymnia nebulosa* (Montagu), *Loimia meduse* (Savigny), *Nicolea gracilibranchis* (Grube)

☆ Family – Sabellidae: *Branciomma nigromaculata* (Baird), *B. serratibranchis* (Grube), *Hypsicomus phaeotaenia* (Schmarda), *Sabella fusca* (Grube), *S. melanostigma* (Schmarda)

☆ Family – Serpulidae: *Spirorbis (Dexiospira) foraminosus* (Busch), *Ditrupa arietina* (Muller), *D. arietina var monilifera* (Fauvel), *Pomatostegus stellatus* (Abildgaard), *Spirobranchus giganteus* (Pallas)

☆ Family – Glyceridae: *Glycera cirrata* (Grube), *G. prashadi* (Fauvel), *G. rouxii* (Audouin and Milne-Edwards), *G. tessellate* (Grube).

Chapter 27
Marine Reptiles

Among reptiles, sea snakes and turtles are important and represented worldwide by 50 and 7 species respectively. These are generally oceanic forms but a majority of them visit the shore at some part of their life. About 26 species of sea snakes belonging to one family, Hydrophiidae, and five species of sea turtles were reported from seas around India. Oceanic islands seem to harbor more reptiles in their marine environment. All the sea snakes and four species of turtles in their marine environment are known from islands of Andaman and Nicobar. Nesting sites of an amphibious snake were reported from the shores of North Andaman Islands (Whitaker, 1985). Turtles visit the shore during breeding time to lay their eggs. The shore visit of these turtles, especially, the olive ridley is a spectacular sight on the sandy beach at Gahirmatha near Bhitarkanika in Orissa. Andaman and Nicobar Islands have the best nesting beaches for the leatherback, the hawksbill and the green turtle in addition to the olive ridley (Baskar, 1993).

Turtles

Five of the seven species of marine turtles are found in Indian coastal waters and atleast four of them are observed at nesting beaches and or feeding areas. These are leatherback turtle, *Dermochelys corincea*, hawksbill turtle, *Eretmochelys imbricate*, loggerhead turtle, *Caretta caretta*, green turtle, *Chelonia mydas* and olive ridley turtle, *Lepidochelys olivacea*. Olive ridleys are considered the most abundant of the turtles. A significant portion of the world's olive ridley turtle population nests at nesting sites along the eastern coast of India. They exhibit the phenomenon of synchronous mass nesting. The main rookeries in Orissa are Gahirmatha, Rushikulya and Devi river mouths of which Gahirmatha is the largest (with over 100000 turtles most often and 600000 of them in some years).

Figure 27.1: Hawksbill Turtle

Turtles are cold blooded creature, that is, their body temperature changes with their surroundings. The body of the turtle is covered with a special bony or cartilaginous shell, which is developed from its ribs.

Figure 27.2: Turtle with Cleaner Fish

Many turtles spend a large period of their lives under water. Yet all turtles breathe air and must surface at regular intervals to refill their lungs. They can also spend much of their lives on dry land. The turtles can take up dissolved oxygen from the water using cloacal papillae, with rich blood supply, in much the same way as fish use gills to respire.

Turtles lay eggs, like other reptiles, which are slightly soft and leathery. The eggs of the largest species are spherical, while the eggs of the rest are elongated. Their albumen is white and contain a different protein than bird eggs, which is such that it will not coagulate when cooked. Turtles eggs, prepared to eat, consist mainly of yolk. In some species, temperature determines, whether an egg develops into a male or female; a higher temperature results in a female and a lower temperature causes a male. Large number of eggs are deposited in holes dug into mud or sand. They are then covered and left to incubate by themselves. When the eggs hatch, they squirm their way to the surface and head towards the water.

Sea turtles lay their eggs on dry sandy beaches. Immature sea turtles are not cared for by the adults. Turtles can take many years to reach breeding age, and in many cases they breed every few years rather than annually.

Morphological characters of hawksbill turtle, showing oval carapace with imbricate scutes and strong horny beak made it easy to identify the turtle as hawksbill. The carapace length was 36 cm and it weighed about 5 kg. Generally along the sandy beaches of Sindhudurg district, olive ridley turtles and their nestings are seen, but the occurrence of hawksbill is rare.

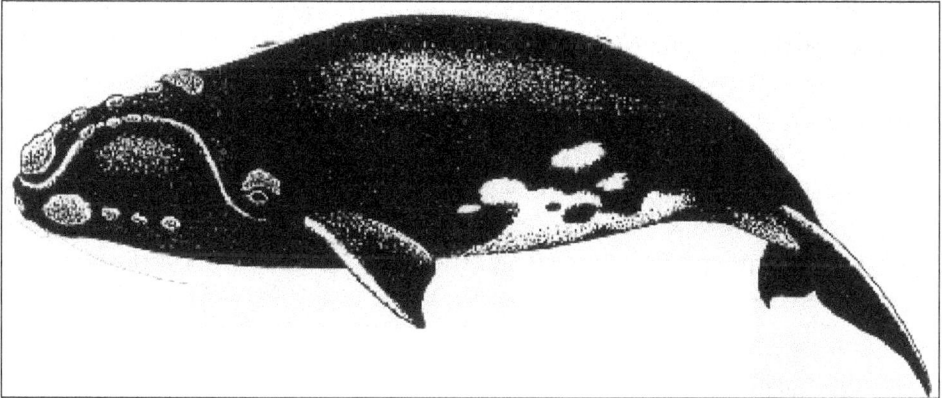

Figure 27.3: A Right Whale by Richard Ellis

Of the five species of turtles occurring in Indian waters, except loggerhead turtle, all the other four species are seen off the coast of Mahasrshtra. The olive ridley is the most common species followed by green turtles, leatherbacks and hawksbill, whose sporadic nesting is reported from all the sites in Maharashtra. Olive ridleys are seen throughout the year in the sea and their sporadic nesting is reported from all the sites in Maharashtra.

The turtles are associated with rocky areas, where they feed on algae.

Leatherback turtle, measuring 195 cm in total length and weighing obout 350 kg was entangled in gill net off Vizhinjam, Kerala. Leatherback is the largest living turtle, which reaches a size of 270 cm in adult stage. The leatherback is composed of a leather like smooth covering. The head of the adult leatherback is small, round and scaleless. Dorsal side is essentially black with scattered white blotches that are usually arranged along the keels and with pinkish blotch on neck and shoulder.

Leatherback turtles routinely travel long distances and are found in the North Atlantic far from the tropical and sub-tropical nesting beaches.

Ocean currents seem to play little part in driving broad scale movements, with turtles swimming against across and with the major current system in the north Atlantic. Periodic residence in specific areas is probably linked to locally enhanced prey availability.

Normally more than half an individual's time is spent diving to depths below 10 m. Dives are generally within epipelagic zone and over 99 per cent of all dives were shallower than 250 m. Turtles occasionally dive very deep, which is sporadic and very rare. Pan-oceanic movements and shallow diving is disadvantageous in that they both increase the interaction of leatherback turtles with long line fishing.

Green sea turtle, *Chelonia mydas* has a map that is partly based on geomagnetic cues. Ocean waves and the Earth's magnetic field, both serve as orientation cues for newly hatched turtles as they migrate to sea for first time. These young turtles migrate towards geographic targets that are no more specific than a vast oceanic region, but the juveniles and adult turtles will return to particular coastal feeding sites with pin point accuracy after displacement and long migrations. This implies that these older turtles are following a map that enables them to establish their position relative to some distant target.

On investigations, it is found that juvenile turtles have a magnetic map sense that helps them to navigate to specific targets.

The precise magnetic features that turtles detect and exactly how the magnetic map is organized remain to be detected. Turtles may possess a map in which magnetic cues provide only one coordinate with another environmental feature providing the second. The turtles may swim along the coast line until they encounter a magnetic parameter that marks a specific coastal location. Alternatively, turtles may detect two magnetic elements, such as, inclination and intensity and rely on bicoordinate magnetic navigation.

As sea turtles mature, they acquire the ability to exploit magnetic information in a more complex way than hatchling using it as a component of a classic navigational map, which permits an assessment of position relative to specific destinations.

The major threat to the turtles is from poaching of eggs and adults and due to the developmental activities along the coast. Feral dogs and wild pigs cause considerable damage to the nests of leatherback, green and hawksbill turtles in Andaman Sea. Trawling has been identified as the biggest cause of turtle mortality throughout the world. The main cause of mortality is illegal gill net and trawl fishing in the offshore waters, where the turtles die as incidental catch.

The endangered species of turtles are also a focus of attention of the international community, which looks upto India to provide safety to the nesting sites and to the turtle population, that seasonally arrive at various sites to perform reproductive functions. The Convention on International Trade in Endangered Species of Flora and Fauna lists Olive Ridleys in Appendix I (prohibition from international trade) and the International Union for the Conservation of Nature currently lists the species as endangered.

The olive ridley (*Lepidochelys olivacea*) is one of the smallest species of sea turtles. They have a high domed shell, with carapace having dark olive green colour with yellowish underside. The carapace margins are smooth and is made up of five pairs of costal scutes, with occurrence of upto 6-9 divisions per side. The ear is large. The ridley turtles have a peculiar habit of mass arrival to the shore (arribada). They make 4 to 7 nests having a large number of eggs per nesting season.

Nesting Sites of Sea Turtles in India

West coast – Gujarat, Maharashtra, Goa, Karnataka and Kerala - Green, Olive Ridley and Leatherback – Nesting sites are Mandvi in Kutch, Sea beach between Okha and Okha Madhi, Bhaidar, Beyt, Nora and Chank Islands. Maharashtra –Near Gorai, Kihim, Manowrie and Versova

East coast – West Bengal, Orissa, Andhra Pradesh and Tamil Nadu –Olive Ridley –Gulf of Mannar, Point Calimere and coastline 50 km south of Chennai in Tamil Nadu. Kakinada coast, Sea beach near the mouth of Godavari and Krishna and near Visakhapatnam in Andhra Pradesh. All along the coast south of Dhamra river mouth, beaches at Gahirmatha and Rushikulya in Orissa. In the sandy beaches of Sunderbans of West Bengal.

Andaman and Nicobar Islands – Olive Ridley, Green, Hawksbill and Leatherback – Great Nicobar, Little Andaman, Rutland, Middle Andaman, Katchal, South Sentinel, South Reef and Teris Island.

Chapter 28

Marine Mammals

Marine mammals belong to three orders, Sirenia, Cetacea and Carnivora. About 120 species are estimated to occur in World Seas and of these 30 are reported from seas around India. But a majority of these is found in oceanic forms and occasionally a few individuals may get stranded on the shore. The sea cow occurs in near shore waters.

The sirenia or sea cows are small group of obscure origin and grotesque form. They inhabited inshore and estuarine areas of Andaman Sea (in Hut Bay of Little Andaman). They feed on marine vegetation. They consisted, until recently of three groups, the Indo-Pacific dugong, the mantee of the tropical Atlantic and much larger northern sea cow of North Pacific. The first two survive, although actively hunted, mainly by native tribes for their flesh and bladder. The sea cow became extinct some thirty years after its discovery in 1741 by the young German naturalist Georg Steller. The destruction of the sea cow is the greatest blow yet dealt by man to the marine fauna, equivalent to that of the equally remarkable, harmless and locally distributed dodo.

The Dugong

The dugong (*Dugong dugong*) is a member of the family Dugongidae in the order Sirenia. Sirenians or sea cows appear to have descended from terrestrial herbivores before the Eocene. Sea cow a close relative of the dugong, was exterminated as late as 1768, within thirty years after its discovery in the Bering Islands. The dugong is listed as vulnerable in the International Union for the Conservation of Nature, Red Data Book.

Average adult dugongs measure from 2.4 to 2.7 m in length and the corresponding weights are 230 to 360 kg. The skin of these fusiform marine animals is thick and smooth, but not shiny. Skin colour in young animals less than 95 cm in length is pale cream, which darkens to pale brown, when the animal reaches approximately 110 cm in length. With maturation, the colour continues to darken and adults are a deep slate colour, dorsally and laterally with a slightly paler ventrally. Short unpigmented hairs are sparsely distributed 3 to 5 cm apart over the body. Hairs are more numerous on the back than on the belly and are virtually absent from the flippers. Numerous cuts, scratches and scars are usually present in the skin as a result of encounters with coral reefs, oyster beds and other sharp objects obscured in the "sea grasses".

The head of the dugong blends into the body with little suggestion of a neck. Dorsally, a shallow furrow in the skin demarcates the head from the rest of the body, and ventrally, the bulbous chin serves the same purposes. Crescent shaped nostrils, approximately 18 mm in diameter and 16 mm apart, are located on the summit of the head. A unique, bulbous muzzle comprises the bulk of the head and slopes forward and strongly downward from the nostrils, before terminating in a short, broad (approximately 17 cm long and 23 cm wide) facial disc, which is directed ventrally. On each side of the disc is a single lateral fold, separated by a deep groove from the central portion of the plate. Short bristles cover the plate, which bears a shallow medial furrow. A large knob of compact fibrous connective tissue, with the appearance of the tip of a tongue, projects between the disc and the mouth (a narrow transverse slit).

The knob is prominent even when the jaws are closed. Long coarse, backward directed bristles project on either side of this knob and interlace with corresponding bristles borne on the fleshy lower lip.

Small round eyes are on the sides of the head, about 18 cm behind the nostrils. The eye is dark because the iris and the choroids layer contain black pigment. Lids are without lashes and close with a spincter action and upper and lower lids are not discernible. Nictitating membranes close over the eye from the front to the rear. The copious flow of viscous tears, evident when a dugong is removed from the water, is caused by secretion from glands under the eyelids. On an average adult, the small (3 mm in diameter) ear openings are about 15 cm posterior to the eyes and are without external pinnae.

The paddle like flippers are composed only of the forearm and the horizontally flattened phalanges, which lack nails. The forelimbs are not used for power stroking and are held loosely at the sides during swimming. However, they do have considerable range of movement and are used for sculling, for slow directional movement during feeding, for sweeping food plants in toward the mouth and for carrying the young. Mammae number two, one teat is located in each akilla behind the flipper. Hind limbs are absent, and the body tapers posteriorly to a horizontally flattened notched tail fluke similar to that of a whale. Locomotion is accomplished by powerful vertical strokes of the posterior part of the body and the tail.

The umbilicus is in the center of the abdomen. Testis are abdominal, and unless the penis is extruded, it is the close proximity of the genital aperture to the umbilicus, which identifies the animal as a male. In females, the genital aperture is located further back than the males, near the anus.

Snow (1970) reported that dugongs are neither present in the Nicobar Islands (contrary to the report of Crusz, 1960), nor in the Barren, Narcodom, Coco or Christmas Islands, although a population persists in the Andaman Sea (near Hutbay, Little Andaman, Dugong Creek).

The tropical and sub-tropical dugong is found in seas, that range from 21-38°Celsius throughout the year. They do not appear to select a particular salinity and not known to ascend rivers. Four basic conditions necessary for proper dugong habitat are; (1) shallow saline water, (2) shelter from high winds and heavy waves, (3) an abundant food source and (4) warm water. Unlike their estuarine relatives, the mantees, dugongs are totally marine and usually are found in coastal waters from 2 to 3 fathoms deep. The sea floor at such a depth is usually sandy with varying amounts of mud and silt and sea-grasses which form the major portion of the dugong diet grow here. These plants are intertidal and subtidal, generally occurring from the lower neap tide level to a depth of about 3 to 4 fathoms Dugongs are occasionally found at depths upto 8 fathoms, either feeding on deepwater plants, or escaping choppy waters at shallower depths. Heavy winds and rough waters, caused by seasonal monsoons combined with exposure to coral reefs during rough weather, may cause appreciable mortality. In 1954, a large number of dead dugongs washed ashore following a cyclone on the south Indian coast and Travis (1967) reported the death of a dugong on the Somali coast, apparently caused by heavy surf on the outer reef.

There appears to be some degree of regular daily movement within the general habitat. Dugongs reportedly spend most of the day light hours in deep water. At dusk or with the rising tide, they move to the shallower waters along established swimways, thus avoiding the coral.

On the basis of earlier reports, Annandale (1905) stated that large herds of dugongs once frequented the shallow waters of gulf of Mannar. The animals no longer inhabit the shallow water and that, whether singly or in herds, they remain in the deeper waters of the gulf. There had been a shift in habitat preference from shallow to deep waters.

Long distance migration is unknown for dugongs, but seasonal changes in their abundance in shallow coastal waters are apparent. These changes are correlated with the changing monsoon seasons and with resulting shifts in abundance of food sources The dugongs may move to inshore waters to feed on *Halophila*, *Zostera* and *Diplanthera*. These grasses are virtually absent during the north east monsoon. With the coming of the north east monsoon, the animals travel to deeper waters further from the shore and appear to be less concentrated.

A similar movement, apparently in response to rough water occurs along Indian coasts. Dugongs move inshore seeking shelter during south west monsoon.

Dugongs are largely herbivorous. Sea grasses are undoubtedly the primary food of dugongs. In Andaman Sea, dugongs feed largely on phanerogamous sea grasses, *Cymodocea*. Large masses of green marine algae, *Caleurpa*, formed the stomach content. The apparent change in diet over the years from algae to sea grasses is a result of change in habitat preference.

Feeding is the prominent activity of the dugong. The interlacing bristles on the lip pads are used for grasping sea-grasses. Evidence for the rooting action by the face and mouth can be seen in the task wear of the males and in females possessing erupted tusks. The heavily worn and some times broken tusks suggest scraping and knocking against a hard substrate. While grazing dugongs move along the bottom by walking on their flippers. Dugongs have also been observed drifting with the flippers dragging against the substrate, which may create the calloused areas. A captive dugong was observed stuffing cut grasses into its mouth with its flippers, demonstrating the flexibility and control of forelimbs. After grasping each mouthful of food, the animal shook its head back and forth in an apparent effort to remove mud and sand. Dugongs favor night feeding and spend day light hours in deeper off shore waters. Dugongs prefer to feed with the rising tide at dawn or dusk and most commonly on overcast days.

Dugongs generally remain submerged from one and half to four minutes. In 108 timed dives of a 1.8 m female dugong, the average submergence time was 3.01 minutes. Traveling dugongs reportedly rise to breathe less frequently than do feeding animals.

When swimming to and from feeding grounds, they travel at about 10 km per hour. But when disturbed, they are capable of greater speeds. Maximum speed of dugong was estimated to be 18.5 km per hour over short distances.

Vocalizations from animals in distress as described as whistling sounds. Calves are said to have a bleating lamb-like cry.

Dugongs swam both within and beyond the reef and were neither elusive nor shy. The young calves left the herd in the afternoon to form a sort of nursery close to he sandy beaches.

Sex and age composition of dugong groups vary. Calves usually accompany their mothers for a period of over a year, forming a stable social unit for that time. Males are not thought to remain with this unit. The animals tame rather quickly and this is taken by some investigators as evidence of some degree of intelligence.

In an attempted mating of a pair of dugongs, the behaviour began early in the morning. The male initiated activities by swimming belly up, towards the female and pushing and biting at her neck. Both animals moved vigorously and splashed and thrashed about until late at night. The penis of the male projected 5 to 8 cm from the body. Similar behaviour continued sporadically the following day, and at late afternoon the female for the first time also turned upside down and both rolled together from side to side. In the wild, it is believed that copulation occurs in sheltered bays.

A single young is usually born, although twins are reported rarely. Newborn calves are about 1.1 m long. Young animals have frequently been reported riding on the backs or shoulders of an adult female, and mothers are known to clasp their young to their side while traveling. Calves as long as 1.83 m have recorded accompanying their mothers.

Occurrence of Mammals along Indian Coasts

On 18th May, 2008, a humpback whale (*Megaptera novaengliae*) was found stranded in decaying condition at Thalikulam landing center in Thrissur district,

Kerala. The animal was having a total length of 9.8 m, width- 3.45 m and flipper length, 2.6 m, fluke- 2.4 m with approximate weight 2 tonnes.

A baleen whale, measuring 12.12 m was found washed ashore on Dona Paula beach, Goa on 18.6.2008. Another baleen whale (*Balaenoptera* sp.) measuring 14.5 m weighing approximately 10 tonnes was found washed ashore at Thalikulam, Trissur, Kerala on 23.4.2009. One baleen whale of 18.3 m in length, 2.4 m body diameter and 10 tonnes weight was found washed ashore at Murudeswara beach, Karnataka on 24.11.2005. Another whale was found washed ashore at Padukere in Udipi district of Karnataka on 2.12.2005. The total length of the whale was 5.5 m.

Five female spinner dolphins, *Stenella longirostris* were accidentally caught off Chennai from 70-75 m depth in the gill net operation, targeted for sharks, rays, tuna and seer fish. The total length of the animals ranged from 149 to 160 cm and weight between 30-40 kg. Spinner dolphin is the most commonly caught species along the Indian coasts in gill net operations.

A female dolphin, *Stenella longirostris* was found stranded in dead condition near Thalikulam landing center, Kerala. The animal weighed about 80 kg and the length of the body is 198 cm.

Chapter 29

Special Traits for Marine Life

Floating Mechanisms of Sea Creatures

Sea water contains about 35 g of salts per kg and has a specific gravity of about 1.026. If an animal could exclude all salts from its body fluids, it could gain about 26 mg lift for every ml of such fluid. The teleosts often have tissue fluids with less than half the salt content of sea water, and while this is unimportant for the buoyancy of the adults, it is probably vital for their eggs. All the common marine fishes except the herring have eggs that float upwards in still water. It seems that the yolk sacs of such eggs are impermeable to water, so that the dilute fluids of a parent can be preserved in its eggs and used to give them buoyancy. The principal advantages to a fish of its dilute body fluids is that of allowing it to lay floating eggs.

The usual disadvantage of having body fluids less salty than the external sea water is that osmotic forces tend to draw water out of the animal, and osmotic work has to be done continuously if the "watery" body fluids are to be maintained. It is found, therefore, that most marine invertebrates have body fluids that are isotonic with sea water and that they can gain lift from these fluids only by changing the kind of solutes that they contain. One of the most effective of such changes is that of replacing some of the usual cations of sea water by ammonium ions.

The osmotic equilibrium of *Noctiluca miliaris* (a very brightly luminescent protozoan of 0.05 cm in size) with the surrounding sea water is effected, that its sap contained an osmotically active substance of lower specific gravity than NaCl. It is more likely that *Noctiluca* gained buoyancy merely by excluding the heavier divalent ions of sea water.

Recently a quantitative study has shown beyond all doubt that the Cranchid squids do use ammonium ions to gain buoyancy. These animals can hang almost motionless in sea water without any apparent effort. When the mantle of such a squid is cut open, an enormous liquid-filled coelomic cavity can be seen. If the membrane surrounding this cavity is punctured, and the fluid it contains drained off, the squid loses its buoyancy and sinks. The coelomic fluids from several species studied had specific gravities between 1.010 and 1.012, and the coelomic fluid accounted for about two-thirds of the weight of each animal. Since the specific gravity of the sea water was 1.026, these animals gained about 16 mg of buoyancy for every ml of coelomic fluid.

Although the coelomic fluids were isotonic with sea water, they nevertheless gave about two-thirds the lift that would be achieved by having pure water as body fluid. Such low specific gravities could not be given by replacing the heavier ions of sea water with lighter ones, such as replacing sulphate ions by chloride ions. They arise from the extraordinary accumulation of ammonium chloride, the ammonium ion accounting for four-fifths of the total cation content. Ammonia occurs as an end product of protein metabolism, and the Cranchidae have evolved the very neat trick of accumulating their excretory ammonia to give themselves lift.

The diatoms form a very important fraction of the tiny drifting plant life on which all life in the sea depends. They can not propel themselves, yet they must remain in the upper sunlit layers of the sea if they are to grow and multiply. The regulation of their buoyancy must be of the first importance. Under the optimal conditions the vegetative cells of this diatom (*Ditylum brightwelli*, size about 0.01 cm) can have the same specific gravity as sea water, while under unfavourable conditions resting spores are formed, with a specific gravity significantly higher than sea water. In the formation of resting spores, the cell sap that fills the bulk of the vegetative cell is expelled, that its specific gravity is such that it will give a lift of 2.5 mg per ml of sea water. The exclusion of the divalent ions from the sap would account for its lower density.

In larger gelatinous planktonic animals, like, medusae (jellyfish), ctenophores (sea gooseberries or comb jellies), heteropod and pteropod mollusks (sea elephants and sea butterflies) and tunicates (or salps), the body fluids had specific gravities less than that ofsea water. These body fluids were isotonic with sea water, but contained the common ions in different relative amounts. Of these changes in composition, by far most important in giving lift is the partial exclusion of the sulphate ion. Although medusae and ctenophores have sulphate concentration as low as 40 per cent of that of sea water, this will give only about 1.5 mg lift for every ml of body fluid and it can buoy up less than 0.5 per cent of its volume of protein.

Although single-celled animals or plants may exist that use fat as their principal buoyancy device, some oceanic copepods have a great deal of fat, and this will undoubtedly bring them much closer to neutral buoyancy than they would otherwise be.

In some fish, such as the mackerel, the amount of fat varies greatly with season, in others, it varies with the stage in the animal's life. Although the basking shark sinks when dead, the very large fatty liver undoubtedly has an important hydrostatic function. The most striking example of the use of fat for buoyancy is provided by some species of sharks of the family Squalidae

The fat of these livers contains very large fractions (sometimes as much as 90 per cent) of the hydrocarbon squalene. The squalene has the low specific gravity of 0.86, a figure that may be compared with the more typical figure of 0.93 for cod liver oil. This difference in specific gravity does not seem very great, but when calculated, the relative lifts given by 1 gram of squalene and 1 gram of cod liver oil, the squalene is found to be about 70 per cent more effective at giving lift. For this special utility of squalene, these sharks accumulate squalene rather than fats in their livers.

The deep diving whales and dolphins, make use of low density fat to give buoyancy. While the shallow diving whales, such as blue whale, have fat densities close to that of cod liver oil, whales that dive deeply, like the sperm whale, have large amounts of low-density fats. Fat gives the blue whale appreciable lift, but since it feeds in the surface waters of sea on tiny animals and does not normally dive very deeply, the gas space of its lungs must be an important variable in determining its specific gravity. The sperm whale on the other hand, dives so very deeply that the gas within its lungs must often be compressed to a volume that allows it to give only a trivial lift; the only effective buoyant component is the whale's fat, and the animal clearly gains great advantage in having those fats which will give the necessary lift in the least volume. The low-density fats used by the whales are not hydrocarbons, like the squalene used by the deep sea sharks, but esters of long-chain aliphatic alcohols with fatty acids.

Under atmospheric pressure, air has a density only about 0.00125 of that of water, and gas spaces offer the most effective way of giving lift. Animals use two principal methods to maintain gas spaces. In the first method, the chamber defining the gas space does not have rigid walls and the gas pressure within the space equals the external hydrostatic pressure. This is the method used in the swimbladders of fish. Since hydrostatic pressure in the sea increases by one atmosphere for every 10 m increase in depth, this may imply large pressures-101 atmospheres at 1000 m depth. The gas pressure will, except at the very surface of the sea, always exceed the combined partial pressures of the common gases dissolved in the sea, so that the gas will have to be actively secreted into the space and some arrangement made to prevent it from simply dissolving in sea water.

The second method is to have a chamber with rigid walls and to pump water out of the chamber. This is the method used in the cuttlebone of the cuttlefish, *Sepia officinalis*. The gas within the chamber can now approach equilibrium with the gases dissolved in the sea and its pressure rises towards 1 atmosphere. The rigid wall has to sustain the difference in pressure between the external hydrostatic pressure and the pressure of gas within the chamber.

Many fish, such as the cod and the perch, bring themselves to neutral buoyancy by having gas space within themselves. This space amounts usually to about 5 per cent of the total volume of a marine fish, and gives a lift that just balances the weight in sea water of the other tissues. The swimbladder wall is not rigid, so that if a fish swims upwards or downwards in the sea and so changes the external pressure acting on it, the gas in the swimbladder either expands or is compressed, and this changes the animal's buoyancy. In response to longer-lasting changes, fish can secrete more gas into the swimbladder or reabsorb gas from the swimbladder so as to restore neutral buoyancy. The swimbladder is used down to astonishing depths, to 2000 m and probably down to 4500 m. at this later depth, the pressure will be 450 atmospheres, and the density of the gas itself will demand that the volume be greatly increased if it is to provide neutral buoyancy.

It has been found that whereas the gas from the swimbladders of fish living near the surface of the sea often contains less oxygen than air, the proportion of oxygen increases with depth and gas taken from fish living at appreciable depths is mostly oxygen. Some swimbladders must be capable of secreting oxygen against very steep pressure-gradients. The gas is secreted into the swimbladder from a special tissue, known as gas gland. This secretion seems to be intimately linked with the structure called the rete mirabile in which arterioles going towards the gas gland break up into capillaries that come into intimate contact with corresponding venous capillaries arising from the gas gland. These capillaries are longest known in nature and deeper the fish lives, the longer are the capillaries. The secretion of gas into the swimbladder is clearly the principal problem, for the mechanism of gas reabsorption are easy to understand. Some of the fish have a valved duct leading to the oesophagus through which they can allow gas to escape. Others have a structure called the oval, in which some region of the swimbladder wall can be either exposed to or occluded from the gases in the swimbladder by the action of a ring muscle round its perimeter. This region is served by a blood supply without a rete, and gas can thereby be carried away from the swimbladder when the oval is open.

The cuttlebone consists of a number of thin chambers laid down one below the other at the rate of about one or two a week as the animal grows. The calcareous walls of these chambers are spaced at about 0.7 mm intervals and are held apart by numerous pillars. The whole cuttlebone accounts for about 9 per cent of a *Sepia's* total volume and contains gas spaces that give it a specific gravity of about 0.6 and so allow it approximately to balance the weight in sea water of the rest of the animal. It has been shown that the cuttlefish can vary its specific gravity and that it does so by varying the proportions of liquid and gas space that the cuttlebone contains.

The gas within the cuttlebone is principally nitrogen, and its average pressure is about 0.8 atm, no matter at what depth the cuttlefish is caught. There can therefore be no question of liquid being pumped out of the cuttlebone by gas pressure. When a new chamber is formed, liquid is actively extracted, leaving behind a space that contains gas under very low pressure. Gas then slowly diffuses into this space until its partial pressure equals that in the surrounding tissues.

It has also been shown that liquid is pumped in and out only through the posterior ends of the chambers and there is a specialized membrane which is thought to do the active pumping. When an animal has been kept in very shallow water, the liquid inside the cuttlebone is isotonic with sea water; in animals just hauled up from the sea bottom it is markedly hypotonic. This suggest that equilibrium is maintained by balancing an osmotic force, between the blood and the cuttlebone liquid, tending to extract liquid from the cuttlebone against the hydrostatic pressure of the sea that tends to push water into the cuttlebone. The calcareous structure must sustain the inevitable difference in pressure between the gas inside the chambers and the hydrostatic pressure of the sea.

It is clearly seen that the elegant buoyancy devices found in higher animals, such as squid and fish have their counterparts in the so-called "simple animals", such as protozoans. The so-called simple animals have not merely the potentialities for evolution into "higher ones", but have already evolved the principal functions that the higher ones possess.

Deep Sea Diving of Marine Animals

The apnoeic or breath-hold dives made by marine mammals is limited in duration by the low oxygen reserves trapped in the diver's lungs. In fact, the large whalebone whales (*Balaena mysticetus*) commonly dive to 200 meters, at which depth they feed on the beds of small crustaceans. Certain toothed whales (*Physeter macrocephalus*) reach depths of more than 400 meters and can remain submergd for more than a hour. On the other hand, the pinnipeds (seals and otaries) do not dive below a few dozen meters and remain submerged for less than a quarter of an hour.

The great depths reached by the cetaceans seem to be attributable to their specialized anatomy; in these animals, all the ribs are floating and there is no sternum. The thoracic cage is therefore not rigid, and can become partially flattened under the effect of pressure. The lungs can be collapsed almost completely; the volume of residual air in deep diving is certainly very small. The respiratory tract is short, wide, and mechanically very tough; it becomes blocked during diving by specific anatomical structures in the blowhole, spiracular chamber and glottis, which prevent any water from entering the respiratory tract. The pinnipeds, which are only moderate divers, do not possess these anatomical peculiarities, which are likewise absent in man. In fact, the maximum depth that can be reached by a breath-hold diver depends on the ratio between the total pulmonary volume and the residual volume.

In cetaceans, the absence of a sternum, and the very oblique positioning of the diaphragm facilitate these mechanisms, so that very deep dives can be performed; the lungs are then totally collapsed, and the air initially present in the alveoli is driven by the ambient pressure into the anatomical dead space. This also explains why the cetaceans can return to the surface after a deep and prolonged dive, no matter what depth had been reached. At great depths, the alveolo-capillary exchanges are zero, and the animal is in no danger of accidents due to gases (nitrogen narcosis, oxygen toxicity) or decompression accidents on surfacing.

Other anatomical adaptations are also favourable to diving in some pinnipeds; for example, the specific vascular networks (*rete mirabile*) distributed along various arteries, and the presence of arterio-venous shunts above certain organs. The vena cava is often enormous, and constitutes a reservoir where blood can accumulate through the action of precardial sphincter which slows down the return of blood to the heart. It is probable that during apnoea, in atleast some of the diving mammals, a large number of organs (such as the muscle) receive no blood supply and function under conditions of total ischaemia. The vascular networks just mentioned, the sphincters and the arterio-venous shunts then ensure that the blood circulates only through the most important organs, that is the nervous system and the heart muscle. However, these anatomical modifications are not found consistently, being present in some species and not in others.

Immersion bradycardia exist in all diving animals (duck, porpoise, seal, dolphin, whale etc) and even in man. This bradycardia (slow beating of the heart) appears in the first few seconds following immersion, and persists until the animal returns to the surface. It is often accompanied by a decreased cardiac output; this decrease may reach as much as 75 per cent in the seal. There seems to be a relationship between the rate of establishment and the intensity of bradycardia on the one hand, and the possibility of prolonged apnoea on the other. For example, in seals, which can remain without breathing for 10 to 12 minutes, bradycardia develops in about 20 seconds and lowers the heart rate to 20 per cent of that at the surface. In porpoises, on the other hand, which are not such good divers and can remain without breathing only for 3 minutes, bradycardia sets in, in 90 seconds and lowers the heart rate by only 50 per cent.

This modification of the heart rate takes place by a vagal cardio-moderator effect. The origin of diving bradycardia is poorly understood. In man and in certain animals, the action of cold water on the region around the mouth seems to be the principal, but not the only factor. Pulmonary distension in certain cases plays a role of some importance. On the other hand, changes in blood gases during apnoea have no marked bradycardia-inducing effect.

The action of diving bradycardia is reinforced in many diving mammals by an intense peripheral constriction of the blood vessels which occurs concomitantly with breath-hold immersion. This vasoconstriction affects mainly the muscle masses, whereas the heart and central nervous system continue to receive a normal blood supply.

As a result of their respiratory adaptations, diving mammals can very rapidly renew alveolar oxygen between two dives, and the tidal volume represents 80 per cent of the lung capacity. The level of blood oxygen uptake is very high, and may reach 10 per cent (as compared with 3 to 4 per cent in terrestrial mammals). The large diving mammals also have a very high tolerance to hypercapnia (excess of carbob di oxide in the lungs and blood) and the expired air may contain up to 10 per cent of carbon di oxide at the end of apnoea. The fixation of oxygen by myglobin could account for 50 per cent of the oxygen reserves available in diving. During apnoea, and because of the muscular vasoconstriction, the muscle uses this source of oxygen exclusively.

All these cardiovascular and respiratory adaptation mechanisms help to prolong apnoea. But a calculation of oxygen balance in diving mammals shows that the reserves carried by the animal are insufficient to maintain an aerobic metabolism comparable to that of terrestrial mammals. Such calculation in fact indicates that apnoea could last only about 5 minutes in the seal, and about 15 minutes in baleen whales, and yet the times attained are respectively 15 and 60 minutes. The explanation for the prolonged periods of apnoea achieved lies firstly in the high tolerance of the nerve centers to hypercapnia, and secondly in the protection of the most vital organs against anoxia. The heart and central nervous system are particularly sensitive to lack of oxygen, and during diving they are isolated from the rest of the circulation, so that a sort of " heart-lung-central nervous system circuit" is set up. The oxygen carried by this reduced circulatory circuit is then used only for these vital organs. There is no doubt that these are the principal functional characteristics of adaptation to diving in diving mammals.

Hydrodynamic Function of Marine Fishes

Fishes in still water, sometimes cease making swimming movement, glides forward by their momentum, as if it were rigid but gradually moves more slowly. This retardation process can be determined by the rate of loss of momentum and hence the force with which the water resists the fish's motion.

This resistance of the water is much the same for a live fish as for a wooden object of the same shape and size. Admittedly, for most fishes, this shape is streamlined and gives rather a low resistance. However, a fish can maintain a steady speed only if it can move its body and fins, so as to produce a net forward force or thrust exactly balancing the resistance of the water.

In the hundreds of millions of years during which fishes have evolved, aspects of swimming have in many environments had particular importance for survival. In the open ocean, the speed may be one such quality. A great many marine fishes live by capturing and eating other organisms, and probably most die either from starvation or by being themselves captured and eaten. Small improvements in speed can reduce

Figure 29.1: Rays in Motion

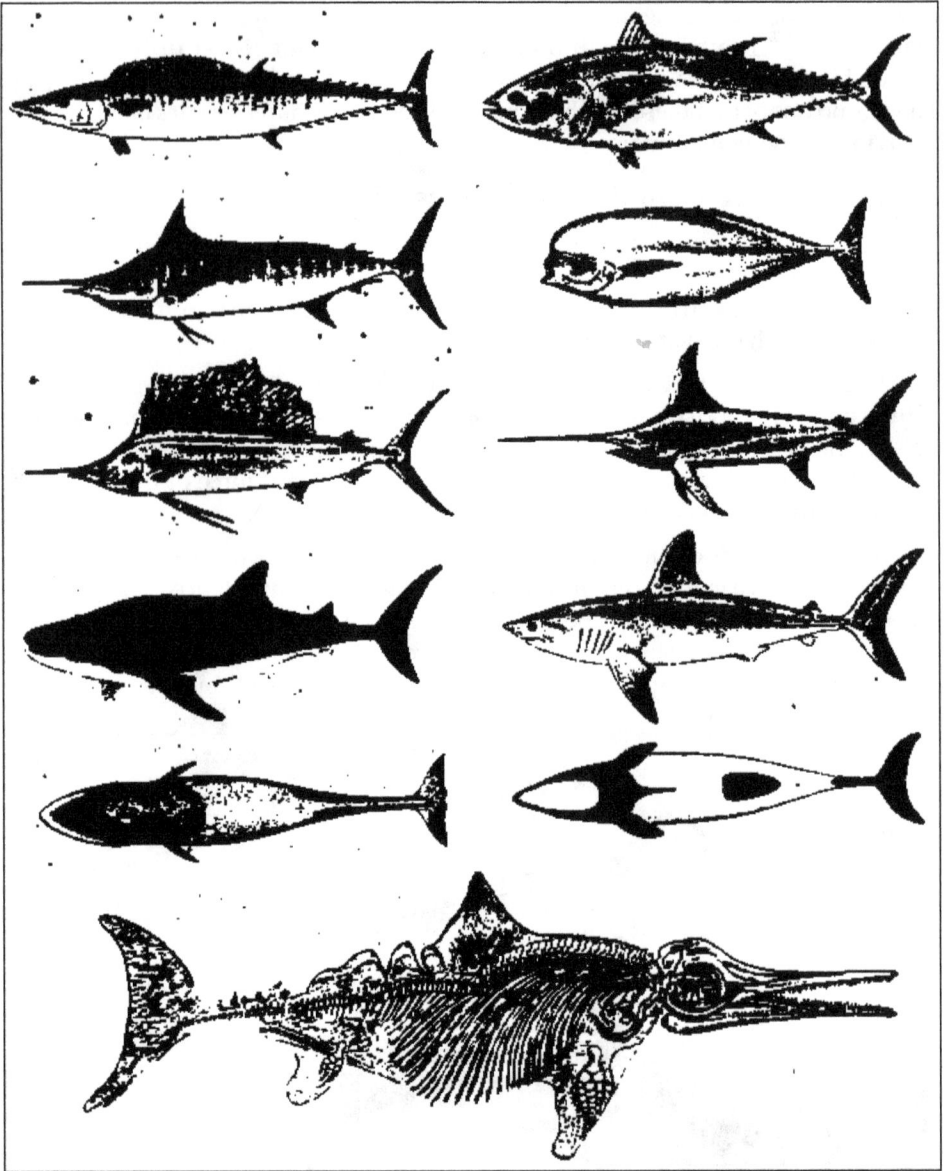

**Figure 29.2: Varients of Lunate Tail in 6 Percomorph Fishes,
2 Sharks, 2 Cetacean Mammals, 1 Extinct Reptile**

the chance of premature death through either of these causes. In consequence, fishes have acquired some remarkable capabilities; for example, a tuna fish about the size of a man can swim ten times as fast as an Olympic champion.

Another quality that has survival value is swimming efficiency. Without this, the fish would too rapidly use up the supply of energy derived from food while moving around to find its next meal. A swimming fish, which produces a thrust

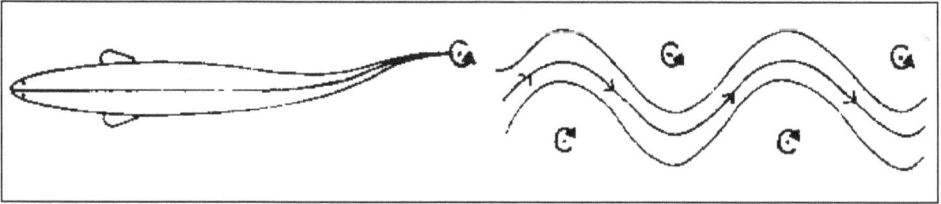

Figure 29.3: Plan View of Vortices, Cast Off by a Lunate Tail as Fish Moves

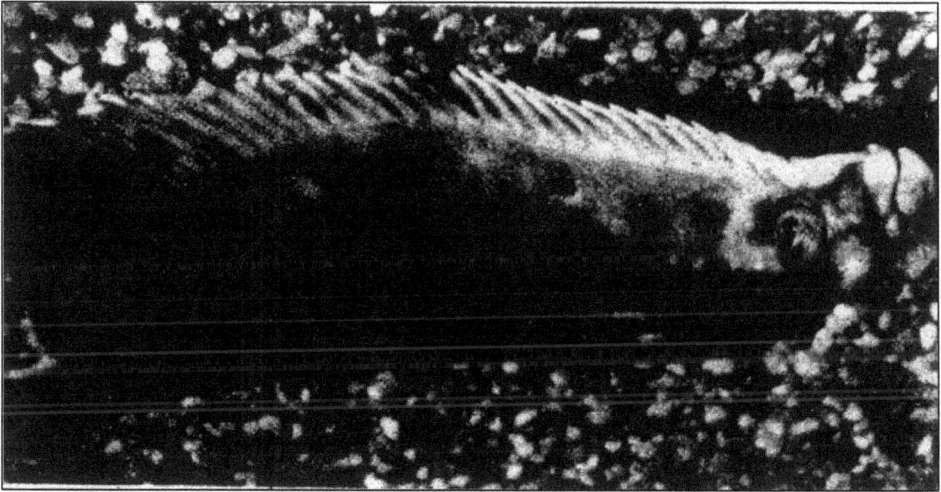

Figure 29.4: The Plaice, *Pleuronectes* spp.

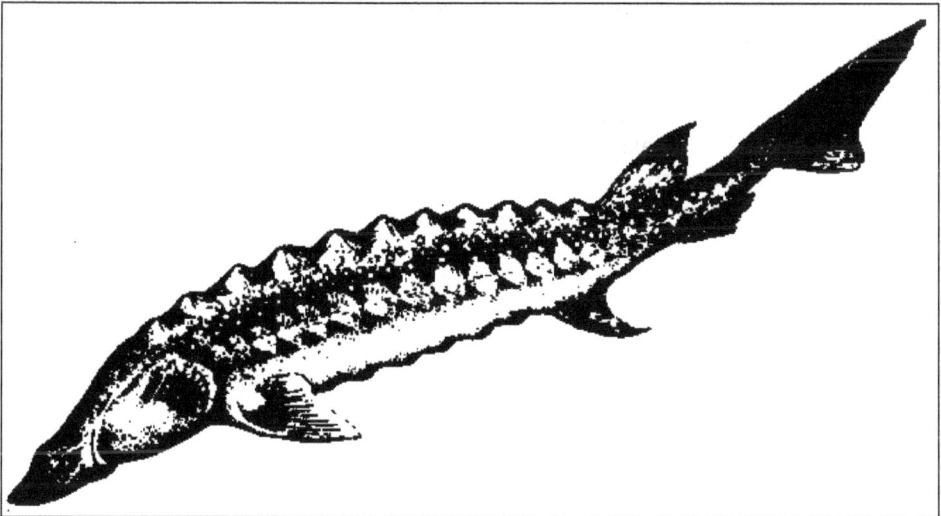

Figure 29.5: The Sturgeon, *Acipenser* spp.

balancing the resistance of the water, is doing work-"useful work" in the engineer's sense- at a rate equal to thrust multiplied by speed. At the same time, its body and fin movements may be wasting energy by churning up the water behind it into a turbulent wake. For efficiency it is important that the rate at which energy is wasted in making this eddying wake is as small a fraction as possible of the rate at which useful work is done.

Of course, there are some environments, such as coral reefs and jungles of marine plant life, where speed and efficiency are not the factors giving the greatest chance of survival; where, for example, body camouflage or precise manoeuvrability may give a net advantage even at the cost of reduced speed and efficiency. In such environments, evolution may proceed in a direction that lowers speed and efficiency, as in the Plectognathi and Solenichthyes

Fishes representing an earlier line of vertebrate development, do not have a bony skeleton, but a cartilaginous one. Fishes with perfected bony skeleton, collectively called teleosts, possess in general a very effective hydrostatic organ, in the shape of a bladder full of gas, the so-called swimbladder and they can control the quantity of gas in the bladder in such a way that their weight is exactly balanced by their buoyancy. The teleosts, can, when they choose, cease swimming movements altogether without any resulting tendency to sink or rise.

Among the swimming methods, one appears to have been the swimming mechanism of the earliest fishes. It depends upon a transverse wave, or side to side undulation, passing down the body. Such an undulatory mode of swimming is found quite commonly among the invertebrates, with the wave passing back from head to tail and increasing somewhat in amplitude as it does so. It occurs in the more efficient form in the vertebrates, including both the more primitive jawless lampreys and the fishes proper. For vertebrates, this lateral undulation is much more effective because they have laterally compressed tails which greatly improve the efficiency of the swimming method.

Such lateral undulation is used by dogfishes and also by some sharks as well as by sturgeons and by lung-fishes. It is also used by certain group of teleosts, notably the eels with their highly extended shape. Fishes of generally similar shape like the ribbon-fishes and unicorn-fishes, as well as by various other fishes including cod.

With teleosts, in which weight is exactly balanced by buoyancy, no special movements are required for maintaining their vertical position, and therefore the undulatory mode is particularly easy to follow. The common eel, *Anguilla vulgaris* having laterally compressed tail, reduced body section, long continuous fins present dorsally and ventrally helps in maintaining the total body depth, although the lateral thickness is enormously reduced. At the posterior end, the tail practically becomes a vertical edge, which an aerodynamicist would call a "trailing edge".

In an eel swimming in a tank, the wave can be seen by progress backwards towards the tail, Positions of greatest curvature appear gradually further back. The speed of the wave down the body is always found to be greater than the resulting fish-speed through the water, so that even relative to the water the wave passes backwards. The amplitude of lateral motion increases as the wave passes from head to tail. This is the typical anguilliform mode of swimming.

Though fish like the cod does not have continuous dorsal and ventral fins, but has three of each, with only short gaps between them, these short gaps become filled up with vortex sheets and behave mechanically almost like a solid fin, while swimming, according to the theory of flow around slender bodies, so that the anguilliform motion of the cod becomes effectively very similar in its mechanics to that of the eel.

In carangiform swimming, the front part of the fish has lost its flexibility and the undulation is confined almost entirely to the rear half, or even third of the body length. A wave that passes backwards can still be discerned, but it is a wave whose amplitude increases rather fast from almost zero at the mid-point of the fish's length to a large value at the tail. As before, the trailing edge's motion lags behind that of front section although now there is almost no motion further forward still. This modified undulatory montion confined to the neighbourhood of the tail is known as carangiform motion (from the family Carangidae).

It is known from engineering experience that this rapid acceleration produces a better result because it keeps the kinetic energy of the water's motion down to the half mw square value, whereas slower acceleration gives time for water motions with extra kinetic energy to appear, due to eddies shed from the tops and bottoms of fine cross-sections.

Carangiform motion is well developed in the salmon family, *Salmo salar*. The front half of the body is not at all flexible, but over the rear half there is a rapid increase in wave amplitude to the large value it reaches at the caudal fin. The fishes, using carangiform method swimming include the horse-mackerels and jacks of the family Carangidae, as well as perches, red-mullets, barracuda and red cardinal-fishes.

The fins in the fish's plane of symmetry, the dorsal, ventral and above all, caudal fins are important for propulsion and for stability against side to side movements. However, paired fins off the plane of symmetry also exist. These are the pelvic fins and more important, the pectoral fins (just behind the head). The paired fins give stability against heaving and pitching motions. Fins in general are important, too, in enabling fishes to make fine adjustments of their position in the water, and in stopping and starting. Fishes are helped to make a rapid start from rest because muscle can exert about four times as much power for a brief period as it can continuously. Fishes therefore can start quickly by making normal swimming motions with extra force. On the other hand, stopping is not so easy because the streamlined shape of the fish enables it to glide forward a considerable distance under its own momentum. For a sudden stop, however, fishes have learnt to use their pectoral fins so as greatly to increase water resistance, much as an aircraft pilot uses his airbrakes.

Fishes which have succeeded best in the struggle for existence in the surface waters of the deep ocean, and in doing so have become the fastest fishes of all. They have achieved this by a major change in the shape of the caudal fin, what an aerodynamicist would call an increase in its aspect ratio, defined as depth square/surface area. There is already an aeronautical appearance about the herring fin, which closely resembles a pair of highly sweptback wings; increasing the aspect ratio can be regarded as making them not too sweptback.

Thrust can be increased if there is an increase in the virtual mass at the trailing edge, and this mass in turn is proportional to the square of the depth of the caudal fin. However,if the caudal fin were made too big, the resistance as its large area was dragged through the water would become excessive, so that speed might not much improve. This indicates the importance of the aspect ratio, which can be increased by reducing the sweepback well below. As a result, thrust is raised without much increase in resistance, and the fish can go faster

The tunas, the striped marlin, the sailfish and swordfish, all are with different ends, but with broadly similar lunate tails. These are all outstandingly fast, active fishes, constantly on move, to such an extent that they have gradually lost all pumping apparatus for bringing water in contact with their gills. It is no longer necessary because their unceasing motion forces water through their mouths and out of their gill slits to a quite sufficient extent.

There is something about the lunate tail that especially fits it for high speed marine propulsion. The method is strictly carangiform, all the propulsive mechanism being in the tail, and the reduction in cross section depth just before the caudal fin is even more extreme than in the salmon. All these fishes possess exceptionally powerful musculature for moving the tail at very high frequency (as much as 10 Hz), and the body temperature is found to be unusually high (almost 30°Celsius).

All the lunate tails show a strong similarity to certain configurations of aeroplane wings; they have good "aerofoil sections", with a nice blunt leading edge. It is in fact possible to analyse the thrust developed by a fish's lunate tail due to its carangiform motion by just the methods used to find the forces sustained by aircraft wings when pitching and heaving.

The lunate tail of tuna, as the fish moves to the left, shows the eddy or vortex cast off at each extreme of the tail's movement and depicts the backward stream of water that these vortices generate between them, indicating that this way of obtaining thrust has something in common with jet propulsion. The character of the vortices at roughly the level of the fish's nose. In three dimensions, taking into account the curved shape of the lunate tail in the vertical, it can be regarded as a horizontal section of a sequence of vertical vortex rings pushed backwards diagonally, alternately to the right and to the left, by the fish. It is well known that smoke rings (which are vortex rings in air) have enormous momentum, and the lunate tail may be so effective propulsively because it can especially well give thrust by the reaction of these backward moving vortex rings, that have so much momentum in relation to their energy.

Fishes heavier than water are not so outstandingly fast or efficient in their swimming, because their weight is not exactly balanced by the buoyancy of a controlled volume of swimbladder gas. These include most of he sharks. The slower sharks and most of the dogfishes have a specially shaped asymmetric (heterocercal) tail, which support their excess weight while swimming. Essentially, the heterocercal tail consists of a single large sweptback wing above and a much smaller one below. As the large wing moves from side to side, it produces thrust at right angles to itself, which therefore, contains a certain vertically upward component, or lift. At the same time the pectoral fin is well developed, and set an angle of incidence to carry the main part of the load, just as does an aeroplane wing. The weight of an aeroplane can be balanced in a stable fashion only if wing lift ahead of the center of gravity takes part of the load while tail lift behind takes the rest. With the shark the arrangement is the same, the pectoral fin is like the wing, and the anguilliform motion of the body and the heterocercal tail gives not only the thrust, but also the necessary tail lift.

The same arrangement is found in the sturgeon, which has the same problem of weight support. All these animals, as a result, have just the dynamical properties of an aeroplane and they can, for example, gracefully "loop the loop".

In skates, rays and allied fishes, there is still more pronounced development of the pectoral fin, initially for weight support, but increasingly contributing to propulsion. Some, like the so-called guitar-fishes, still have effective caudal fins.

The bottom living skates and rays, have in fact become almost all pectoral fin, the two fins together approximate to a square, with an extremely thin body and tail down the diagonal. They swim very beautifully by passing backwards over the pectoral fins an undulation very like the basic undulatory mode with which it was started, but now with up-and-down motions instead of side to side. These fish live near the bottom and when not actively swimming, their wing-like shape, with aspect ratio of about 2, enables them to achieve a conveniently low angle of glide down to the bottom.

On the other hand, eagle rays have developed a mode of swimming even closer to the flight of a bird or devil rays swim in a rather similar manner. Their pectoral fins have developed into wings having an aspect ratio of around 4 and much less than a whole wavelength of undulation is visible at any one time. They use in fact a strong downstroke and a rather more feathered upstroke, like many birds.

The bottom living teleosts, like plaice have lost their swimbladder, so that when they are inactive they glide down on to the bottom, where their excellent protective colouring gives them some security. However, when very young, they are fishes of normal shape, swimming in anguilliform motion, and it is only at a certain age that they turn over on their sides and the lower eye comes round to the top. Then their lateral anguilliform motion, turned through 90°, becomes an up-and-down one, and their final shape and mode of swimming are not unlike those of a skate, although the skate reached this condition quite differently, by an enormous development of the pectoral fin.

Colouration and Light Emission in Marine Animals

Marine animals, especially shallow-living tropical fish are well known to have a great variety of strikingly brilliant colours, shades and patterns. But even animals living in the profound depths of cold waters, show a great variety of colours when brought to the surface. At these depths, where sunlight does not penetrate, the only light available is that produced by the animals themselves. This light is of varying colour and intensity. The greater part of the volume of the deep ocean is pitch black.

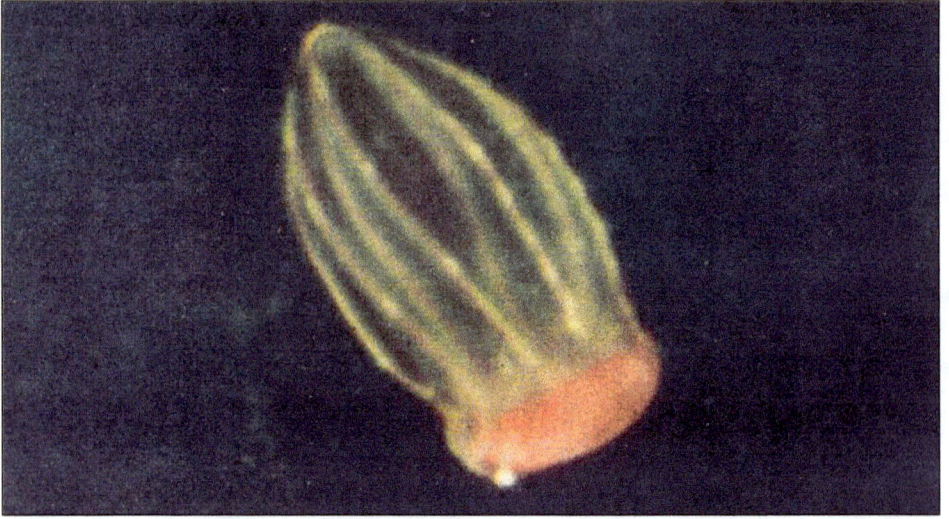

Figure 29.6: Sea-gooseberry, *Bcroe forskali*

1cm

Figure 29.7: Cydippid Larva of Ctenophore

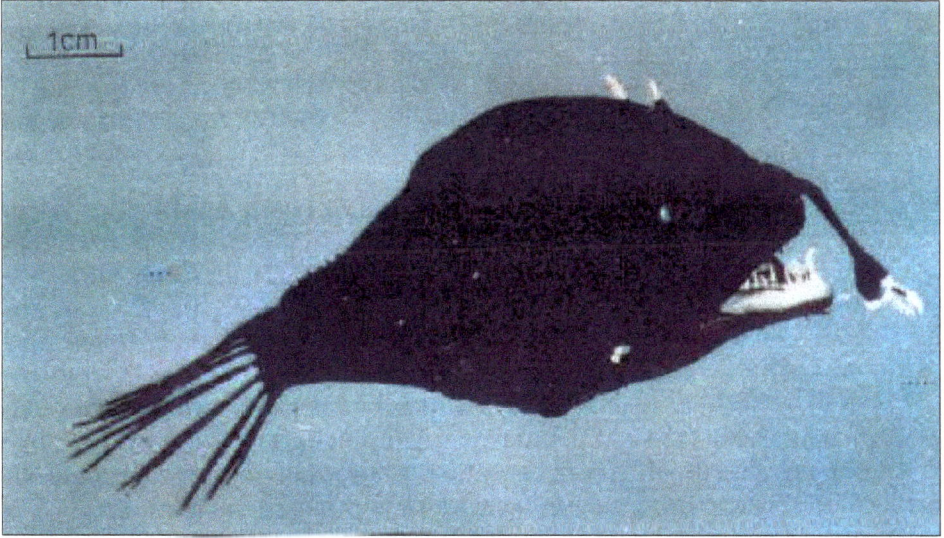

Figure 29.8: A Certoid Angler Fish *Melanocetus*

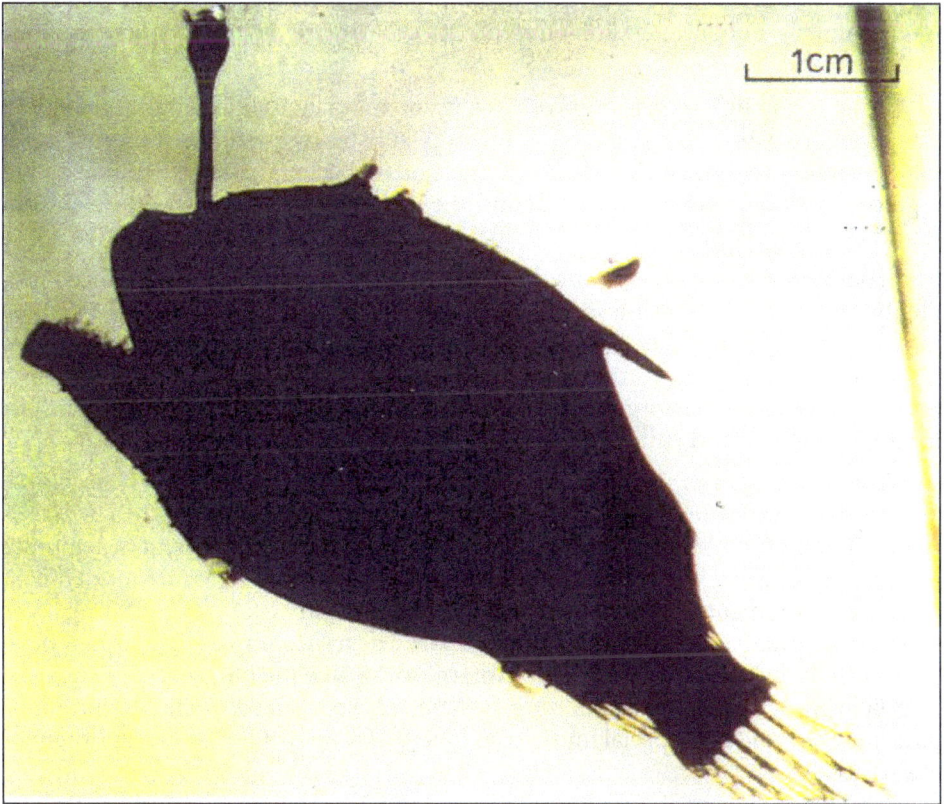

Figure 29.9: Angler Fish Hoisting Light Organ

Figure 29.10: Deep Sea Angler Fish, *Astronesthes*

As light enters the ocean it is both scattered and absorbed. The surface water most strongly absorbs the red and ultraviolet light and it is thus green and blue light that penetrates most deeply. The light emitted by marine animals can come from any part of the visible spectrum, but it is usually blue or green. This light is probably that to which their eyes are most sensitive. Light production is common amongst marine animal than terrestrial ones and is exceedindly rare in fresh water animals. Light is produced by animals both with and without eyes and by sessile benthic as well as motile animals. The simplest known forms of light-producing life are found amongst bacteria.

Most of the light-emission processes involve the luciferin-luciferase reaction. The reactions vary from the simple enzyme-substrate system in the crustacean *Cypridina*, the pyridine-nucleotide linked reactions of bacteria, the adenine-nucleotide type of reaction as occurs in *Renilla* and the sea pansy, to the peroxidase system found in *Balanoglossus*.

There are three main ways by which marine animals are able to produce light. Firstly it may be produced intracellularly in special cells that are sometimes organized into special structures, such as lanterns and light organs. Sometimes the light is produced extracellularly and can be discharged into the surrounding sea as a luminous cloud. Other organisms, instead of making light themselves, harbour luminous bacteria and cultivate them.

The special light-producing organs are called photophores, and vary from simple to very elaborate structures. In any one species there may be from one to a dozen or more different types of photophore, the whole complex being arranged in a definite pattern. The simpler structures are no more than a luminous part surrounded by a cup of black pigment cells. In the more elaborate structures there is a reflecting layer between the pigment cup and the luminous part. The most complex light organs have a lens resembling that of a bull's-eye lantern, some, like the hatchet fish *Polyipnus*, even being provided with a colour filter. In extremely specialized species the luminous parts may be moved by special muscles and there may be an adjustable diaphragm of pigment cells.

Direct control of light emission, as in the pelagic gastropod *Phyllirrhoe* is rare, where some of the glandular cells produce light and have swollen nerve endings abutting against them, presumably the neuro-effector junctions. Even in those photophores that have been shown to be innervated it is not known whether the initiation of luminescence depends on neuronal control. Indirect neuronal control of luminescence is well known from many cases. It is manifested either as the control of the diameter and therefore of the flow of blood in vessels supplying oxygen to the photophores, or, as the control of the muscles that squeeze out the luminous secretions from the glands, as in the squid *Heteroteuthis*. Similarly, in both the squid *Watasenia scintillans* and the lantern eye fish *Photoblepharon*, rapid obscuring of their light emission is achieved by drawing an opaque pigmented fold of skin in front of the light organ. In the closely related fish, *Anomalops* muscles rotate the light organ so that the light-emitting part is buried in the animal's opaque tissues.

Electrical stimulation in many animals, such as the sessile pennatulids results in a wave of luminescence passing outwards over the colonies from the point stimulated. Mechanical disturbances can produce similar effects. Exposure to light too has an effect upon luminescence; it usually reduces the ability of the animal to luminesce and it needs a long period in the dark before it can recover its full output. If only small segments of the animal are illuminated, then only these areas are affected in this way.

Some of the transparent pelagic animals, such as small crustacean sometimes appear luminous as a result of eating phosphorescent food. However the functions of some light organs such as luminous lures are apparent; presumably prey is attracted to the lure and then captured.

Some fishes have bright light organs that illuminate the visual fields of their owners and act like search lights. It is thought that some of these lights are strong enough to be reflected by the reflectors within the light organs of other fish and to illuminate them, despite the victim having extinguished its light and contracted the pigment cover. Light organs may act as recognition signals; in some fishes there are differences in the photophore patterns between the sexes; in a similar way they may act as a basis for schooling responses. The idea that they may have some defensive function is well established, and it is suggested that changing patterns of flashing might confuse enemies. Indeed the deep sea squid *Heteroteuthis dispar* secretes from glands near its mouth a substance that on contact with sea water produces a cloud of luminous sparks. Surprisingly, many of the fixed benthic organisms, such as sea pen, *Pennatula* glow with brilliant phosphorescent light. One would have thought this could only make them conspicuous to predators.

Animals living at any depth may possess light organs and although the ability to produce light presumably most useful to those that live in the inky blackness of the ocean depths, it is the population that lives between 100-500 m, that is frequently characterized by luminescence. Although nearly all these species exist below 500 m their highest concentrations are found above this level.

Colourations

There is a recognizable pattern in the colours of deep sea life and this is organized on a vertical scale. The animals that inhabit the surface down to about 150 m are either transparent or blue. Below this depth and down to about 500 m the inhabitants are mainly silvery and greyish. Below this depth again the population consists of dark-coloured fish and red prawns. Many of the deep sea cephalopods, such as, *Mastiogteuthesis* and *Calliteuthis* are also red. The development of colour may have a concealing effect, since at depths where there is no red light, creatures appear black. This colouration is peculiar to salt-water animals that dwell in the dark regions, whereas cave dwelling animals are usually a pallid white. The colours of deep sea animals may of course have no ecological significance, and they could represent stores of waste products, or of metabolically useful intermediate products. It is known that starvation causes the depletion of carotenoids from the ink and liver of *Octopus bimaculatus*

In animals, such as the euphausiids, their red colour renders them invisible in the twilight zone of the mid-ocean waters where red light does not penetrate, and they are not rendered visible by the light usually emitted by predators. But some of the fish that feed upon them emit red light from large luminous organs. It may be that the colours of some fishes have been influenced not only by incident light from above, but also by the need to avoid reflections from light emitted by predators.

Besides colours, many marine animals, especially those of the twilight zone, have very efficient reflecting surfaces. These surfaces reflect incident light falling upon them so effectively that they cause the fish to merge into the background and become difficult to detect. Because the light falling upon these animals has different intensities in different directions, many of them supplement their reflecting surfaces with organs that emit light so that the animal's own light output matches that of its surroundings. This is most useful to the animal that is trying to disguise itself from predators that are below it and are attempting to view it as a dark object silhouetted against the lighter water above. The ventrally situated light organs can destroy this silhouette and in order to use them in this way a very fine technique is used; the animal has a light organ that shines into its own eye. It appears that by matching the intensity of the output of this organ with that of the background, the animal can determine the output required by its other light organ to match the background illumination.

Luminescence is found in most groups of deep-sea animals. Very often each individual photophore lights up only momentarily, producing sparkling effect, although the photophores near the eyes usually glow with a steady light.

References

Ansari, Z.A and A.H. Parulekar. Meiobenthos of Andaman Sea, Tech.Rep., 02/80, NIO, Goa, India, 1980.

Appukuttan, K.K. On the occurrence of green mussel, *Perna viridis* in Andaman Island, Indian, J of Fisheries, Vol. 24, Nos. 1 and 2, 1977.

Alagarswami, K *et al*. Pearl oyster resources of India, CMFRI Bulletin No. 39, 1987.

Asha, P.S and K.Diwakar A note on exploitation of star fish, *Protoreaster lincki* in Tuticorin, Mar.Fish.Inf. Serv. T and E Ser. No 187, 2006.

Anil, M.K. *et al*. A note on leatherback turtle, *Dermochelys coriacea* rescued at Vizhinjam, Kerala, Mar.Fish.Inf.Serv., T and E Ser, No. 200, 2009.

Ahmed Basheer. Recovery of an injured hawksbill turtle in Sindhudurg, Ratnagiri, Mar.Fish.Inf.Serv Adam Shiledar T and E Ser, No. 200, 2009.

Bak, R.P.M. Ecological variables including physiognomic-structural attributes and classification G.D.E. Povel of Indonesian coral reefs, Netherlands Journal of Sea Research 23(2), 1989.

Brewer, P.G., D.W. Spencer Anomalous fluoride concentrations in the North Atlantic. Deep Sea Research and P.E. Wilkniss 17, 1970.

Bhattathiri, P.M.A. and V.P. Devassy. Primary production of the Andaman Sea, Tech. Rep. 02/80, NIO, Goa, India, 1980.

Baskarudoss, K.P. *et al*. Cobia (*Rachycentron canadum*) culture, Fishing Chimes, Vol 30, No 8, 2010.

Baskaradoss, K and Pawan Kumar. Endangered species of Sea Horse, conservation and management, Fishing Chimes, Vol. 28, No. 9, 2008.

Batcha, H. *et al*. Rare occurrence of diamond back squid, *Thysanoteuthis rhombus* off Chennai Coast, Mar.Fish.Inf.Serv., T and E Ser No.201, 2009.

Babu, D.E. Promotion of mud crab culturing, need for national level planning, Fishing Chimes Vol. 29, No. 1, 2009.

Baby, K.J. Occurrence of humpback whale at Tuticorin, Mar.Fish.Inf.Serv, T and E Ser, No. 200, 2009.

Baby, K.G. Unusual occurrence of *Pempheris moluca* at Azhikode, Kerala, Mar. Fish. Inf. Serv., T and E Ser, No. 201, 2009.

Baby, K.G.and Sethy Prakasham. Baleen whale washed ashore, Mar.Fish.Inf.Serv., T and E Ser. No. 201, 2009.

Brad, A. Seibal *et al*. Post spawning egg care by a squid, Nature, Vol. 426(6), 2003.

Bhaumik, Utpal and A.P. Sarma. The fishery of Indian Shad (*Tenualosa ilisha*) in the Bhagirathi-Hoogly river system, Fishing Chimes, Vol. 31, No. 8, 2011.

Biju Kumar, A and R.B. Pramod Kiran. The Indian Spiny Turbot (*Psettodes erumei*), alleged extinction status along Kerala coast needs probe, Fishing Chimes, Vol. 29, No. 12, 2010.

Biswas, G. *et al*. Rare occurrence of slender sunfish off Sagar Island coast, West Bengal, Fishing Chimes, Vol. 30, No. 5, 2011.

Biswas, K.P. Corals of Tropical Oceans, Daya Publishing House, Delhi, 2008.

Biswas, K.P. Fishes around Indian Ocean, Daya Publishing House, Delhi, 2009.

Biswas, K.P. Marine prawns and shrimps, Daya Publishing House, Delhi, 2011.

Biswas, K.P. Advancement in Fish, Fisheries and Technology, Narendra Publishing House, Delhi, 2012.

Banerjee, P. *et al*. Occurrence of Pinjalo snapper (*Pinjalo pinjalo*) in the fishery of Maharashtra. Mar.Fish.Inf.Serv., T and E Ser.No. 186, 2005.

Culkin, F and R.A.Cox. Sodium, potassium, magnesium, calcium and strontium in sea water, Deep Sea Research, 13,1966.

Colborn J.G. The thermal structure of Indian Ocean, IIOE Monograph No. 2, The Univ. of Hawaii Press Honolulu, 1975.

Chaniyappa, M Bumper landings of giant sea catfish, *Arius thalassinus* by purse seiners at Malpe Fisheries Horbour, Karnataka, Mar.Fish.Inf.Serv, T and E Ser, No 200,2009.

Carina Dennis. Close encounters of the jelly kind, Nature, Vol. 126(6), 2003.

Chidambaram, L. Unusual heavy landings of juvenile catfishes (*Arius caelatus*) at Pondicherry Fishing harbour, Mar.Fish.Inf.Serv, T and E Ser. No. 186, 2005.

Chavan, B.B and S. Sundaram. On the occurrence of juvenile grouper in Dol net landings at New Ferry Wharf Mumbai, Mar.Fish.Inf.Serv, T and E Ser. No. 186, 2005.

Chouteau, J and J.H. Corriol. Physiological aspects of deep sea diving, Endeavour, Vol. XXX. No. 110, 1971.

Draper, L Proceeding No. 10, Coastal Engineering, Tokyo, ASCE, 1960.

Dhanesh, K.V. *et al*. Tuna fishery in Lakshadweep, current status, exploitation and need for Augmentation, J. Indian Ocean Studies, Vol. 19, No. 3, 2011.

Devassy, V.P. and P.M.A. Bhattathiri. Distribution of phytoplankton and chlorophyll *a* around little Andaman Island Tech. Rep. No. 02/80, NIO, Goa, India, 1980.

Daw, Rosamund Give a shell a break, Nature, Vol. 427, 2004.

Dharmaraj, S.K. and T.S. Velayudhan. Gastropod predation on the sacred chank, *Xancus pyrum* in the Gulf of Mannar Mar.Fish.Inf.Serv, T and E Ser, No 186, 2005.

Das Thakur and Sujit Sundaram. Sucker fish (Spearfish Remora) *Remora brachyptera*-rare occurrence in Mumbai waters, Fishing Chimes, Vol. 29, No. 10, 2010.

Dhaneesh, K.V. *et al*. Unusual mass landings of jelly fish at Cuddalore district, Tamil Nadu, Fishing Chimes, Vol. 29, No. 6, 2009.

Devanathan, K and M. Srinivasan. Gastropod, *Babylonia spirata* from Cuddalore coastal waters, Fishing Chimes. Vol. 29, No. 11, 2010.

Dineshbabu, A.P. *et al*. Targeted trawl fishery for moontail bullseye, *Priacanthus hamrur*, off Mangalore for surumi production, Mar.Fish.Inf.Serv, T and E Ser, No. 200, 2009.

Das Thakur *et al*. Bumper catch of sea bass, *Lates calcarifer* by gill netters in Mumbai waters, Mar. Fish.Inf.Serv, T and E Ser, No. 200, 2009.

Das Thakur *et al*. Observations on the fecundity of *Rhynchobatus djiddansis*, Fishing Chimes, Vol. 31, No. 8, 2011.

Das Thakur *et al*. Accidental capture and landing of whale shark, *Rhincodon typus* and the tiger shark *Galeocerado cuvier* by trawlers at Mumbai, Mar.Fish.Inf.Serv, T and E Ser. 205 2010.

Das Madhumita. Unusual landings of *Trachino cephalus* by trawlers at Visakhapatnam fishing harbour, Mar.Fish.Inf.Serv, T and E Ser.No 204, 2010.

Deshmukh V.D. Purse seine brings huge catch of red snapper, *Lutjanus argentimaculatus* at Mumbai, Cadalmin, No 125, 2010.

Dilly, P.N. The enigma of colouration and light emission in deep sea animals, Endeavour, Vol XXXII, No 115, 1973.

Denton, E.J. Buoyancy mechanisms of sea creatures, Endeavour, Vol XXII, No 85, 1963.

Delong, E.F. and David M Karl. Genomic perspectives in microbial oceanography, Nature, Vol 437/15, 2005.

Gouveia, A.D. *et al*. Wave characteristics in Andaman Sea, Tech. Rep. No. 02/80, NIO, Goa, India, 1980.

Guptha, S.M.V. Nanoplankton from recent sediments off Andaman Island, Tech. Rep. No 02/80, NIO, Goa, India, 1980.

George, M.J. *et al*. Indian Journal of Marine Science, Vol. 4, 1975.

Ghosh, S. *et al*. Sea anemones-perfect symbiotic animals in the coral reef ecosystem, Fishing Chimes, Vol. 24, No. 3, 2009.

Ganguli, S. *et al*. Jelly fish, an important bio-technological tool, Fishing Chimes, Vol.30, No5, 2010.

Giraspy Beni Sea cucumber farming in Australia, paving the way for a sustainable sea cucumber industry, Fishing Chimes, Vol. 28, No 9, 2008.

Ghosh, S. *et al*. Record size landing of black banded trevally, *Seriolina nigrofasciata* at Verabal, Mar.Fish.Inf.Serv, T and E Ser, No 201, 2009.

Gopkumar, G. *et al*. Breeding and seed production of silver pompano for the first time in India, Cadalmin, Vol 130, 2011.

Giovannoni, S.J, Molecular diversity and ecology of microbial plankton, Nature, Vol 437/15, 2005.

Hogg, N.G. Steady flow past an Island with applications to Bermuda, Geophys, Fluid Dyn, Vol 4, 1972.

Hoeksema, B.W. and Willem Moka. Species assemblages and phenotypes of mushroom corals (Fungiidae) related to coral reef habitats in the Flores Sea, Netherland J. of Sea Research, 23(2), 1989.

Hays, G.E. *et al*. Pan-Atlantic leatherback turtle movements, Nature, Vol 429, 2004.

James, D.B. Marine poisonous Echinoderms, Fishing Chimes, Vol 30, No 1, 2010.

James, D.B. *et al*. New records of echinoderms from the west coast of India, Fishing Chimes, Vol 28 No. 9, 2008.

Jacob, A.Y. and V.A. Narayanankutty. Report on the first occurrence of *Bathynomus giganteus*, a deep sea isopod fromthe west coast of India, Mar.Fish.Inf.Serv, T and E Ser, No 187, 2006.

Kamal Sarma. *et al*. Reef ecosystems and possible threats, management and conservation strategies, Fishing Chimes, Vol 31, No 9, 2011.

Krishnamurthy, V. Fisheries around Andaman and Nicobar Islands, Fishing Chimes, Vol 31, No 1, 2011.

Kripa, V. *et al*. Mabe pearl production- a technology for Andaman and Nicobar Islands Development, Fishing Chimes, Vol 30, No 3, 2010.

Khan, S.A. Management of spiny lobster fishery resources, National Biodiversity Authority 2006.

Kathirvel, M. Marine poisonous brachyuran crabs of India, Fishing Chimes, Vol 29, No 9,2009.

Kaladharan, V.S.P. and P. Rohit. Pycnogonids, the sea spiders, their role in marine ecology, Fishing Chimes, Vol. 30, No 9, 2010.

Kakati, V.S. *et al.* Occurrence of spider crab with sea anemone on its carapace at Mandapam waters. Cadalmin, No. 127, 2010.

Livingstone, D.A. Chemical compositions of rivers and lakes, U.S.Geol. Serv. Pap. 440-G, 1963.

Lyman, J. Oceans and Seas, Encyclopedia of Science and Technology, Mc Grahill Book Co, New York, 1966.

Laxmilatha, P. *et al.* Biology of *Mactra violacea* from Kerala, south west coast of India, Mar.Fish.Inf. Serv, T and E Ser, No 203, 2010.

Lohmann, K.J. *et al.* Geomagnetic map used in sea turtle navigation, Nature, Vol. 428, 2004.

Lighthill, M.J. How do fishes swim?, Endeavour, Vol XXIX, No 107, 1970.

Murty, C.S. *et al.* On some physical aspects of the properties of the surface waters around the Little Andaman Island, Tech. Rep. No 02/80, NIO, Goa, India, 1980.

Mondal Tamal, *et al.* Occurrence of seven Scleractinian corals in Ritchie's Archipelago, Andaman Islands of India, Proc. Zool. Soc, 64(1), 2011.

Mack Maurice Coccoliths, Endeavour, 24 (131), 1965.

Movachan, O.A. Soviet Fisheries Investigations, 1973.

Madhupratap, M. *et al.* Zooplankton abundance in the Andaman Sea, Tech. Rep. 02/80, NIO, Goa, India 1980.

Madhupratap, M. *et al.* Distribution of zooplankton in relation to thermocline of the Andaman Sea, Tech. Rep. No. 02/80, NIO, Goa, India, 1980.

Madhupratap, M. *et al.* Composition of major crustacean groups and total zooplankton diversity around Andaman-Nicobar, Tech. Rep. No. 02/80, NIO, Goa, India, 1980.

Murugan, A. *et al.* Assessment of Syngnathid distribution and its biological status along Tamil Nadu Coast, Fishing Chimes, Vol 29, No.10, 2010.

Miara, K.C. An aid to the identification of the fishes of India, Burma and Ceylon, *Elasmobranchii* and *Holocephali*, Records of the Indian Museum, Vol. XLIX, Part-I, pp 89-137, 1952.

Manickraja, M. *et al.* Occurrence of deep sea crab, *Thalanita crenata* in shallow water gill net operation at Tharuvaikulam, north of Tuticorin, Mar.Fish.Inf.Serv, T and E Ser, 201, 2009.

Makadia, B.V. Unusual heavy landings of *Otolithoides biauritus* and *Protonibea diacanthus* at Salaya landing center, Jamnagar, Gujarat, Mar.Fish.Inf.Serv, T and E Ser, No 200, 2009.

Mani, P.T. Heavy landings of juvenile groupers in trawl at Neendakara, Kollam, Mar.Fish.Inf Serv, T and E Ser, No. 187, 2008.

Manojkumar, P.P. Spotfin porcupine fish spotted at Malabar coast, Cadalmin, No.130, 2011.

Mohammed G. *et al.* Unusual landings of *Protonibea diacanthus* at Okha, Gujarat, Mar.Fish.Inf.Serv, T and E Ser, No. 205, 2010.

Manojkumar, P.P. First record of hound shark, *Mustelus mosis* from Calicut, Cadalmin,No125, 2010.

Manojkumar, P.P. First record of pelican flounder, *Chascanopsetta lugubris* from Malabar region, Cadalmin, No. 126, 2010.

Mohan, S. *et al.* Heavy landings of *Upeneus taeniopterus* along Chennai coast, Mar.Fish.Inf.Serv, T and E Ser, No 186, 2005.

Murugan, M. *et al.* Marine ornamental fishes of the Kittle Andaman waters, Fishing Chimes, Vol 28, No. 8, 2008.

Mehta, S.R. and Raptorial jaws in the throat help moray eels swallow large prey, Nature, Vol. 499/6 P.C. Wainwright 2007.

Naqvi, S.W.A and C.V.G. Reddy. On the variation in calcium content of waters in the Laccadives (Arabian Sea), Marine Chem. 8, 1978.

Nornha, R.J. *et al.* Calcium, magnesium and fluoride concentrations in the Andaman Sea, Tech. Rep. No. 02/80, NIO, Goa, India, 1980.

Naqvi, S.W.A. *et al.* Mahasagar, Bull. Natl. Inst. Ocenogr., 12, 1979.

Nakhawa, A.D. Mangrove biodiversity, Fishing Chimes, Vol. 29, No. 3, 2009.

Norris E. Richard Unarmoured marine dinoflagellates, Endeavour, 28(184), 1969.

Nayar, K.N. and S. Mahadevan. Ecology of pearl oyster bed, CMFRI Bulletin, No. 39, 1987.

Narasimham, K.A. *et al.* Laboratory breeding of clams, CMFRI News Letter, No. 39, 1988.

Nagabhushanam, R. *et al.* Reproductive biology of Indian rock oyster, *Crassostrea cuculata*, Indian, J. Fisheries, Vol 24, No 1 and 2, 1977.

Naomi, T.S. *et al*, Re-occurrence of oilfish in the landings of the south west coast of India, Mar. Fish.Inf.Serv, T and E Ser, No. 201,2009.

Naomi, T.S. *et al.* New distributional record of insular shelf beauty, *Symphysanodon typus* from Indian waters, Mar.Fish.Inf.Serv, T and E Ser, No. 204, 2010.

Oommen, V.P. Two octopods new to Arabian Sea, Indian J. Fisheries, Vol. 24, Nos 1 and 2, 1977.

Parulekar, A.H. and Z.A. Ansari. Benthic macrofauna of the Andaman Sea, Tech. Rep. No. 02/80, NIO, Goa, India.

Phadke, G.G. *et al.* Conservation of ecologically sensitive coral, Fishing Chimes, Vol. 31, No.4, 2011.

Pant, A. Extracellular production by phytoplankton in the Andaman Sea, Tech. Rep. 02/80 NIO, Goa, India.

Prabhu Matondkar, S.G. Microbiological studies of the sediments of Andaman Sea, Tech. Rep. 02/80, NIO, Goa, India.

Packard Andrew and Geoffrey Sanders. What the octopus shows to the world, Endeavour, Vol 28, No 104, 1969.

Prakash, S. *et al*. Scope for aquaculture of marine ornamental shrimps in Lakshadweep, a new prospect in India, Fishing Chimes, Vol. 31, No. 11, 2012.

Pillai, S.L. and P. Thirumilu. Exploitation of sand crabs (*Emertia asiatica* and *Alubunea symnista*) as a source of income during post-tsunami period along Chennai coast, Tamil Nadu, Mar.Fish. Inf. Serv, T and E Ser, No. 186, 2005.

Paul, Sijo. Unique landing of *Sardinella sirm* at Neendakara, Kollam, Mar.Fish.Inf.Serv, T and E. Ser., No. 201, 2009.

Pillai, N.G.K. *et al*. Lantern fishes (Myctophids) by catch of deep sea shrimp trawlers operated off Kollam, south west coast of India, Mar.Fish.Inf.Serv, T and E Ser No 202, 2009.

Patel, P.P. *et al*. Grey mullet, *Mugil cephalus* in Okhamandal region, maturity and biometric study, Fishing Chimes, Vol. 30, No. 12, 2011.

Qasim, S.Z. and Z.A. Ansari. Detrital content of the Andaman Sea, Tech. Rep. 02/80, NIO, Goa, India.

Qasim, S.Z. Indian Journal of Marine Science, 1979.

Rama Raju, D.V. *et al*. Some physical characters of the Andaman Sea, Tech.Rep.No. 02/80, NIO, Goa, India.

Ramesh Babu, V. *et al*. Hydrography of Andaman Sea, during late winter, IJMS, Vol.5, 1996.

Riley, J.P. *et al*. The major cation chlorinity ratio in sea water, Chem.Geol. 2, 1967.

Rodolfo, U.S. "Encyclopaedia of Oceangraphy", Van Nostrand Reinhold Company, 1966.

Ramesh Babu, V. *et al*. Indian Journal of Marine Science, Vol. 5, 1976.

Rao, C.V. *et al*. Journal of Marine Biological Association of India, 17, 1975.

Rao, G. Chandrasekhara. On the zoogeography of the interstitial meiofauna of the Andaman and Nicobar Islands, Indian Ocean, Rec.Zool.Surv.India, 77, 1980.

Rasal, K. and A. Rasal. Indian turtles, its status and conservation, Fishing Chimes, Vol 31, No. 8, 2011.

Rao, R.B. A report on the Olive Ridely turtle eggs found in Janjira region of Raigad, Maharashtra, Mar.Fish.Inf.Serv, T and E Ser, No. 201, 2009.

Ramani, K. *et al*. An overview of marine fisheries in India during 2007, Mar.Fish.Inf.Serv, T and E Ser, No. 203, 2010.

Rao, N.R. and Rohit, P. First record of threadfin bream, *Nemipterus zysron* from Andhra Pradesh coast, Mar.Fish.Inf.Serv, T and E Ser, No.204, 2010.

Rajabaikavim, S. *et al*. Heavy landings of mullet, *Mugil cephalus* by bag net at Chennai Fishing Harbour, Mar.Fish.Inf.Serv, T and E Ser, No. 186, 2005.

Sewell, R.B.S. Geographic and oceanographic research in Indian waters, Mem. Asiatic Soc. of Bengal, Vol. IX, No. 1, 1925.

Sharma G.S. Transequatorial movement water masses in the Indian Ocean, J.M.R, Vol.34(2), 1976.

Stequert, B. and F. Marsac Tropical tuna-surface fisheries in the Indian Ocean, FAO Fish. Tech Paper, 282, 1989.

Sreekanth G.B. *et al*. Yellowfin and Bigeye tuna of Indian EEZ, Fishing Chimes, Vol.30, No.4, 2010.

Sandra, L.H. A review of the literature of the dugong (*Dugong dugong*), Wildlife Research Report No.4, Fish and Wildlife Service, Washington D.C., 1975.

Sengupta, R. *et al*. A study of fluoride, calcium, magnesium in the Northern Indian Ocean, Mar. Chem. 6, 1976.

Sengupta, R. *et al*. Chemical characteristics of the Andaman Sea, Tech. Rep. 02/80, NIO, Goa, India.

Sulochanan Bindu, *et al*. *Entiglus acoroides* fruits observed in Gulf of Mannar, Mar.Fish.Inf.Serv, T and E Ser No. 200, 2009.

Sundaram, Sujit. Octopus fishery off north west (Maharashtra) coast, Fishing Chimes, Vol 30 No. 8, 2010.

Silas, E.G. *et al*. Identity of common species of cephalopods of India, CMFRI Bull. 37, 1985.

Sarma Kamal, *et al*. Tiger shrimp brooders of Andaman waters, Fishing Chimes, Vol 29, No10, 2010.

Sulochanan, B. *Enhalus acoroides* (Lf) Royle fruits observed in Gulf of Mannar, Mar.Fish.Inf. L.S. korabu Serv, T and E Ser, No 200, 2009.

Sudarsan, Sujit. Occurrence of *Octopus dollfusi* in Maharashtra waters, Mar.Fish.Inf.Serv, T and E A.D.Sawant Ser, No. 205, 2010.

Sundaram, Sujit. The various uses of cephalopods, Fishing Chimes, Vol 29, No.8, 2009.

Sreeram, M.P. *et al*. Unusual trend in landings of *Portunus sanguinolentus* in Mumbai, Mar.Fish.Inf. Serv, T and E Ser, No 186, 2005.

Sundaram, Sujit, *et al*. Trends of landings of *Sepia aculeate* and *Sepia prashadi* at Mumbai, Maharashtra Mar.Fish.Inf.Serv, T and E Ser No 187, 2006.

Sundaram, S. and M.Z. Khan. Record of *Octopus membranaceous* in Maharashtra waters, Mar.Fish.Inf.Serv, T and E Ser. No 200, 2009.

Sanaye, S.V. and R.M.Tibile. Status of turtle conservation in Maharashtra, Fishing Chimes, Vol. 29, No. 1, 2009.

Seetha, P.K. Heavy landings of juvenile lizard fish and silverbellies at Neendakara, Mar.Fish. Inf. Serv, T and E Ser No. 201, 2009.

Sukumaran, S and C. Kasinathan. A note on the landing of a slender sunfish near Rameswaram, Mar.Fish.Inf.Serv, T and E Ser, No 187, 2006.

Salvi, P.S. and V.D. Deshmukh Landing of "flying gurnard" in Mumbai, Mar.Fish.Inf.Serv, T and E Ser, No 205, 2010.

Sundaram S and M.Z.Khan Record of cuttle fish, *Sepia elliptica* off Maharashtra coast, Mar.Fish.Inf. Serv, T and E Ser, No 204, 2010.

Sreeram, M.P. *et al*. Record of rare serranid fish, *Sacura boutengeri* from Mumbai waters, Mar.Fish. Inf.Serv, T and E Ser, No 206, 2010.

Sarang J.D. and Sujit, Sundaram. Emergence of oil sardine fishery as alternative resource for dol netters at Arnala Mar.Fish.Inf.Serv, T and E Ser, No 206, 2010.

Sreeram, M.P. *et al*. Bumper landing of dolphin fish at Sasson Dock, Mumbai, Mar.Fish.Inf.Serv, T and E Ser, No 186, 2005.

Smetacek, Victor. The Ocean's veil, Nature, Vol 419, 2002.

Thomas, P.A. Sponge generated bioerosion in Lakshadweep, Mar.Fish.Inf.Ser. No 86, 1988.

Talbot, F.H. and M.S. Talbot. The crown of thorn starfish (*Acanthaster*) and the Great Barrier Reef, Endeavour, 1970.

Thangavelu, R. *et al*. Mussel resources in Andaman Islands, Mar.Fish.Inf.Serv, T and E Ser No 201,2009.

Thomas, M.M. Decapod crustaceans new to Andaman and Nicobar Islands, Indian J.Fish. Vol 24, No. 1 and 2, 1977.

Thangavelu, R. *et al*. Record landings of cephalopods by trawlers at Verabal during first quarter of 2009, Mar.Fish.Inf.Serv, T and E Ser, No 205, 2010.

Thangavelu, R. *et al*. By catch of gastropod, *Tibia* spp. in gill nets, operating along Gujarat coast, Mar. Fish.Inf.Serv, T and E Ser. No. 204, 2010.

Thomas, S. *et al*. Pufferfish, *Lagocephalus inermis*, an emerging fishery along Mangalore coast of Karnataka, Mar.Fish.Inf.Serv,T and E Ser. No 200, 2009.

Thangavelu, R. *et al*. Unusual heavy landings of catfish, *Arius dussumieri* in Rajapara Bay of Gujarat Coast, Mar.Fish.Inf.Serv, T and E Ser No 206, 2010.

Thangavelu, R. *et al*. Occurrence of a large cornetfish, *Fistularia patimba* in trawl catches of Verabal Coast, Gujarat, Mar.Fish.Inf.Serv, T and E Ser. No 202, 2009.

Thangavelu, R. *et al*. Rare occurrence of bramble shark, *Echinorhinus brucus*, along the Verabal Coast, Mar.Fish.Inf.Serv, T and E Ser No 202, 2009.

Venkataramanujam, K. *et al*. Gorgonian resources of Tamil Nadu, Sea Food Export Journal, March, 1982.

Venkataraman, K. Coral reefs in India, National Biodiversity Authority, Chennai, India, 2006.

Venkataraman, K. Natural Aquatic Ecosystems of India, NBSAP, Zool. Surv.India, 2003.

Vijayalakshmi, R.N. *et al.* Chaetognatha of the Andaman Sea, Tech.Rep. 02/80, NIO, Goa, India.

Venkatesan, C.K. and A. Shanmugavelu. On the first record of a rare marine ornamental fish from the Gulf of Mannar, Mar.Fish.Inf.Serv, T and E Ser No 187, 2006.

Vijayaraghavan, P. Life history and feeding habits of spotted seer, *Scomberomorus guttatus* Indian J. Fisheries, Vol. II (2), 1955.

Vaidya, N.G. *et al.* Shore seine operations during monsoon at Karwar of Karnataka, Mar.Fish. Inf.Serv, T and E Ser No. 206, 2010.

Vinoth, R. *et al.* The first time record of the dusky grouper, *Epinephelus marginatus* in Vellar Estuary, south east coast of India, Fishing Chimes, Vol.30, No. 12, 2011.

Wyrtki, Klaus. Physical oceanography of the South East Asian waters, NAGA Report, Vol 2 1961.

Waghmare, K.B. *et al.* An emerging commercial fishery of *Rachycentron canadum* at New Ferry Wharf, Mumbai, Mar.Fish.Inf.Serv, T and E Ser.No.201, 2009.

Yogesh, C. Captive breeding of sea horse, Success in NIO, Goa, Fishing Chimes, Vol.28 No. 10/11, 2009.

Zernova, V.V. and Ju. A. Ivanov. Trud. Inst. Okeanol 64, 1964.

Zaneveld, J.S. Economic marine algae of tropical South East Asia and their utilization, IPFC, Spl. Pub. No.3, Bangkok, 1955.

Zacharia, P.U. *et al.* Landing of the dog cohelk, *Nassaria nivea* and the beak shell *Tibia fusus* by Trawlers at Tuticorin, 2009, Mar.Fish.Inf.Serv, T and E Ser. No.203, 2010.

Zacharia, P.U. Unusual landings of rays and skates at Tuticorin Fisheries Harbour, Mar.Fish. P. Kandan Inf. Serv, T and E Ser. No. 205, 2010.

Index

Figure 1.2: Vertical zonation of near shore sea

Figure 4.1: Marine Micro-world

Figure 7.1: *Cheilosporum spectabile* 1, Red algae

Figure 7.2: *Halymenia floresia*

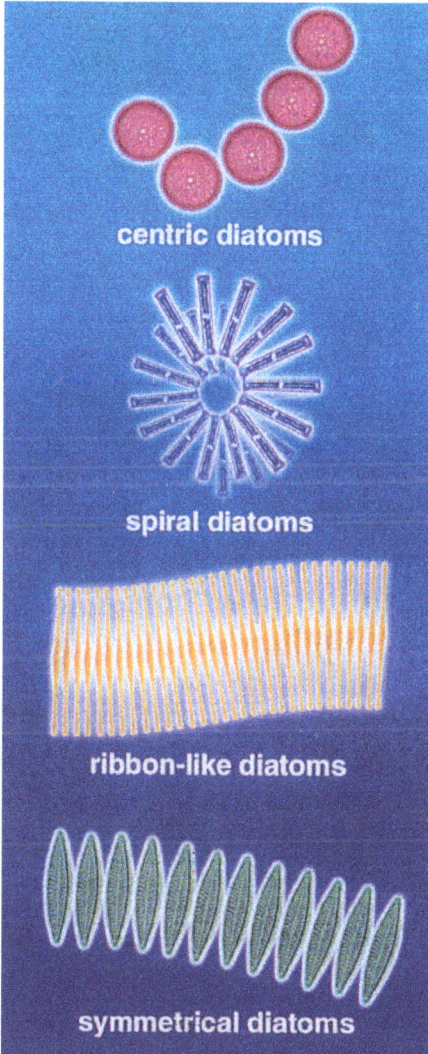

Figure 8.2: Phytoplankton Communities of the Sea, Diatoms

centric diatoms

spiral diatoms

ribbon-like diatoms

symmetrical diatoms

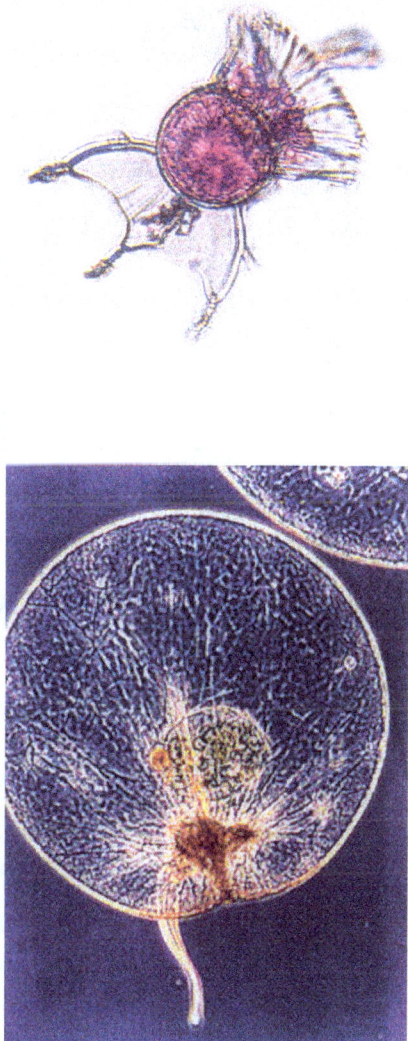

Figure 8.3: *Noctiluca* spp.

Figure 9.1: Coral Polyps

Figure 9.2: Orange Cup Corals

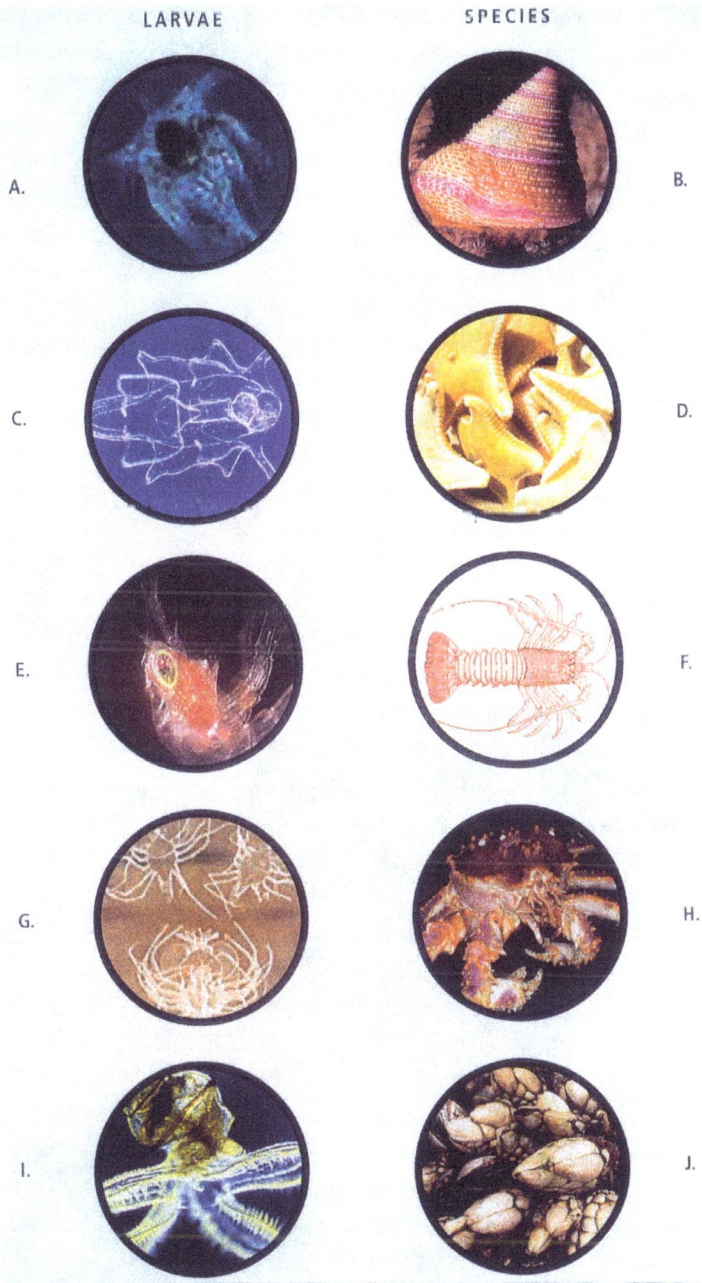

LARVAE   SPECIES

A.

B.

C.

D.

E.

F.

G.

H.

I.

J.

Figure 9.3: A: Nauplius larva of a barnacle; B: Blue-ring top snail; C: Starfish larva;
D: Mud sea star; E: Blue king crab zoea larva; F: Spiny rock lobster;
G: New Zealand red rock lobster larva; H: Blue king crab;
I: Larva of the gastropod; J: Barnacle

Figure 9.4: Nudibranchs

Figure 16.1: *Porpita porpita*

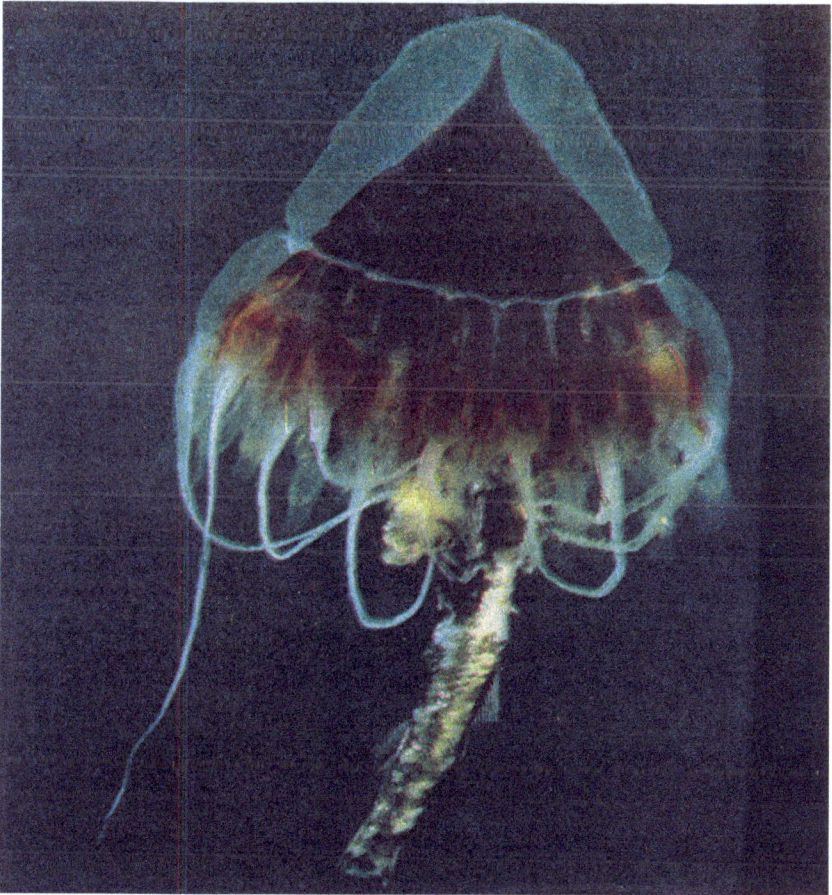

Figure 16.2: Deep Sea Jellyfish, *Periphylla*

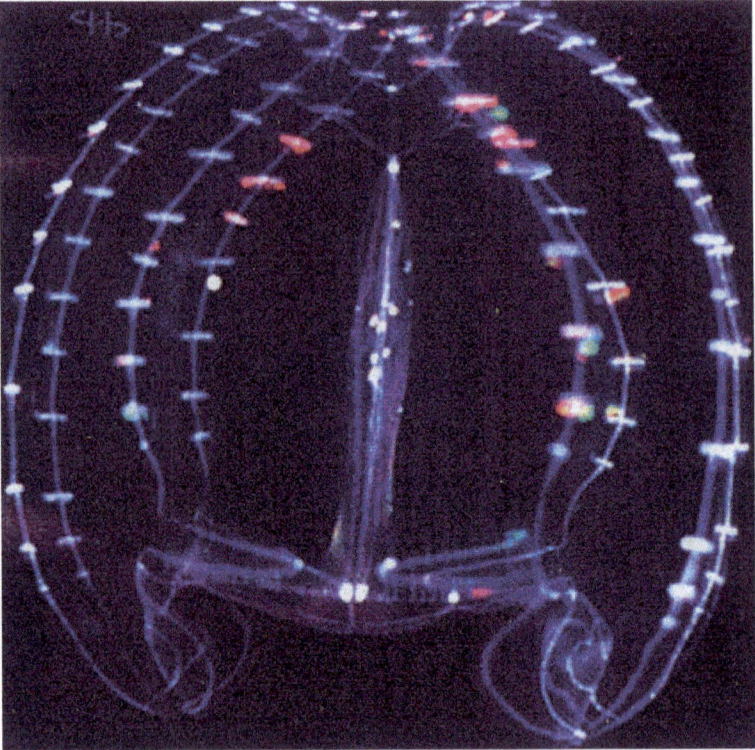

Figure 16.3: Polyp (Free swimming)

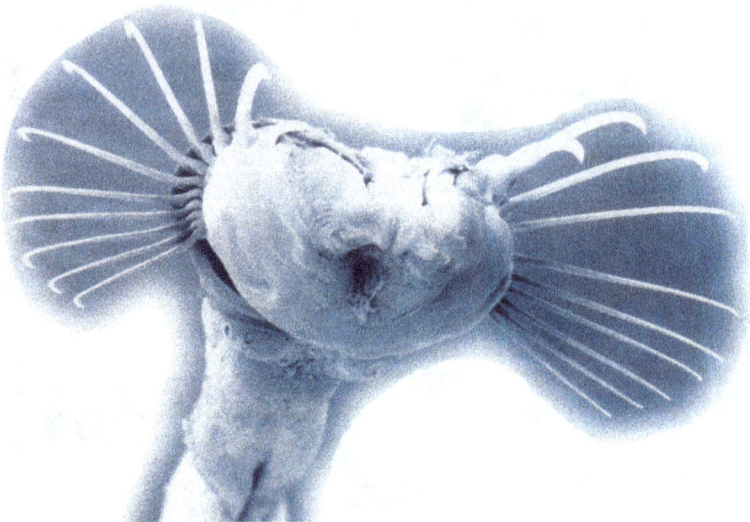

Figure 19.1: Head of *Sagitta* sp.

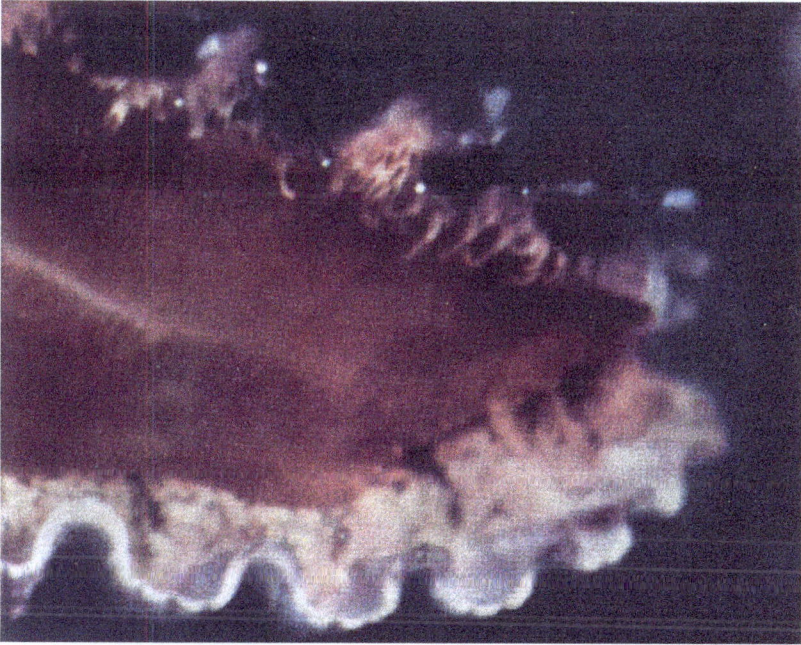

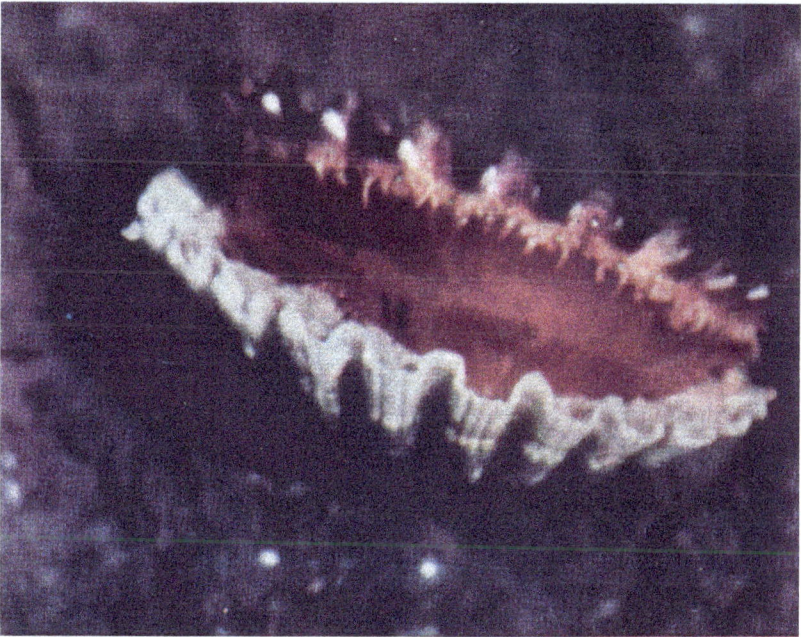

Figure 21.3: Scallop, *Pecten jacobaeus*

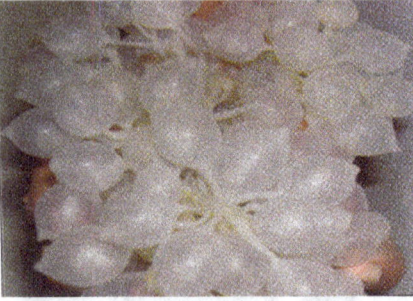

Figure 22.2: Egg Cluster of *Sepia pharaonis*

Figure 22.3: Sepia Young Ones

Figure 22.4: Cuttle Fish, *Sepia* spp.

Figure 22.5: Cuttle Fish, *Sepia elliptica*

Figure 22.6: The Copulating Cuttle Fish, *Sepia officinalis*

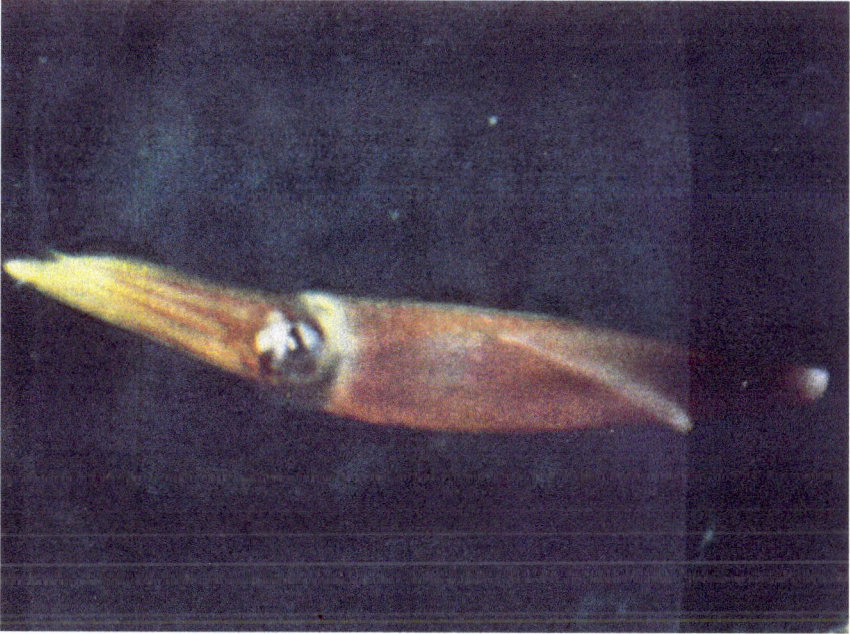

Figure 22.7: The Squid, *Loligo vulgaris*

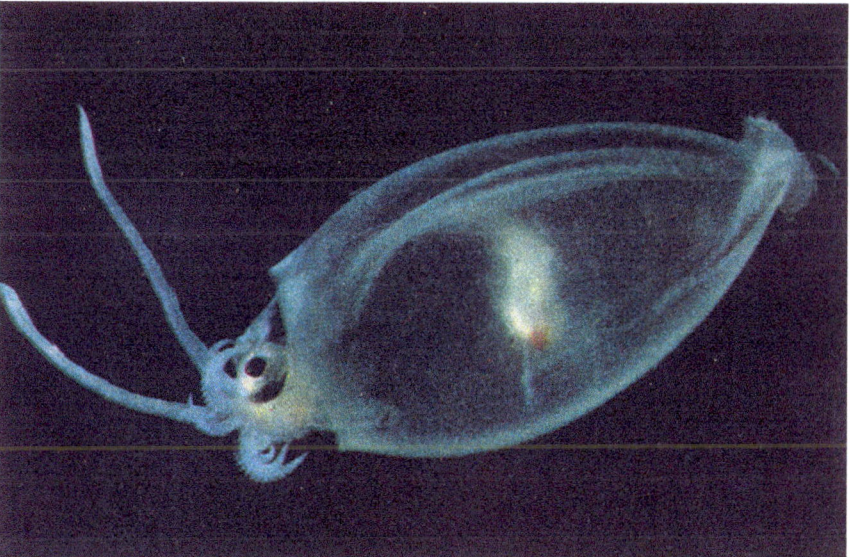

Figure 22.8: Cranchid Squid

Figure 22.9: Squid, *Loligo* spp.

Figure 22.10: A Cranchid Squid, *Cranchia scabra*

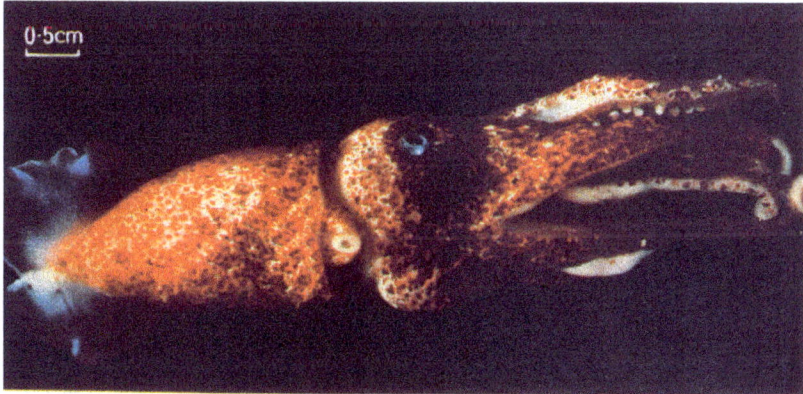
Figure 22.11: Benthic Crinoid, *Pentacrinus*

Figure 22.12: Crinoid

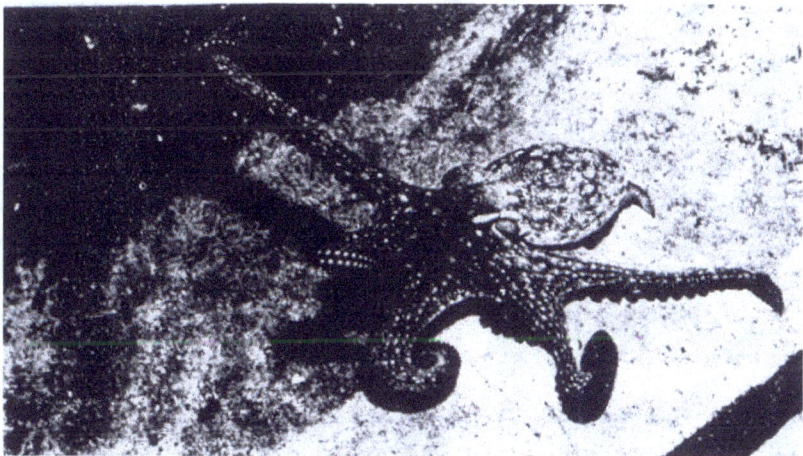
Figure 22.13: Cephalopods Regular Components of Patterning

Figure 22.14: Octopus, *Octopus* spp.

Figure 22.15: *Octopus macropus*

Figure 23.2: Deep Sea Shrimp, *Gnatheuphausia*

Figure 23.3: Squat Lobster, *Galathea strigosa*

Figure 23.4: Rock Lobster,
Panulirus spp.

Figure 23.5: Sand Lobster,
Thenus orientalis

Figure 23.6: Spiny Rock Lobster

Figure 23.7: Dorsal View of
Scyllarides tridacnophaga

Figure 23.8: Hermit Crab, *Pagurus arrosor*

Figure 23.9: Blue Crab

Figure 23.10: Sea Crab,
Portunus sanguinolentus

Figure 23.11: Sea Crab, *Charybdis* spp.

Figure 23.12: Sea Crab,
Portunus pelagicus

Figure 23.13: Mud Crab, *Scylla serrata*

Figure 23.14: Berried Female of *Portunus sanguinolentus*

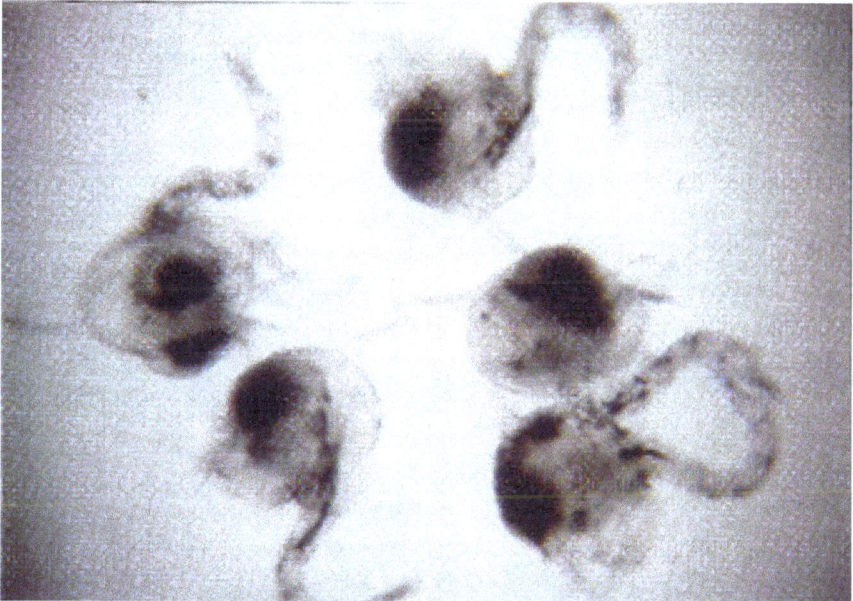

Figure 23.15: Zoe of *Portunus sanguinolentus*

Figure 25.1: *Psudanthias* sp.

Figure 25.2: *Chelidoperca investigatoris* (Alcock, 1895)

Figure 25.3: *Lutjanus argentimaculatus*

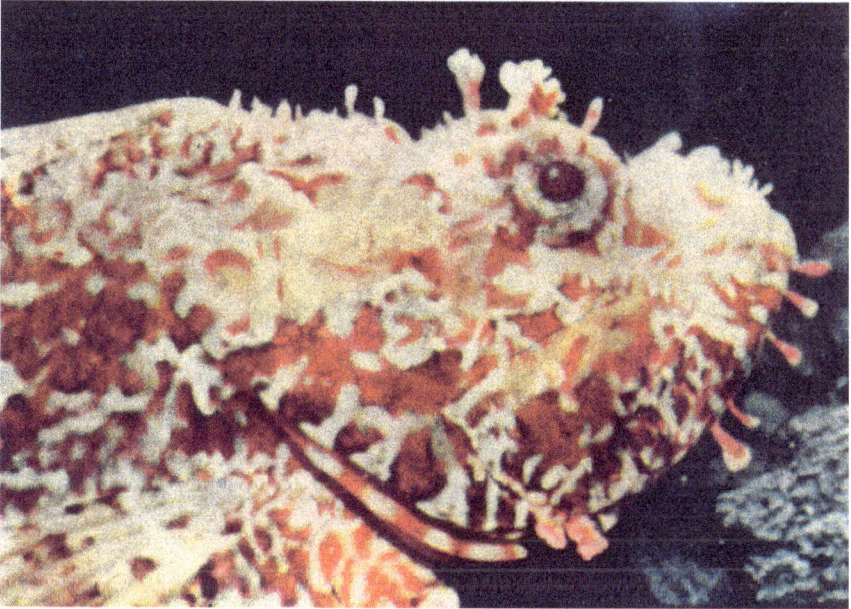

Figure 25.4: Scorpion fish, *Scorpaena scropha*

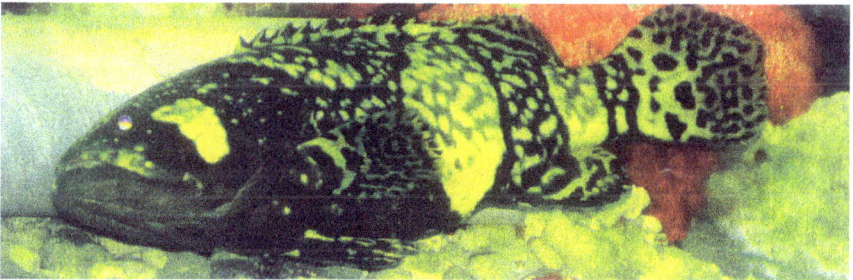

**Figure 25.5: Lateral View of Giant Grouper *Epinephelus lanceolatus*
Maintained in the Marine Research Aquarium**

Figure 25.8: Quagga Shark, *Halaelurus quagga*

Figure 25.9: *Pseudanthias* sp.

Figure 25.11: Adult Mullet FIsh

Figure 25.12: Dissected Out Matured Mullet Fish
Showing Ripened Ovary in the Abdomen

Figure 25.13: Iced Mullet Fish Ready to Transport

Figure 25.14: Harvested Mullet Fish

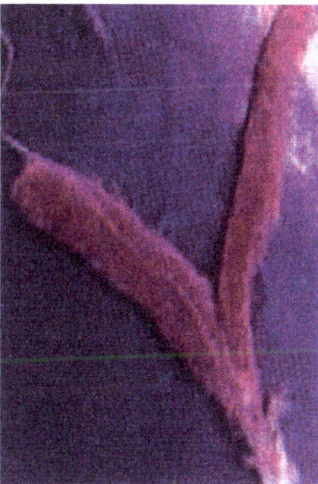

Figure 25.15: Developing Ovary

Figure 25.16: Ripe Ovary

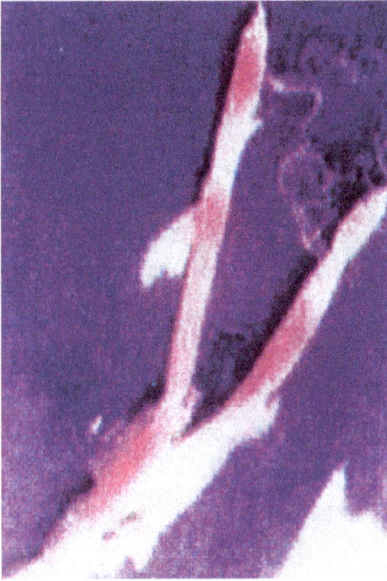

Figure 25.17: Spend and on Recovery

Figure 25.18: Testis

Figure 25.20: Seer Fish, *Scomberomorus linceolatus*

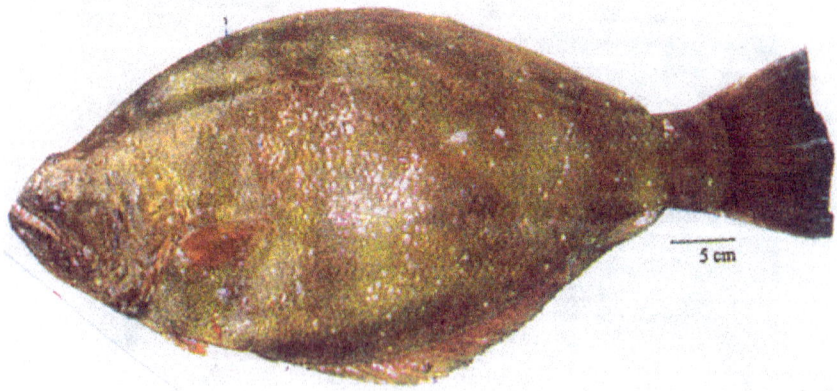

Figure 25.21: Flat Fish, *Psettodes erumei*

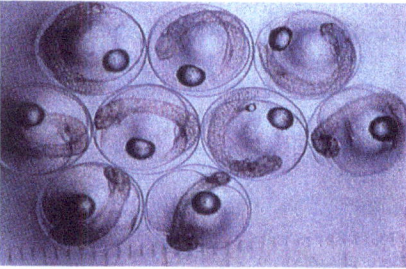

Embryonic Development

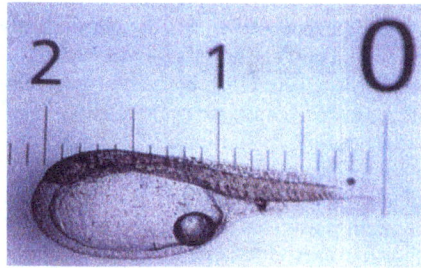

Newly Hatched Larva

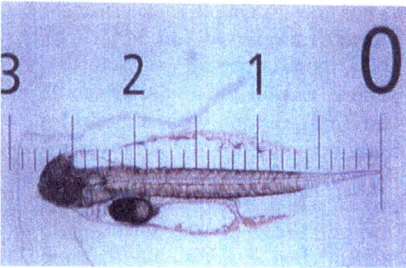

Larva on 2 dph

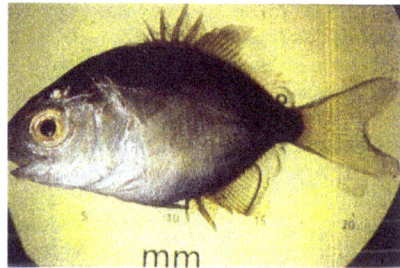

Fingerling on 28 dph

Figure 25.24: Larval Development of Silver Pompano

Figure 25.25: Silver Pompano, *Trachynotus blochii*

Figure 25.26: *Chascanopestta lugubris*

Figure 25.28: Flying Gurnard *(Dactyloptena peterseni)*

Figure 25.29: *Roa jayakari* (Norman, 1939)

Figure 27.2: Turtle with Cleaner Fish

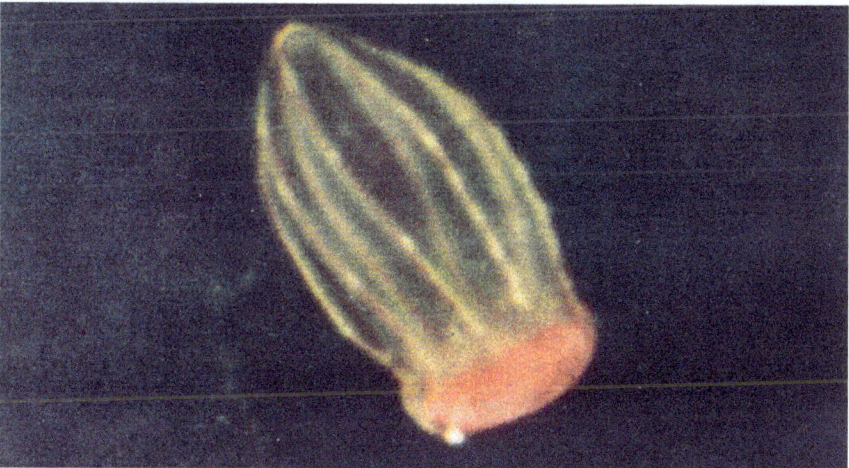

Figure 29.6: Sea-gooseberry, *Bcroe forskali*

Figure 29.7: Cydippid Larva of Ctenophore

Figure 29.8: A Certoid Angler Fish
Melanocetus

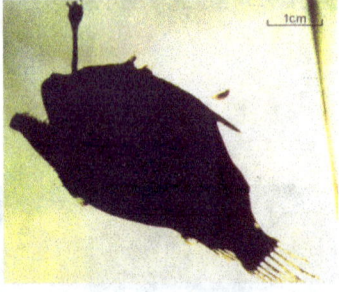

Figure 29.9: Angler Fish Hoisting Light Organ

Figure 29.10: Deep Sea Angler Fish, *Astronesthes*

www.ingramcontent.com/pod-product-compliance
Lightning Source LLC
Chambersburg PA
CBHW050127240326

41458CB00124B/1458